The Estimation and Tracking of Frequency

CAMBRIDGE SERIES IN STATISTICAL AND PROBABILISTIC MATHEMATICS

Editorial Board

R. Gill, *Department of Mathematics, Utrecht University*
B.D. Ripley, *Department of Statistics, University of Oxford*
S. Ross, *Department of Industrial Engineering, University of California, Berkeley*
M. Stein, *Department of Statistics, University of Chicago*
D. Williams, *School of Mathematical Sciences, University of Bath*

This series of high quality upper-division textbooks and expository monographs covers all aspects of stochastic applicable mathematics. The topics range from pure and applied statistics to probability theory, operations research, optimization and mathematical programming. The books contain clear presentations of new developments in the field and also of the state of the art in classical methods. While emphasizing rigorous treatment of theoretical methods, the books also contain applications and discussions of new techniques made possible by advances in computational practice.

Already published
1. *Bootstrap Methods and Their Application,* by A. C. Davison and D. V. Hinkley
2. *Markov Chains,* by J. Norris
3. *Asymptotic Statistics,* by A. W. van der Vaart
4. *Wavelet Methods for Time Series Analysis,* by Donald B. Percival and Andrew T. Walden
5. *Bayesian Methods: An Analysis for Statisticians and Interdisciplinary Researchers,* by Thomas Leonard and John S. J. Hsu
6. *Empirical Processes in M-Estimation,* by Sara van de Geer
7. *Numerical Methods of Statistics,* by John Monahan
8. *A User's Guide to Measure-Theoretic Probability,* by David Pollard

The Estimation and Tracking of Frequency

B. G. Quinn

University of Manchester Institute of Science and Technology

E. J. Hannan

Late, Australian National University

PUBLISHED BY THE PRESS SYNDICATE OF THE UNIVERSITY OF CAMBRIDGE
The Pitt Building, Trumpington Street, Cambridge, United Kingdom

CAMBRIDGE UNIVERSITY PRESS
The Edinburgh Building, Cambridge CB2 2RU, UK
40 West 20th Street, New York, NY 10011-4211, USA
10 Stamford Road, Oakleigh, VIC 3166, Australia
Ruiz de Alarcón 13, 28014 Madrid, Spain
Dock House, The Waterfront, Cape Town 8001, South Africa

http://www.cambridge.org

First published 2001

Printed in the United States of America

Typeface Computer Modern 11/13 pt.   *System* LATEX [AU]

A catalog record for this book is available from the British Library.

Library of Congress Cataloging in Publication data
Quinn, Barry G.
The estimation and tracking of frequency / B.G. Quinn, E.J. Hannan.
p. cm. — (Cambridge series in statistical and probabilistic mathematics)
Includes bibliographical references and indexes.
ISBN 0-521-80446-9
1. Signal processing—Mathematics. 2. Signal processing—Data processing. 3.
Modulation (Electronics)—Mathematical models. 4. Time-series analysis. 5.
Approximation theory. 6. Radio frequency. 7. Audio frequency. 8. MATLAB. I.
Hannan, E. J. (Edward James), 1921–   II. Title.   III. Cambridge series on statistical and
probabilistic mathematics.
TK5102.9 .Q85 2001
621.382′2—dc21   00-051944

ISBN 0 521 80446 9 hardback

To Margie, Tim, Jamie, Nicky and Robbie
and to the memory of Ted.

Contents

vii

Preface

In late 1982, Ted Hannan discussed with me a question he had been asked by some astronomers – how could you estimate the two frequencies in two sinusoids when the frequencies were so close together that you could not tell, by looking at the periodogram, that there *were* two frequencies? He asked me if I would like to work with him on the problem and gave me a reprint of his paper (Hannan 1973) on the estimation of frequency. Together we wrote a paper (Hannan and Quinn 1989) which derived the regression sum of squares estimators of the frequencies, and showed that the estimators were strongly consistent and satisfied a central limit theorem. It was clear that there were no problems asymptotically if the two frequencies were fixed, so Ted's idea was to fix one frequency, and let the other converge to it at a certain rate, in much the same way as the alternative hypothesis is constructed to calculate the asymptotic power of a test. Since then, I have devoted much of my research to sinusoidal models. In particular, I have spent a lot of time constructing algorithms for the estimation of parameters in these models, to implementing the algorithms in practice and, for me perhaps the most challenging, establishing the asymptotic (large sample) properties of the estimators.

We commenced writing this book in 1992, while I was working for the Australian Department of Defence as a scientist specialising in underwater signal processing. Our aim was to provide a comprehensive examination of various techniques for estimating fixed and varying frequency without dwelling too much on the minutiae of the theory. We hoped that the book would be accessible to both statisticians and engineers. I wrote what I thought would be the most straightforward chapter, on my work with Jose Fernandes, who had been a PhD student in the Electrical Engineering Department at the University of Newcastle, NSW, Australia. In late 1993, Ted wrote to me, concerned that the book was moving too slowly, and worried that it might never be completed. He died on January 7th, 1994, having completed several drafts of Chapters 2 and 3. Although I have corrected and added to these chapters, I have adjusted neither the style in which Ted wrote, nor his turn of phrase. It has taken me more than 6 years to complete and revise the book, for a number of reasons, among which have been several changes of location, from the Defence Science and Technology Organisation (DSTO) in Salisbury, Adelaide, Australia to Goldsmiths' College, University of London, and to the University of Manchester Institute of Science and Technology (UMIST).

This book is reasonably self-contained. We begin by motivating the problems by appealing to a small number of physical examples. For this reason, and since some of the important examples are not mentioned elsewhere, it is recommended that all readers consider the Introduction first. While there is also a fair amount of theory there, most theoretical issues are left until Chapter 2, which contains the statistical and probability theory needed for an understanding of the remainder of the book. Chapter 3 is concerned with the inference for fixed frequencies, including the problem of estimating two close frequencies and the estimation of the number of sinusoids. In Chapter 4, we present the Quinn and Fernandes (1991) technique. Unlike most techniques in the book, which involve Fourier transforms, this technique uses sequences of linear filters. In Chapter 5, we consider several popular techniques based on autocovariances, discussing their asymptotic properties and conditions under which they exhibit reasonable behaviour. In Chapter 6 we look at techniques which use only the Fourier transform of a series at the Fourier frequencies (the so-called Fourier coefficients) to estimate the system parameters. Finally, in Chapter 7, we treat the problem of changing frequency – a class of hidden Markov Model (HMM) procedures for tracking frequency as it changes slowly in very low signal-to-noise environments. Readers looking for a quick introduction to the estimation and related problems, and further applications are referred to the excellent expository article by Brillinger (1987).

I would like to thank the many people who have directly or indirectly encouraged our writing of this book. In particular, I would like to thank Dawei Huang (Queensland University of Technology), Bob Bitmead, Peter Hall and Joe Gani (Australian National University), Ross Barrett, Vaughan Clarkson, David Liebing and Darrell McMahon (DSTO, Salisbury), Doug Gray (DSTO and University of Adelaide), Brian Ferguson (DSTO, Sydney), Peter Kootsookos (Cooperative Research Centre for Robust and Adaptive Systems and e-Muse Corporation, Dublin, Ireland), Stephen Searle (Cooperative Research Centre for Robust and Adaptive Systems), Peter Thomson

(Statistics Research Associates Limited, Wellington, N.Z.) and T. Subba Rao and Maurice Priestley (UMIST). I would also like to thank Roy Streit, of the US Naval Undersea Warfare Center, for insights which led to the formulation of our HMM frequency tracker.

The research was partly supported by grants from the Australian Research Council, and also by the Australian Government, under the Cooperative Research Centre scheme.

I am indebted to my editor, David Tranah, who did a thorough and speedy job of editing the final versions, and made many invaluable suggestions concerning improvements to the layout of the book.

Finally, I would like to acknowledge my debt to Ted Hannan, who, after supervising my PhD, introduced me to the topic of the book and encouraged me in every aspect of my early career. I have missed him greatly, both personally and professionally, and I am sure that he would not mind much my dedication of this book to his memory. An account of Ted's life may be found at http://www.asap.unimelb.edu.au/bsparcs/aasmemoirs/hannan.htm, written by his good friend Joe Gani.

Manchester, U.K.   Barry Quinn
October, 2000.

1

Introduction

1.1 Periodic functions

We encounter periodic phenomena every day of our lives. Those of us who still use analogue clocks are acutely aware of the 60 second, 60 minute and 12 hour periods associated with the sweeps of the second, minute and hour hands. We are conscious of the fact that the Earth rotates on its axis roughly every 24 hours and that it completes a revolution of the Sun roughly every 365 days. These periodicities are reasonably accurate. The quantities we are interested in measuring are not precisely periodic and there will also be error associated with their measurement. Indeed, some phenomena only *seem* periodic. For example, some biological population sizes appear to fluctuate regularly over a long period of time, but it is hard to justify using common sense any periodicity other than that associated with the annual cycle. It has been argued in the past that some cycles occur because of predator-prey interaction, while in other cases there is no obvious reason. On the other hand, the sound associated with musical instruments can reasonably be thought of as periodic, locally in time, since musical notes are produced by regular vibration and propagated through the air via the regular compression and expansion of the air. The 'signal' will not be *exactly* periodic, since there are errors associated with the production of the sound, with its transmission through the air (since the air is not a uniform medium) and because the ear is not a perfect receiver. The word 'periodic' will therefore be associated with models for phenomena which are approximately periodic, or which would be periodic under perfect conditions. While the techniques discussed in this book can be applied to *any* time series, it should be understood that they have been derived for random sequences constructed from regularly sampled periodic functions and additive noise.

The simplest periodic function that models nature is the sinusoid. Con-

sider a particle which rotates in a plane around some origin at a uniform
speed of λ radians per second. The y-coordinate of the particle at time t
(secs) is then of the form

$$y(t) = A\cos(\lambda t + \phi), \tag{1.1}$$

where A is the (constant) distance between the origin and the particle and
ϕ is another constant which indicates the relative position of the cycle at
time 0. The graph of $y(t)$ for $\lambda = 2\pi\frac{15}{512}$ and $\phi = 0$ is shown in Figure 1.1.
Where a frequency is quoted, it will be understood from now on that it is
measured in radians per unit time, unless otherwise stated.

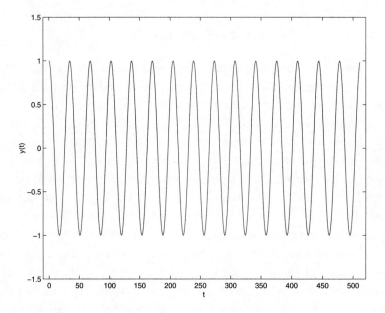

Fig. 1.1. Sine wave, frequency 0.1841

The period of $y(t)$ is the shortest time taken for $y(t)$ to repeat itself,
and this is obviously $2\pi/\lambda$. This function is also the general solution of the
differential equation which describes simple harmonic motion

$$\ddot{y}(t) = -\lambda^2 y(t).$$

In general, if a well-behaved function $y(t)$ is periodic, with period $2\pi/\lambda$,
then it can be written in the form

$$y(t) = c_0 + \sum_{k=1}^{\infty} c_k \cos(k\lambda t) + \sum_{k=1}^{\infty} s_k \sin(k\lambda t),$$

which is called the Fourier expansion of $y(t)$. The coefficients may be calculated using the equations

$$
\int_0^{2\pi/\lambda} \cos(j\lambda t)\, y(t)\, dt \;=\; c_j \int_0^{2\pi/\lambda} \cos^2(j\lambda t)\, dt
$$

$$
=\; \left\{ \begin{array}{ll} c_j \frac{\pi}{\lambda} & ; \quad j \geq 1 \\[2mm] c_0 \frac{2\pi}{\lambda} & ; \quad j = 0 \end{array} \right.
$$

and

$$
\int_0^{2\pi/\lambda} \sin(j\lambda t)\, y(t)\, dt \;=\; s_j \int_0^{2\pi/\lambda} \sin^2(j\lambda t)\, dt
$$

$$
=\; s_j \frac{\pi}{\lambda}.
$$

Note that the range of integration may be replaced by any interval of length $2\pi/\lambda$.

The 'square wave' function, depicted in Figure 1.2, again for the case where $\lambda = 2\pi \frac{15}{512}$ and $\phi = 0$, is periodic and has a Fourier expansion, even though the latter has many discontinuities.

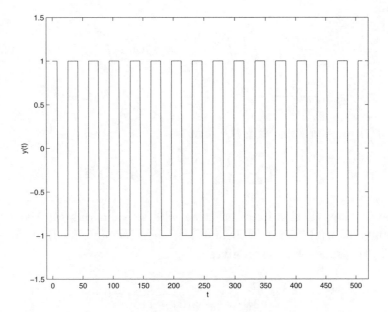

Fig. 1.2. Square wave, frequency 0.1841

Here

$$y\left(t\right) = \begin{cases} 1 & ; \quad -\pi/\left(2\lambda\right) < t \leq \pi/\left(2\lambda\right) \\ -1 & ; \quad \pi/\left(2\lambda\right) < t \leq 3\pi/\left(2\lambda\right), \end{cases}$$

and the coefficients in the Fourier expansion of $y\left(t\right)$ are

$$
\begin{aligned}
s_j &= \int_{-\pi/(2\lambda)}^{\pi/(2\lambda)} \sin\left(j\lambda t\right) dt - \int_{\pi/(2\lambda)}^{3\pi/(2\lambda)} \sin\left(j\lambda t\right) dt \\
&= \frac{\lambda}{j\pi\lambda} \left(\int_{-j\pi/2}^{j\pi/2} \sin x\, dx - \int_{j\pi/2}^{3j\pi/2} \sin x\, dx \right) \\
&= 0, \\
c_0 &= \frac{\lambda}{2\pi} \left(\int_{-\pi/(2\lambda)}^{\pi/(2\lambda)} dt - \int_{\pi/(2\lambda)}^{3\pi/(2\lambda)} dt \right) \\
&= 0, \\
c_{2j} &= \frac{\lambda}{\pi} \left\{ \int_{-\pi/(2\lambda)}^{\pi/(2\lambda)} \cos\left(2j\lambda t\right) dt - \int_{\pi/(2\lambda)}^{3\pi/(2\lambda)} \cos\left(2j\lambda t\right) dt \right\} \\
&= \frac{\lambda}{2j\pi\lambda} \left(\int_{-j\pi}^{j\pi} \cos x\, dx - \int_{j\pi}^{3j\pi} \cos x\, dx \right) \\
&= 0, \ j \geq 1
\end{aligned}
$$

and, for $j \geq 0$,

$$
\begin{aligned}
c_{2j+1} &= \frac{\lambda}{\left(2j+1\right)\pi\lambda} \left(\int_{-j\pi-\pi/2}^{j\pi+\pi/2} \cos x\, dx - \int_{j\pi+\pi/2}^{3j\pi+3\pi/2} \cos x\, dx \right) \\
&= \frac{4\left(-1\right)^j}{\left(2j+1\right)\pi}.
\end{aligned}
$$

Thus $y\left(t\right)$ may be written in the form

$$y\left(t\right) = \sum_{j=0}^{\infty} \frac{4\left(-1\right)^j}{\left(2j+1\right)\pi} \cos\left\{\left(2j+1\right)\lambda t\right\}.$$

The expansion, in this extreme case, will be valid only at the points of continuity of $y\left(t\right)$. This is most easily seen by noticing that, when $t = \pi/\left(2\lambda\right)$, the above expansion gives

$$y\left(\frac{\pi}{2\lambda}\right) = \sum_{j=0}^{\infty} \frac{4\left(-1\right)^j 0}{\left(2j+1\right)\pi} = 0,$$

which is midway between the true values just before and just after $\pi/\left(2\lambda\right)$.

Note that the expansion of $y(t)$ has no sine terms and that the even-indexed cosine terms are also missing.

The frequencies and periods of the sine wave and square wave can easily be determined by inspection of the above pictures. However, more often than not, when data is measured, it comes with noise, which can often be assumed to be additive. A model for the received data might therefore be

$$y(t) = A\cos(\lambda t + \phi) + x(t), \qquad (1.2)$$

rather than equation (1.1), where $x(t)$ is the noise associated with the measurement at time t. It may then not be possible to tell from the data that $y(t)$ has a periodic deterministic component. For example, the Figures 1.3 and 1.4 depict the noisy sinusoid and square waves corresponding to the previous two figures, with $A = 1$ and noise standard deviation 2.

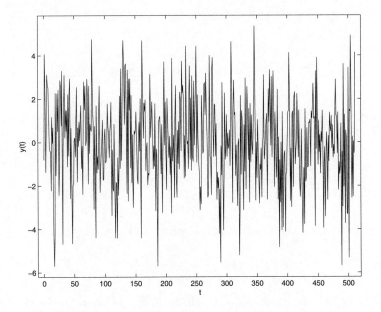

Fig. 1.3. Sine wave + noise, frequency 0.1841

If the deterministic function is more general, but is still periodic, we could consider the model

$$y(t) = \mu + \sum_{j=1}^{\infty} A_j \cos(j\lambda t + \phi_j) + x(t),$$

or a version with only a finite number of terms, since an infinite number

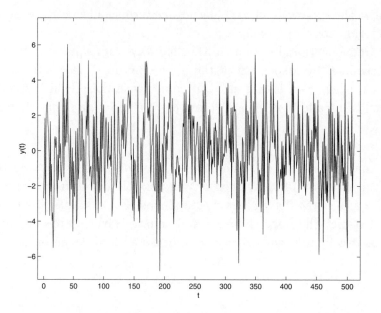

Fig. 1.4. Square wave + noise, frequency 0.1841

of parameters would in practice be impossible to estimate. More generally still, we shall consider in this book the model

$$y\left(t\right) = \mu + \sum_{j=1}^{r} A_j \cos\left(\lambda_j t + \phi_j\right) + x\left(t\right),$$

where r is the number of sinusoidal terms, and the frequencies

$$\{\lambda_j; j = 1, \ldots, r\}$$

may or may not be harmonically related. The term μ is called the overall mean or 'DC' term, and A_j, λ_j and ϕ_j are the amplitude, frequency and (initial) phase of the jth sinusoid. We shall be interested in estimating several (fundamental) frequencies and possibly their harmonics.

Even when there is no noise, it may be difficult to identify the frequencies. For example, if $r = 2$, and the two frequencies λ_1 and λ_2 are close together, the phenomenon known as 'beats' arises. Figure 1.5 was generated using $\lambda_1 = 2\pi 15/512 \sim 0.1841$ and $\lambda_2 = 0.2$. The amplitudes in both cases were 1 and the phases were 0. Although the data seem periodic, there appear to be a frequency near the two, and another much lower frequency associated with the evolution of the 'envelope'. This is because

$$\cos\left(\lambda_1 t\right) + \cos\left(\lambda_2 t\right) = 2\cos\left(\frac{\lambda_1 + \lambda_2}{2}t\right)\cos\left(\frac{\lambda_1 - \lambda_2}{2}t\right),$$

which is the product of two sine waves. We can think of the above either as an amplitude-modulated sinusoid with frequency $\frac{\lambda_1 + \lambda_2}{2}$, or an amplitude-modulated sinusoid with frequency $\frac{\lambda_1 - \lambda_2}{2}$.

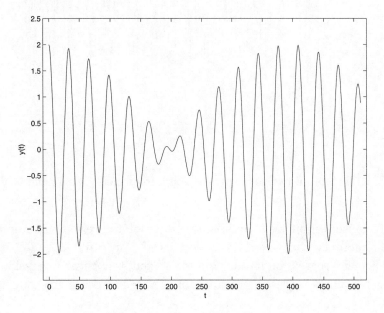

Fig. 1.5. Sum of two sine waves, frequencies 0.1841 and 0.2000

It is possible to estimate the two frequencies by observation, but, again, if there is noise, it may not be possible to tell from a graph that there are any periodicities present. Interestingly enough, it is possible to 'hear' the periodicities when listening to the noisy data depicted in Figure 1.6, even though the two sinusoids cannot be seen.

1.2 Sinusoidal regression and the periodogram

We consider now the simple sinusoidal model but with an additive constant term, and rewritten in the slightly different, but equivalent, form

$$y\left(t\right) = \mu + \alpha\cos\left(\lambda t\right) + \beta\sin\left(\lambda t\right) + x\left(t\right).$$

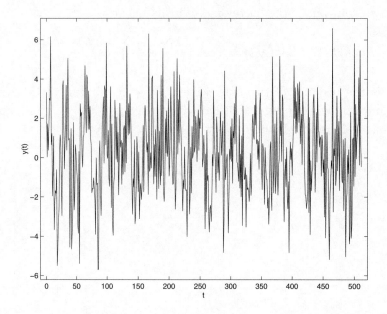

Fig. 1.6. Sum of two sine waves + noise, frequencies 0.1841 and 0.2000

As long as we sample at T equidistant time-points, we may assume without loss of generality that the data are observed at $t = 0, 1, 2, \ldots, T - 1$. The model is just a regression model, and for λ fixed, is a *linear* regression model. The least-squares regression estimators of μ, α and β are given by

$$
\begin{bmatrix} \widehat{\mu} \\ \widehat{\alpha} \\ \widehat{\beta} \end{bmatrix} = D^{-1}(\lambda) E(\lambda),
$$

where

$$
D(\lambda) = \begin{bmatrix} T & \sum_{t=0}^{T-1} \cos(\lambda t) & \sum_{t=0}^{T-1} \sin(\lambda t) \\ \sum_{t=0}^{T-1} \cos(\lambda t) & \sum_{t=0}^{T-1} \cos^2(\lambda t) & \sum_{t=0}^{T-1} \sin(\lambda t) \cos(\lambda t) \\ \sum_{t=0}^{T-1} \sin(\lambda t) & \sum_{t=0}^{T-1} \sin(\lambda t) \cos(\lambda t) & \sum_{t=0}^{T-1} \sin^2(\lambda t) \end{bmatrix}
$$

and

$$
E(\lambda) = \begin{bmatrix} \sum_{t=0}^{T-1} y(t) \\ \sum_{t=0}^{T-1} y(t) \cos(\lambda t) \\ \sum_{t=0}^{T-1} y(t) \sin(\lambda t) \end{bmatrix}.
$$

The residual sum of squares, for fixed λ, is then given by

$$
\begin{aligned}
\mathrm{SS}\,(\lambda) &= \sum_{t=0}^{T-1} y^2\,(t) - E'\,(\lambda)\,D^{-1}\,(\lambda)\,E\,(\lambda) \\
&= \sum_{t=0}^{T-1} \{y\,(t) - \bar{y}\}^2 - \left\{E'\,(\lambda)\,D^{-1}\,(\lambda)\,E\,(\lambda) - \bar{y}^2\right\},
\end{aligned}
$$

where $\bar{y} = T^{-1}\sum_{t=0}^{T-1} y\,(t)$. We may thus estimate the parameter λ, which makes the model a *nonlinear* regression model, by maximising with respect to λ the regression sum of squares

$$
E'\,(\lambda)\,D^{-1}\,(\lambda)\,E\,(\lambda) - \bar{y}^2. \tag{1.3}
$$

This expression is simple enough to compute exactly, but the computations will be onerous if T is large. However, many of the calculations can be simplified by using the fact that

$$
\sum_{t=0}^{T-1} e^{i\lambda t} = \left\{
\begin{array}{ll}
\frac{e^{i\lambda T}-1}{e^{i\lambda}-1} & ;\quad e^{i\lambda} \neq 1 \\
T & ;\quad e^{i\lambda} = 1
\end{array}
\right. .
$$

In fact, for large T, we have the approximations

$$
D\,(\lambda) = \left[
\begin{array}{ccc}
T & O\,(1) & O\,(1) \\
O\,(1) & \frac{T}{2}+O\,(1) & O\,(1) \\
O\,(1) & O\,(1) & \frac{T}{2}+O\,(1)
\end{array}
\right],
$$

and

$$
D^{-1}\,(\lambda) = T^{-1} \left[
\begin{array}{ccc}
1 & o\,(1) & o\,(1) \\
o\,(1) & 2+o\,(1) & o\,(1) \\
o\,(1) & o\,(1) & 2+o\,(1)
\end{array}
\right],
$$

where $o\,(1)$ and $O\,(1)$ denote terms which converge to zero and are bounded, respectively. Hence, the regression sum of squares is, from (1.3), approximately

$$
\begin{aligned}
I_y\,(\lambda) &= \frac{2}{T}\left\{\sum_{t=0}^{T-1} y\,(t)\cos\,(\lambda t)\right\}^2 + \frac{2}{T}\left\{\sum_{t=0}^{T-1} y\,(t)\sin\,(\lambda t)\right\}^2 \\
&= \frac{2}{T}\left|\sum_{t=0}^{T-1} y\,(t)\,e^{i\lambda t}\right|^2,
\end{aligned}
$$

which is called the *periodogram* of $\{y\,(t)\,; t = 0, 1, \ldots, T-1\}$. This leads us to consider a second method of estimating λ; namely, maximising $I_y\,(\lambda)$ with respect to λ. If the model is correct, the two methods lead to equivalent estimators.

Figures 1.7 and 1.8 depict the periodograms of the noiseless sine and square waves, measured at the times $t = 0, 1, \ldots, 511$. The periodograms have been calculated using the finite Fourier transform, at the so-called Fourier frequencies

$$\left\{ 2\pi \frac{j}{T}; j = 0, 1, \ldots, T - 1 \right\}.$$

The Fourier transform is calculated at these frequencies because the the 'fast' Fourier transform algorithm of Cooley and Tukey (1965), and later more general versions, which are available in packages such as MATLAB^{TM}, enable the T complex coefficients, equispaced in frequency, to be calculated in $O\left(T \log T\right)$ operations rather than the expected $O\left(T^2\right)$. It will be seen later, furthermore, that the joint statistical properties of these Fourier coefficients are simpler than those at other frequencies.

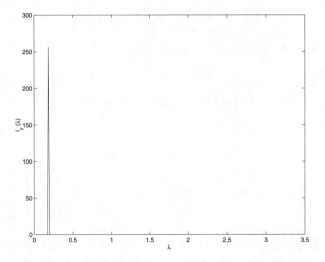

Fig. 1.7. Periodogram of sine wave, frequency 0.1841

Note that the periodogram of the sine wave has a very pronounced peak, near the true frequency, and that the periodogram of the square wave has pronounced peaks at the odd harmonics of the fundamental frequency. The latter might seem a little surprising, since periodogram maximisation, as a frequency estimation technique, has been motivated only by the single frequency case. It is shown in Chapter 3 that when there are r sinusoids, the regression sum of squares, as a function of the r frequencies, is approximated

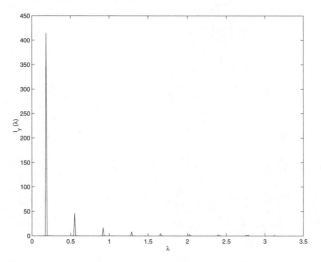

Fig. 1.8. Periodogram of square wave, frequency 0.1841

by the sum of the periodograms at the r frequencies, with the approximation becoming more accurate with increasing T when the frequencies are fixed. However, the approximation breaks down if we consider T fixed and observe what happens as the frequencies coalesce. For example, the periodogram in Figure 1.9 of the sum of the two sinusoids of Figure 1.5 exhibits only one peak. Even if the frequencies are further apart, provided that the amplitudes are quite different, the periodogram may show only the more dominant frequency. These problems are discussed further in Chapter 3, where the full regression procedure is described and analysed.

Now consider the case where we have additive noise. Recall that it was very difficult to see periodic behaviour when noise with standard deviation 2 was added to the sine and square waves. Figures 1.10 and 1.11 represent the periodograms of the noisy sine and square waves shown in Figures 1.3 and 1.4. Note that the peak near the true frequency in the periodogram of the sine wave is still pronounced, but several spurious spikes have appeared. Note as well that only the fundamental and third harmonic of the square wave are now visible, the other peaks having been obscured by 'noise'.

The periodogram is nevertheless a visually powerful tool for locating a frequency. It will be seen in Chapter 3 that the maximiser of the periodogram over all frequencies cannot be bettered, in terms of asymptotic variance, by any other technique, without extensive knowledge of the distribution of the noise. In fact, although it is often assumed that the noise process is

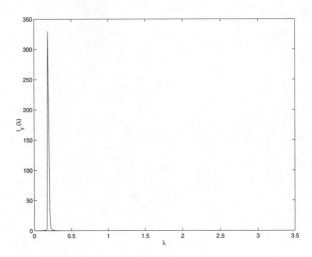

Fig. 1.9. Periodogram of sum of two sine waves, frequencies 0.1841 and 0.2000

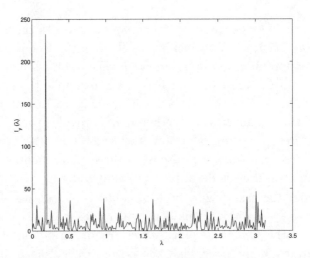

Fig. 1.10. Periodogram of sine wave + noise, frequency 0.1841

white, and some authors will even assume Gaussianity, it turns out that the regression sum of squares and periodogram maximisers have excellent properties even when the noise process is neither white nor Gaussian. Moreover, even if it were known that the noise was Gaussian, the resulting maximum likelihood estimator would not have smaller asymptotic variance.

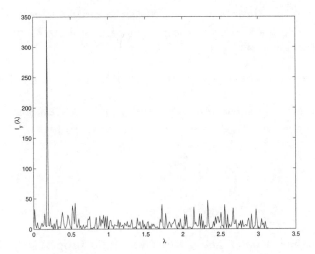

Fig. 1.11. Periodogram of square wave + noise, frequency 0.1841

1.3 Testing for the presence of a sinusoid

We note here that the periodogram of a white noise sequence, since it is, after all, a trigonometric polynomial with many local maxima and minima, will have peaks which could mistakenly indicate the presence of a sinusoid. Figure 1.12 represents the periodogram of 512 pseudo-Gaussian random numbers.

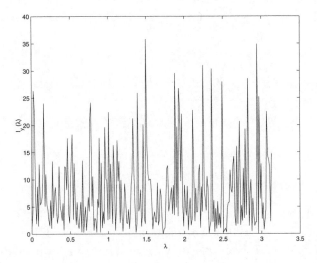

Fig. 1.12. Periodogram of noise

There are quite a few peaks here which could be present because of a sinusoid in the series. To test whether there is a sinusoid or not, consider the subset of sinusoidal models

$$y\left(t\right) = \mu + A\cos\left(\lambda_j t + \phi\right) + x\left(t\right),\ t = 0, 1, \ldots, T - 1$$

where $\lambda_j = 2\pi j/T$ but j is unknown and $\{x\left(t\right)\}$ is Gaussian and an independent sequence, and therefore 'white'. We wish to test

$$H_0 : A = 0$$

against

$$H_A : A > 0.$$

A test which has usually has good asymptotic properties and is usually simple to derive is the likelihood ratio test, which rejects the null on large values of the ratio of the maximised likelihood under H_A to the maximised likelihood under H_0. The former is just

$$-\frac{T}{2}\log\left(2\pi\widehat{\sigma}_A^2\right) - \frac{T}{2},$$

while the latter is

$$-\frac{T}{2}\log\left(2\pi\widehat{\sigma}_0^2\right) - \frac{T}{2},$$

where

$$\widehat{\sigma}_0^2 = \frac{1}{T}\sum_{t=0}^{T-1}\left\{y\left(t\right) - \overline{y}\right\}^2,$$

$$\widehat{\sigma}_A^2 = \frac{1}{T}\sum_{t=0}^{T-1}\left\{y\left(t\right) - \overline{y}\right\}^2 - \max_{1\leq j\leq n} I_y\left(\lambda_j\right)$$

and $n = \lfloor (T-1)/2 \rfloor$. We thus reject H_0 if $\widehat{\sigma}_A^2/\widehat{\sigma}_0^2$ is too small, or, equivalently, if

$$g = \frac{\max_{1\leq j\leq n} I_y\left(\lambda_j\right)}{\sum_{t=0}^{T-1}\left\{y\left(t\right) - \overline{y}\right\}^2}$$

is too large. The statistic is known as Fisher's g (Fisher 1929) and its exact distribution under H_0 is known, although somewhat difficult to calculate directly, especially if T is moderately large. However its probability density function and distribution function are related to certain spline functions, and may thus be calculated relatively quickly and accurately using recursion.

More often than not, the asymptotic approximation will be used: when H_0 is true, the terms

$$\sum_{t=0}^{T-1} y(t) \cos(\lambda_j t)$$

and

$$\sum_{t=0}^{T-1} y(t) \sin(\lambda_j t)$$

are, for $j = 1, \ldots, n$, independent and identically distributed normally with common means zero and common variances $T\sigma^2/2$, where σ^2 is the variance of $x(t)$. Thus the $I_y(\lambda_j)/\sigma^2$ are independent and identically distributed as χ^2 with 2 degrees of freedom. Since it can be shown that

$$\frac{1}{T} \sum_{t=0}^{T-1} \{y(t) - \bar{y}\}^2$$

converges almost surely to σ^2, it follows that Tg is asymptotically distributed as the maximum of $X_1, X_2, \ldots, X_n$, n independent and identically distributed χ^2 random variables with 2 degrees of freedom. Now

$$\Pr\left\{ \max_{1 \le j \le n} X_j \le 2\log n + 2x \right\} = \left\{ 1 - e^{-\frac{1}{2}(2\log n + 2x)} \right\}^n$$

$$= \left(1 - \frac{1}{n} e^{-x} \right)^n$$

$$\to e^{-e^{-x}}.$$

Thus

$$\Pr\{Tg > 2\log(T/2) + 2x\} \to 1 - e^{-e^{-x}},$$

which enables an asymptotic critical region of any size to be calculated.

There are similar techniques for tests concerning the numbers of sinusoids. In Chapter 3 we examine the likelihood ratio test of

$$H_0 : r = r_0$$

against

$$H_A : r = r_A$$

where r_A is some given number greater than r_0. We obtain exact, approximate and asymptotic null distributions of the test statistic. The test has limitations for the same reasons as the previous test based on Fisher's g-statistic, in that the frequencies are limited to the Fourier frequencies and the noise is assumed to be Gaussian and white.

A related problem is that of determining, for fixed T, how large the signal-to-noise ratio (SNR) of a sinusoid should be in order that it be detected, since if it is too small, the largest periodogram value may not occur near the correct frequency. This problem is discussed in Rife and Boorstyn (1974) and Quinn and Kootsookos (1994), but will not be dealt with in this book, as even the analysis for the case of Gaussian white noise and Fourier frequencies is very difficult.

Instead of approaching the problem of inference for the number r of sinusoids from the testing point of view, it is possible to derive an estimator of r using information theoretic ideas which have provided estimation techniques for system order for other statistical and, in particular, time series models. This is done in Section 3.6. The idea is to balance, in an automatic way, the *decrease* in the residual sum of squares as a function of the number of sinusoids, with the *increase* in complexity of the model. Similar ideas have been applied successfully in the estimation of the orders of ARMA models.

The periodogram is used in sonar signal processing, speech processing and many other areas to detect periodicities via the persistence through *time* of 'energy' near a particular frequency. The peaks in periodograms of white noise from consecutive blocks of time-samples will occur uniformly in frequency, while even if buried in noise, a genuine frequency will persist in time and be visible in an image display such as a spectrogram (also known as lofargram or sonogram). Thus, if we increase the standard deviation of the noise used to generate the periodogram in Figure 1.10 from 2 to 5, we obtain the one displayed in Figure 1.13, whose periodicity is not as pronounced. In fact, in other realisations with these values, the peak near the true frequency disappears altogether.

However, Figure 1.14 demonstrates that the periodicity will be detected by the spectrogram, an intensity plot of consecutive periodograms. Finally, as expected, the spectrogram of pure noise shown in Figure 1.15 would not lead anyone to suspect that there was a periodicity.

1.4 Frequency estimation and tracking

This book is primarily concerned with the estimation of the frequencies $\lambda_1, \ldots, \lambda_r$. It is not, therefore, a book on spectral analysis or the spectral theory of stationary processes. There is often confusion between the two. Many techniques have been derived for the estimation of the spectral density

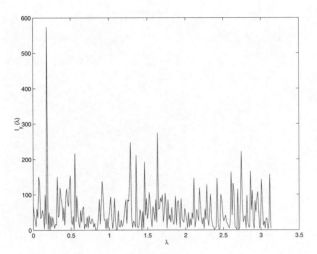

Fig. 1.13. Periodogram of sine wave + noise, frequency 0.1841, std. dev 5.0

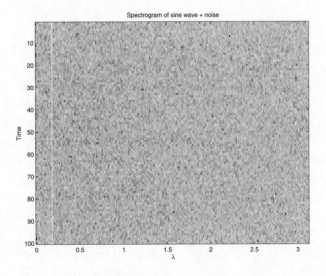

Fig. 1.14. Spectrogram of sine wave + noise

function. Often, however, the first application is to the case of a sinusoid + noise, or several sinusoids + noise (see Kay and Marple 1981 and Marple 1987 for examples of this, and for an excellent engineering perspective on these and related problems). We have previously mentioned two estimation methods: the maximisation of the periodogram and the maximisation of the residual sum of squares. The statistical theory for these estimation

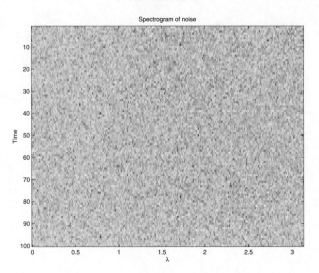

Fig. 1.15. Spectrogram of noise

techniques will depend heavily on the assumptions made concerning the
noise process $\{x(t)\}$. If it were, for example, an independent process with
distribution known apart from a finite number of constants, then standard
statistical theory could provide information concerning the lowest variances
possible for unbiased estimators of the frequencies, the so-called Cramér–
Rao (lower) bounds. However, even when these assumptions are not made,
there is much which can be said about the statistical properties of the estimators, especially in an *asymptotic* or large sample sense. In fact, even
under the simplest assumptions possible – white Gaussianity – neither the
exact variance nor the exact distribution of any estimator of frequency has
been calculated. We shall thus restrict our consideration of statistical properties to asymptotics. Chapter 2 lays the theoretical foundations for the
rest of the book. Discussed there are stochastic models for $\{x(t)\}$, maximum likelihood and the Cramér–Rao bound, and the limit theorems which
will be needed later on. Throughout the text we shall use the least restrictive
assumptions needed to obtain decent behaviour of the estimators. Thus, for
example, for most of the techniques we shall assume that the noise process
is stationary and ergodic, together with a few extra inoffensive conditions.
For others, we shall also have to assume that the noise is white. Often we
shall *derive* techniques using white and/or Gaussian assumptions, but will
be able to show that the resulting technique has very good properties even
when these assumptions are false. Thus, for example, the periodogram max-

imiser is equivalent to the maximum likelihood estimator of frequency under white Gaussian assumptions and therefore attains the Cramér–Rao bound asymptotically. However, the bound is *also* attained by the periodogram maximiser when the noise is coloured and Gaussian, and has excellent properties even when the assumption of Gaussianity is dropped.

In Chapter 3 we examine in detail the statistical theory for the (Gaussian) maximum likelihood estimators of the frequencies. In particular, we look at the cases where there are two frequencies very close together and where a single frequency is close to 0. We also examine the problem of estimating the number of frequencies.

The maximum likelihood estimators and periodogram maximisers are extremely difficult to calculate numerically, mainly since the periodogram has many local extrema. Thus the search has been on for equivalent but computationally efficient estimation techniques. Autoregressive models have been used for many years, and certainly since Yule (1927) modelled the pseudo-periodic Wolfer sunspot data using a second order autoregression. Indeed, if the zeros of the auxiliary polynomial for an autoregression of order 2 are chosen to form a complex conjugate pair, and have moduli close to the unit circle, then the spectral density function of the process will be peaked near the common absolute argument of the complex zeros. It is hardly surprising then that there has been considerable interest in estimating frequency using ARMA estimation techniques. Chan, Lavoie and Plant (1981) have suggested a closed-form technique based on the method of instrumental variables, but interest has focused on adaptive versions of techniques derived from 'comb' or 'notch' filters. The common idea behind the latter techniques is the construction of sequences of Butterworth filters, with poles and bandwidth parameters adjusted iteratively (see Nehorai and Porat 1986 and Fernandes, Goodwin and de Souza 1987, for example). It is very difficult to obtain the statistical properties of such techniques, partly because there are usually tuning parameters involved, but mainly because the poles of the filters are kept away from the unit circle. In Chapter 4 we examine the Quinn and Fernandes (1991) technique which from the outset places the poles on the unit circle, and iteratively achieves an estimator whose asymptotic properties are the same as those of the periodogram maximiser. The technique is similar, and asymptotically equivalent, to that of Truong-Van (1990), who used the fact that the solution to the difference equation

$$x_t - 2\cos\lambda x_{t-1} + x_{t-2} = \cos(\lambda t)$$

involves the terms $t\cos(\lambda t)$ and $t\sin(\lambda t)$ as well as $\cos(\lambda t)$ and $\sin(\lambda t)$ and thus a filter of the type $\left(1 - 2z\cos\lambda' + z^2\right)^{-1}$ applied to a sinusoid makes it 'ring' when the frequency λ' is near the true frequency. In fact, both techniques are designed to find the fixed point of the same function.

In Chapter 5, we look at a number of popular frequency estimation techniques. The autoregressive, Pisarenko and MUSIC techniques all use sample autocovariances, while Kay's two techniques are constructed from the raw (complex) time series. The autoregressive and Kay's first estimator are asymptotically biased (inconsistent), while the others, although consistent when the noise is white, have asymptotic variances of order T^{-1}, and are thus not statistically efficient, since the Cramér–Rao bound is of order T^{-3}. They are nevertheless interesting and the ideas may be used in conjunction with other techniques presented in this book. In particular, Kay's first method may be used with the Fourier coefficients at a single frequency from successive time blocks to yield a fast and accurate estimator (see Quinn 1992).

In many areas of application, and particularly in real-time signal processing, the data which are often most accessible are the Fourier coefficients – the Fourier transform of short lengths of data calculated at the Fourier frequencies $\{2\pi j/T;\ j = 0, 1, \ldots, T-1\}$. While these can be transformed back to the original data, it is fairly obvious that a small number of these coefficients near the true frequency will contain most of the statistical information relevant to that frequency. It is thus interesting to examine the estimation of the unknown parameters in a sinusoid when all that is available is a small number of Fourier coefficients. The PIE (phase interpolation estimation) technique of McMahon and Barrett (1986) uses the relationship between the Fourier coefficients at the same frequency in successive time periods, and, particularly, the difference between their arguments. A similar idea is used in the method of secondary analysis (see Brunt 1917, or Priestley 1981 for a more modern description). The technique is related to the complex demodulation technique of Tukey (1961), in which a signal is multiplied by a complex sinusoid at a frequency of interest, and filtered by a low-pass filter, leaving a very low frequency noisy complex sinusoidal signal, from which the frequency may be estimated by linear regression on the filtered signal's argument.

In Chapter 6 we consider the general problem, and obtain a closed form estimator of frequency for a single time series using only three Fourier coefficients at successive frequencies. Other estimators based on the use of Fourier coefficients alone have been considered in the past. Bartlett (1967) constructed an estimator which used a number of Fourier coefficients. Rife and Vincent (1970) constructed an estimator using only the moduli of three Fourier coefficients. Their technique is shown here to have deficiencies.

MacLeod (1991, 1998) has developed alternative techniques, the first of which is asymptotically equivalent to the FTI (Fourier transform interpolator) technique described by Algorithm 4. The second technique shown here, described in Algorithm 5, is asymptotically superior. Finally, we describe the asymptotic behaviour of techniques which use more than one time-block and/or more than three Fourier coefficients. Such techniques have also been considered by Quinn (1992) and Daley (2000).

Although locally in time a signal may be approximated as a sinusoid with fixed frequency, it may be that a more plausible model for the deterministic component is

$$y(t) = A \cos \{\Lambda(t)\}$$

where the derivative of $\Lambda(t)$, $\Lambda'(t)$, varies slowly in time. For example, it may be that the signal is a constant pitched tone emanating from an object which is moving at constant velocity relative to a fixed receiver. The function $\Lambda(t)$, which would be linear in t if the object were motionless, is in this case not linear, and, in fact, $\Lambda'(t)$ steadily decreases from a maximum to a minimum value, both determined by the relative speed, the closest distance between object and receiver, the time at which this occurs, and the 'rest' frequency. This is the well-known Doppler effect. In this particular case, the dynamics may be used to model the 'total phase' $\Lambda(t)$ and the 'instantaneous frequency' $\Lambda'(t)$. The received signal can be broken into small pieces, the frequency and amplitude estimated from each piece and then used to estimate the unknown speed, distance, time and frequency. The amplitude is, of course, not fixed throughout, but cannot be modelled accurately. As the problem is physically very simple, we shall present a derivation here for the case of a moving source and stationary receiver. The case of a moving receiver and stationary source is even simpler.

Suppose that at time τ, measured in seconds, the (unsigned) amplitude, i.e. the acoustic signal emanating from the source, is, say, $\rho \cos(2\pi f \tau + \phi)$, where f is the frequency in Hz of the sinusoid. If the distance between the source and receiver at time τ is r_τ, and the speed of sound, c, is constant in the medium, then the signal is received r_τ/c seconds later. The amplitude of the signal is, however, not constant, but is some function, ρ, of τ. There are various mathematical models for the amplitude function. For example, cylindrical and spherical spreading wave propagation models have the received amplitude inversely proportional to r_τ and $r_\tau^{1/2}$ respectively. We shall not make any such assumptions, but shall use short-term estimates of amplitude instead. The received noiseless signal at time $\tau + r_\tau/c$ will be

$$y\left(\tau + r_\tau/c\right) = \rho\left(\tau\right)\cos\left(2\pi f\tau + \phi\right).$$

Suppose now that the source is moving in a constant direction at constant speed v, and that it reaches its closest distance R from the receiver at time t_c. Then

$$r_\tau = \left\{R^2 + v^2\left(\tau - t_c\right)^2\right\}^{1/2}.$$

Put $t = \tau + r_\tau/c$. We need to describe τ as a function of t in order to express $y\left(t\right)$ in its own time-frame. Now

$$\begin{aligned} c^2\left(t - \tau\right)^2 &= r_\tau^2 \\ &= R^2 + v^2\left(t_c - \tau\right)^2 \\ &= R^2 + v^2\left\{\left(t_c - t\right)^2 + 2\left(t_c - t\right)\left(t - \tau\right) + \left(t - \tau\right)^2\right\}. \end{aligned}$$

Thus

$$\left(c^2 - v^2\right)\left(\tau - t\right)^2 - 2v^2\left(t - t_c\right)\left(\tau - t\right) - \left\{R^2 + v^2\left(t - t_c\right)^2\right\} = 0$$

and

$$\tau - t = \frac{v^2\left(t - t_c\right) \pm \sqrt{v^4\left(t - t_c\right)^2 + \left(c^2 - v^2\right)\left\{R^2 + v^2\left(t - t_c\right)^2\right\}}}{c^2 - v^2}.$$

Since we must have $t > \tau$, we take the negative sign in the above. Hence

$$\begin{aligned} \tau\left(t\right) &= \frac{c^2 t - v^2 t_c - \sqrt{v^4\left(t - t_c\right)^2 + \left(c^2 - v^2\right)\left\{R^2 + v^2\left(t - t_c\right)^2\right\}}}{c^2 - v^2} \\ &= \frac{c^2 t - v^2 t_c - \sqrt{R^2\left(c^2 - v^2\right) + v^2 c^2\left(t - t_c\right)^2}}{c^2 - v^2} \end{aligned}$$

and

$$y\left(t\right) = \rho\left(\tau\left(t\right)\right)\cos\left\{2\pi f\tau\left(t\right) + \phi\right\}.$$

Now, for fixed t and small δ,

$$\begin{aligned} & y\left(t + \delta\right) \\ & \sim \ \rho\left(\tau\left(t\right) + \tau'\left(t\right)\delta\right)\cos\left\{2\pi f\tau'\left(t\right)\delta + 2\pi f\tau\left(t\right) + \phi\right\} \\ & \sim \ \left\{\rho\left(\tau\left(t\right)\right) + \rho'\left(\tau'\left(t\right)\delta\right)\tau'\left(t\right)\delta\right\}\cos\left\{2\pi f\tau'\left(t\right)\delta + 2\pi f\tau\left(t\right) + \phi\right\}. \end{aligned}$$

Thus, locally in time, assuming the amplitude variation to be very small, the signal appears to be sinusoidal with frequency

$$2\pi f \tau'(t) = \frac{2\pi f c^2}{c^2 - v^2} \left\{ 1 - \frac{v^2 (t - t_c)}{\sqrt{R^2 (c^2 - v^2) + v^2 c^2 (t - t_c)^2}} \right\}. \qquad (1.4)$$

The ratio of the 'received' to 'transmitted' *instantaneous* frequencies at time t is thus

$$\tau'(t) = \frac{c^2}{c^2 - v^2} \left\{ 1 - \frac{v^2 (t - t_c)}{\sqrt{R^2 (c^2 - v^2) + v^2 c^2 (t - t_c)^2}} \right\},$$

which varies as t changes from $-\infty$ to ∞ between the asymptotes

$$\frac{c^2}{c^2 - v^2} \left(1 \pm \frac{v}{c} \right) = \frac{c}{c \mp v}.$$

The problem has been considered in Ferguson and Quinn (1994) and Quinn (1995), where estimation procedures for the unknown parameters f, R, v and t_c are described and evaluated. The signal collected at the receiver is sampled at a high sample frequency and divided into a number of subsets, each of which has a large number of samples. Over each subset, which is of short duration, the frequency and amplitude are estimated using any of the high accuracy techniques described in this book. Since these procedures all yield frequency estimators which have asymptotic variance inversely proportional to the square of the amplitude, a reasonable approach to estimating the unknown parameters is to use weighted least squares, with the true amplitudes replaced by their estimators. Assuming for simplicity that each subset is of the same size, and that the midpoints of the subsets are at times $t_1, t_2, \ldots$, we are thus led to minimise

$$\sum_i \widehat{\rho}_{t_i}^2 \left[\widehat{f}_{t_i} - \frac{f c^2}{c^2 - v^2} \left\{ 1 - \frac{v^2 (t_i - t_c)}{\sqrt{R^2 (c^2 - v^2) + v^2 c^2 (t_i - t_c)^2}} \right\} \right]^2$$

with respect to f, v, R and t_c. This can be simplified by putting

$$\alpha = \frac{f c^2}{c^2 - v^2}, \quad \beta = -\frac{f c v}{c^2 - v^2} \quad s = \frac{R \sqrt{c^2 - v^2}}{v c}$$

and minimising

$$\sum_i \left\{ \widehat{\rho}_{t_i} \widehat{f}_{t_i} - \alpha \widehat{\rho}_{t_i} - \beta \widehat{\rho}_{t_i} \frac{t_i - t_c}{\sqrt{s^2 + (t_i - t_c)^2}} \right\}^2.$$

The problem is now an ordinary nonlinear least squares problem, with two parameters, α and β entering linearly and two, t_c and s, entering nonlinearly. The only remaining problem is, therefore, one of finding good initial estimates for t_c and s.

The case of a moving receiver and stationary source is even more straightforward, for then the signal received at time t is just the signal at the source at time $t - r_t/c$. The instantaneous frequency at time t is thus

$$
\begin{aligned}
2\pi f \left(1 - c^{-1} \frac{dr_t}{dt} \right) &= 2\pi f \left\{ 1 - \frac{v^2 (t - t_c)}{c\sqrt{R^2 + v^2 (t - t_c)^2}} \right\} \\
&= 2\pi \left\{ \alpha + \beta \frac{t - t_c}{\sqrt{s^2 + (t - t_c)^2}} \right\}
\end{aligned}
$$

where now

$$
\alpha = f, \quad \beta = -\frac{fv}{c} \quad s = \frac{R}{v}.
$$

Thus the same algorithm for estimating the unknown four parameters can be used in this case. The reader is referred to Ferguson and Quinn (1994) and Quinn (1995) for further details. Figure 1.16 shows the results of a single simulation, where the second FTI technique and the Quinn and Fernandes techniques were both used to estimate frequency and phase. As the highest SNR throughout the simulation was -15 db, the frequency and amplitude estimates are fairly good, indeed more than adequate to estimate the 'true' dynamical parameters accurately.

More often, however, we do not know the dynamics of the change in frequency. Hannan and Huang (1993) have developed an adaptive version of the Quinn and Fernandes algorithm and analysed its performance when the frequency is fixed. The technique must be re-initialised regularly, although the decisions about when to do this may be automated. The Kalman filter is often successful in adapting to change in system parameters. Anderson and Moore (1979) have developed an *extended* Kalman filter for tracking frequency, since the ordinary Kalman filter cannot be applied to the nonlinear problem of frequency estimation. Their EKF is constructed using the raw data, and like all tracking techniques will break down if the SNR is too low. More recently, La Scala, Bitmead and Quinn (1996) have developed an EKF which is suitable at lower SNRs and uses several Fourier coefficients instead of the raw data. The tracker has also been designed to be more robust than usual, at the expense of a slight drop in accuracy.

The techniques which have recently shown the most promise at low SNRs

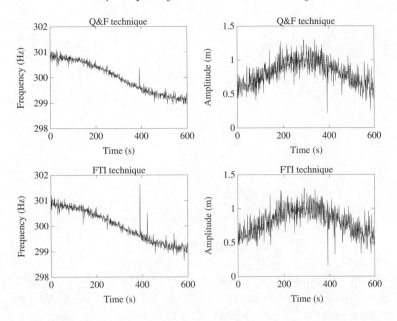

Fig. 1.16. Q&F and FTI frequency and amplitude estimates for a moving acoustic source.

have been based on Hidden Markov Models (HMMs). Streit and Barrett (1990) developed a tracker which used the moduli of the Fourier coefficients as data, and assumed that the frequencies formed an unobservable realisation of a discrete state space Markov chain with states coinciding with the Fourier frequencies. The tracker, although of low resolution, tracks well at much lower SNR than might be expected. Barrett and Holdsworth (1993) and Quinn, Barrett and Searle (1994) refined the procedure. Chapter 7 contains a complete development of the methodology needed for the implementation of HMM techniques, together with a frequency tracking implementation which uses a number of Fourier coefficients at different frequencies and in different time blocks, and where the frequencies, although discrete, are not restricted to coincide with Fourier frequencies.

2

***

Statistical and Probabilistic Methods

2.1 Introduction

We introduce in this chapter those statistical and probability techniques that underlie what is presented later. Few proofs will be given because a complete treatment of even a small part of what is dealt with here would require a book in itself. We do not intend to bother the reader with too formal a presentation. We shall be concerned with a sample space, Ω, which can be thought of as the set of all conceivable realisations of the random processes with which we are concerned. If A is a subset of Ω then $P(A)$ is the probability that the realisation is in A. *Because we deal with discrete time series almost exclusively, questions of 'measurability', i.e. to which sets A can $P(\cdot)$ be applied, do not arise and will never be mentioned.* We say this once and for all so that the text will not be filled with requirements that this or that set be measurable or that this or that function be a measurable function. Of course we shall see only (part of) one realisation, $\{x(t), t = 0, \pm 1, \pm 2, \ldots\}$ and are calling into being in our mind's eye, so to say, a whole family of such realisations. Thus we might write $x(t; \omega)$ where $\omega \in \Omega$ is the point corresponding to a particular realisation and, as ω varies for given t, we get a random variable, i.e. function defined on the sample space Ω. *However, usually and somewhat loosely, but we hope without confusion, we shall write $x(t)$ both for the actual value observed at time t and for the random variable which could have assumed any one of, say, a continuum of values.* Some of the formalism we wish to avoid can be eliminated by the use of the notion of 'expectation'. Thus, for example, we write $E\{x^2(t)\}$ for the expectation of $x^2(t)$ and not $\int x^2(t; \omega)\, dP(\omega)$. This expectation is the average value that $x^2(t; \omega)$ will take over all realisations. The estimation of such a quantity from the data will be discussed later, for example when ergodicity is treated.

The reader may ask why the subject cannot be treated without introducing Ω, referring only to the sequence $\{x(t)\}$ itself, imagined as prolonged for indefinitely large values of t, perhaps for sequences with some special properties. Gardner (1988) takes this viewpoint, for material closely connected with that of this book. However it would be impossible to develop much, indeed most, of what we seek to here in that way and any attempt to do so would lead us back to the beginning. An example is the central limit theorem, which will be important later. In any case, a connection with this viewpoint is obtained through the ergodic theorem discussed in Section 2.2.

We shall speak of the sequence of random variables,

$$\{x(t)\} = \{x(t;\omega) ; t = 0, \pm 1, \pm 2, \ldots\}$$

as a stochastic (i.e. probabilistic) process. We can, of course, consider some subset, say $\{x(0), x(1), \ldots, x(T-1)\}$, and can derive the 'joint probability distribution' of that subset, which prescribes the probability that this set of T random variables will take on values in a set in T-dimensional space. This probability distribution can be defined by a function

$$F(x_0, x_1, \ldots, x_{T-1}) = P\{x(0) \le x_0, x(1) \le x_1, \ldots, x(T-1) \le x_{T-1}\}.$$

Of course we might further derive some probability distribution, $F(\cdot)$, of some scalar function of $x(0), \ldots, x(T-1)$. Often when that is done our conditions will ensure that this distribution is absolutely continuous so that we can equivalently consider its derivative, $f(x) = dF(x)/dx$. One might ask why, if only $x(0), \ldots, x(T-1)$ are observed, we do not commence from the distribution of these random variables. There are many reasons: for example, the understanding that at future time points $x(t)$ will be observable; the fact that only special functions

$$F(x_0, \ldots, x_{T-1})$$

could in this way be considered whereas as T increases the dependence on such speciality will disappear. Moreover, the complexity of many of the functions of $x(0), \ldots, x(T-1)$ to be considered makes it almost impossible to derive $F(x)$, whereas as T increases, the limiting theory becomes tractable.

2.2 Stationary processes, ergodicity and convergence concepts

We say that $\{x(t)\}$ is *stationary* if for each n, the distributions of

$$\left[\begin{array}{ccc} x(0) & \cdots & x(n-1) \end{array}\right]'$$

and

$$\left[\ x\left(t\right)\quad \cdots\quad x\left(t+n-1\right)\ \right]'$$

are the same for all t. Thus the stochastic mechanism generating the $x\left(t\right)$ is not changing through time. This will imply that

$$E\left\{x\left(s\right)x\left(s+t\right)\right\}=E\left\{x\left(0\right)x\left(t\right)\right\},$$

which depends only on t. Also

$$E\left\{x\left(t\right)\right\}=E\left\{x\left(0\right)\right\}=\mu,$$

say. We put

$$\gamma\left(t\right)=E\left\{x\left(s\right)x\left(s+t\right)\right\}-\mu^2$$

for the covariance function so that $\gamma\left(0\right)$ is the variance of $x\left(t\right)$. *We have implicitly assumed that this is finite and shall always do that.* We return to $\gamma\left(t\right)$ later but first discuss ergodicity.

We may construct new stationary processes from $\left\{x\left(t\right)\right\}$ by taking some function of the realisation, say

$$f\left(\ldots,x\left(-2\right),x\left(-1\right),x\left(0\right),x\left(1\right),\ldots\right).$$

Call this $z\left(0\right)$. Then $z\left(t\right)$ is obtained by translating the sequence, that is the argument of $f\left(\cdot\right)$, through t time periods. For example, we could take $z\left(0\right)$ to be $x\left(0\right)x\left(s\right)$ or $x\left(0\right)x\left(s\right)-\mu^2$ so that $z\left(t\right)$ is $x\left(t\right)x\left(t+s\right)$ or $x\left(t\right)x\left(t+s\right)-\mu^2$. We shall say that $z\left(t\right)$ is obtained from $x\left(t\right)$ by a filter. We do not mean here that this is necessarily a linear filter, and shall discuss this later. What is special about the concept is that the translation $x\left(0\right)\rightarrow x\left(t\right)$ commutes with forming $z\left(0\right)$, i.e. gives the same result as $z\left(0\right)\rightarrow z\left(t\right)$. The following result is known as the ergodic theorem.

Theorem 1 *If $\left\{z\left(t\right)\right\}$ is obtained from the stationary process $\left\{x\left(t\right)\right\}$ by a filter and $E\left|z\left(t\right)\right|<\infty$ then*

$$\lim_{T\to\infty}T^{-1}\sum_{t=0}^{T-1}z\left(t\right) \qquad (2.1)$$

exists, almost surely.

By the phrase 'exists, almost surely' we mean that the sequence

$$\left\{T^{-1}\sum_{t=0}^{T-1}z\left(t\right)\right\}$$

converges to a limit except possibly at a set of points, i.e. realisations, which has probability zero. *In future, unless emphasis is needed, we shall omit the qualification almost surely* (which is customarily abbreviated to a.s.) since for all intents and purposes this mode of convergence is just the one we are familiar with from classical analysis.

It is possible that the limit of the sequence (2.1) might depend on ω, i.e. on the particular realisation. *If that is not so, no matter what the filter, i.e. $f(\cdot)$ may be, then we say that $\{x(t)\}$ is ergodic.* It then must be true that this limit is $E\{z(t)\} \equiv E\{z(0)\}$. It is almost obvious that if, as will be so in the cases that interest us, we only ever see one realisation (indeed part of one) then ergodicity must be a costless assumption since we are merely treating as constants, and not as functions of ω (i.e. of the realisation), those quantities that are constant for this realisation, which is all we shall see. It is clear also that it is scientifically meaningless to query ergodicity in our situation since there is no way of telling whether the process is ergodic or not from one realisation. *Thus we always assume $\{x(t)\}$ to be ergodic.* The quantities whose constancy is being asserted, i.e. the limits in Theorem 1, are all invariant functions of the realisation $x(t;\omega)$, i.e. are not changed when the sequence $x(t;\omega)$ is replaced by $x(t+s;\omega)$ for any s. It is only these that are being required to be constant and not quantities such as $x(t)$ itself or, say,

$$\max_{t \le s \le t+n} |x(s)|,$$

which will certainly not be invariant under translation.

Consider some event

$$A = \left\{\omega : \left[\ x(0;\omega) \quad \cdots \quad x(n-1;\omega)\ \right]' \in B\right\}$$

where B is a set in n-dimensional space. Let $I_A(0;\omega)$ be the indicator function of A, which is unity when $\left[\ x(0;\omega) \quad \cdots \quad x(n-1;\omega)\ \right]' \in B$ and zero otherwise. Similarly we may form $I_A(t;\omega)$ from

$$\{x(t), \ldots, x(t+n-1)\}$$

and the same B. Then, using $I_A(t;\omega)$ as our $z(t)$, the mean value in (2.1) converges to $E\{I_A(0;\omega)\} = P(A)$. Thus from an ergodic stationary process we can 'eventually', i.e. given an indefinitely long period of observation, find all of the probability structure of $x(t)$. This links with the brief discussion in the penultimate paragraph of Section 2.1 for, subject to this single, costless, assumption of ergodicity, one can find all about the probability structure of our process from one, full, realisation. We have thus lost nothing by

introducing the notion of a stochastic process but have gained a structure
which enables a very subtle analysis and statistical treatment of the phenomenon. Of course stationarity is involved and there may be situations
where it might not be realistic to impose that condition and where some
weaker requirement such as the convergence of

$$T^{-1} \sum_{s=0}^{T-1} x(s)$$

and

$$T^{-1} \sum_{s=0}^{T-1} x(s)x(s+t)$$

might suffice. However, to repeat, for the developments of this work it is
clearly best to build the theory as we have done.

The mode of convergence with which we have dealt here, i.e. almost
sure convergence, is often referred to as *strong convergence*. A weaker form
is *convergence in probability*. Thus a sequence $\{x_n\}$ of random variables
converges in probability to a random variable x if, for all $\varepsilon > 0$ then

$$P\{|x_n - x| > \varepsilon\} \to 0. \qquad (2.2)$$

Strong convergence implies convergence in probability but not conversely.
We give an example of the latter phenomenon later. Convergence in probability is also implied by *convergence in mean square* i.e.

$$E\left\{(x_n - x)^2\right\} \to 0. \qquad (2.3)$$

Neither strong convergence nor mean square convergence implies the other.

An even weaker form of convergence is *convergence in distribution*, also
called *convergence in law*. This holds if for 'almost all' c

$$P\{x_n < c\} \to P\{x < c\}.$$

This is the same as saying that if $F_n(\cdot)$ is the distribution function of x_n and
$F(\cdot)$ that of x then $F_n(c) \to F(c)$ at all points of continuity of the latter
function. (The 'almost all' refers to this last qualification.) In connection
with convergence in distribution, the limit distribution that we shall mainly
be concerned with is, of course, the normal or Gaussian distribution which
has distribution function

$$\frac{1}{\sqrt{2\pi\sigma^2}} \int_{-\infty}^{x} e^{-\frac{(u-\mu)^2}{2\sigma^2}} \, du, \qquad (2.4)$$

where μ and σ^2 are the mean and variance of the random variable. Of more importance is the multivariate normal distribution for which the density function, corresponding to the integrand in (2.4), is

$$(2\pi)^{-n/2} \left(\det \Gamma\right)^{-1/2} e^{-1/2(x-\mu)'\Gamma^{-1}(x-\mu)},$$

where x is a vector of, say, n components. Here Γ is the covariance matrix of a random vector, x, with this distribution and μ is its mean i.e.

$$E\left\{(x-\mu)(x-\mu)'\right\} = \Gamma, \ E(x) = \mu.$$

All concepts of this section generalise fairly obviously to random vectors and sequences of such and we shall not explicitly deal with that. If distribution functions of a sequence of random vectors converge to a Gaussian distribution, we say that the *Central Limit Theorem* (CLT) holds.

There is a further kind of result to which we shall refer below, again a strong convergence result but of an especially sharp kind. This is the *law of the iterated logarithm* (LIL). Put $\bar{x}_T = T^{-1} \sum_{t=0}^{T-1} x(t)$. We know that this converges to μ from Theorem 1, but how fast does that happen? One way to describe this is to find a positive, deterministic (i.e. non-stochastic) sequence, $g(T)$ say, so that $|\bar{x}_T - \mu|/g(T)$ stays bounded. In fact under fairly general conditions it can be shown that, taking $x(t)$ as scalar for simplicity,

$$\limsup_{T \to \infty} |\bar{x}_T - \mu|/g(T) = 1, \quad \text{a.s.} \tag{2.5}$$

for

$$g(T) = \{2V(T) \log \log T\}^{1/2}, \tag{2.6}$$

where $V(T)$ is the variance of $\bar{x}_T$. In the standard situation $TV(T)$ will converge to a limit V_0 as $T \to \infty$ so we may replace $g(T)$ by

$$\{2V_0 \left(\log \log T\right)/T\}^{1/2}.$$

We shall indicate the value of V_0 for a stationary process in the concluding part of this section.

Since $T^{1/2}(\bar{x}_T - \mu)$ does not converge to zero, even in probability, we see that $g(T)$ must decrease to zero more slowly than $T^{-1/2}$. Of course it is $T^{1/2}(\bar{x}_T - \mu)$ that will obey the CLT. The adjunction of the factor $(\log \log T)^{1/2}$ just suffices to handle this. However $(\bar{x}_T - \mu)/g(T)$ with $g(T) = (2V_0 \log \log T/T)^{1/2}$ would converge in probability to zero. Thus the result (2.5) provides an example of a sequence converging in probability to zero but not almost surely. In fact $(\bar{x}_T - \mu)/g(T)$ would also converge in mean square to zero.

The LIL is not as important as the CLT, for the latter enables useful bounds to be put upon the likely deviation of $\overline{x}_T$ from μ whereas the $2\log\log T$ factor tends to be too large for useful results from the LIL. There is also the question of the speed with which the LIL becomes relevant compared with the CLT. The LIL is nevertheless of some consequence in giving an idea of the speed of strong convergence. We do not wish to imply that it is only for the mean that LIL results hold and indeed shall later give a result for a frequency estimator, which is very nonlinear. We have not indicated conditions under which the CLT or LIL hold but they certainly do hold if the stationary process $\{x(t)\}$ has finite variance and is a process of independent random variables. However the scope of these theorems is very much wider than that, as we indicate in Section 2.4. First however we deal with the spectral theory of stationary processes.

2.3 The spectral theory for stationary processes

Our treatment here will necessarily be brief, for the subject is a large one. We need to give some definitions, in any case, and take the opportunity to give explanations at the same time. There are many books to consult, for example Brillinger (1974), Hannan (1970) and Priestley (1981).

The autocovariance function $\gamma(t)$ of a stationary stochastic process with finite variance is a positive function in the sense that the matrix Γ_n, that has entries $\gamma(j-k), j,k=1,\ldots,n$, in row j column k, is non-negative definite for any n. Because of this,

$$\gamma(t) = \int_{-\pi}^{\pi} e^{it\lambda} dF(\lambda), \qquad (2.7)$$

where $F(\lambda)$ is a non-decreasing function. We shall be concerned only with the case where (2.7) can be written as

$$\gamma(t) = \int_{-\pi}^{\pi} e^{it\lambda} f(\lambda)\, d\lambda \qquad (2.8)$$

because we shall not be concerned with jumps in $F(\lambda)$ since these correspond to sinusoidal components in $x(t)$ (see below) and we shall model these, so that $x(t)$ will correspond to the remainder. Even then, when jumps are ruled out, $F(\lambda)$ can contain a 'purely singular' component, which prevents (2.7) being be replaced by (2.8). However, this component appears to have no physical meaning and so will be treated as nonexistent. The function $f(\lambda)$ is an even, positive function i.e.

$$f(\lambda) \geq 0, \; f(\lambda) = f(-\lambda).$$

The function $f(\lambda)$ is often referred to as the 'spectrum' of $x(t)$, though more properly the spectrum is the support of that function. It is also called the *power spectrum* or the *spectral density*. Of course λ has the interpretation of a frequency as will become apparent (see (2.10) for example). It is often referred to as an 'angular frequency' or frequency in radians per unit time, as distinct from $f = \lambda/(2\pi)$ which is called the frequency in oscillations or cycles per unit time. (We are taking the unit time as the interval between observations.) The term $\lambda/(2\pi)$ has a direct physical meaning and λ is used only to eliminate a factor 2π which otherwise would keep occurring in formulae such as

$$\gamma(t) = 2\pi \int_{-1/2}^{1/2} e^{i2\pi tx} f(2\pi x)\, dx.$$

The connection between $\gamma(t)$ and $f(\lambda)$ can be exhibited in another way. Introduce the unitary matrix U_T with entry $T^{-1/2} e^{-ik\omega_j}$ in row j column $k, 0 \leq j, k \leq T-1$, where $\omega_j = 2\pi j/T$ and $-\pi < \omega_j \leq \pi$. This is unitary because it is easily seen that

$$T^{-1} \sum_{k=0}^{T-1} e^{-ik(\omega_u - \omega_v)} = \delta_{u,v},$$

where $\delta_{u,v}$ is zero for $u \neq v$ and unity for $u = v$. Assuming that $f(\lambda)$ is continuous, then, putting U_T^* for the complex conjugate transpose of U_T, we have

$$U_T \Gamma_T U_T^* = \text{diag}\{2\pi f(\omega_j)\} + o(1), \tag{2.9}$$

where by this we mean that the first matrix on the right side is diagonal with jth diagonal entry $2\pi f(\omega_j)$ and the $o(1)$ term stands for a $T \times T$ matrix whose entries converge, uniformly over the matrix, to zero as $T \to \infty$. If stronger conditions are imposed on $f(\lambda)$, e.g. Lipshitz conditions or differentiability, then the second term on the right in (2.9) will decrease faster. In any case (2.9) shows the connection between $f(\lambda)$ and the eigenvalues of Γ_T and also the fact that Γ_T can almost be diagonalised by a *known* unitary transformation.

The other thing that needs to be done is to connect (2.8) with $x(t)$ itself.

This is done via the theorem that provides the decomposition

$$x\left(t\right) = \int_{-\pi}^{\pi} e^{it\lambda} d\zeta\left(\lambda\right), \tag{2.10}$$

where $\zeta\left(\lambda\right)$ is a complex-valued random (stochastic) function on $[-\pi, \pi]$ with the property that

$$\zeta\left(\lambda_1\right) - \zeta\left(\lambda_2\right) = \overline{\zeta\left(-\lambda_2\right) - \zeta\left(-\lambda_1\right)}, \quad \lambda_2 \leq \lambda_1,$$

$$E\left[\left\{\zeta\left(\lambda_1\right) - \zeta\left(\lambda_2\right)\right\} \overline{\zeta\left(\lambda_3\right) - \zeta\left(\lambda_4\right)}\right] = 0, \quad \lambda_4 \leq \lambda_3 < \lambda_2 \leq \lambda_1$$

and

$$E\left|\zeta\left(\lambda_1\right) - \zeta\left(\lambda_2\right)\right|^2 = \int_{\lambda_2}^{\lambda_1} f\left(\lambda\right) d\lambda, \quad \lambda_2 \leq \lambda_1.$$

Alternatively we can rewrite (2.10) as

$$x\left(t\right) = \int_0^{\pi} \left\{\cos\left(t\lambda\right) d\xi\left(\lambda\right) + \sin\left(t\lambda\right) d\eta\left(\lambda\right)\right\} \tag{2.11}$$

where

$$\zeta\left(\lambda\right) = \left\{\xi\left(\lambda\right) - i\eta\left(\lambda\right)\right\}/2, \quad 0 < \lambda < \pi; \quad \zeta\left(0\right) = \xi\left(0\right), \quad \zeta\left(\pi\right) = \xi\left(\pi\right).$$

Thus $x\left(t\right)$ is being represented as a continuous linear superposition of sinusoidal oscillations of all angular frequencies from 0 to π, with amplitude and phase determined by the differentials $d\xi\left(\lambda\right)$ and $d\eta\left(\lambda\right)$. The relations below (2.10) can be conveniently summarised as

$$E\left\{d\zeta\left(\lambda_1\right) \overline{d\zeta\left(\lambda_2\right)}\right\} = \delta_{\lambda_1, \lambda_2} f\left(\lambda_1\right) d\lambda_1.$$

Correspondingly, $\xi\left(\lambda\right)$ and $\eta\left(\lambda\right)$ satisfy

$$E\left\{d\xi\left(\lambda_1\right) d\xi\left(\lambda_2\right)\right\} = E\left\{d\eta\left(\lambda_1\right) d\eta\left(\lambda_2\right)\right\} = 0, \quad \lambda_1 \neq \lambda_2,$$

$$E\left\{d\xi\left(\lambda_1\right) d\eta\left(\lambda_2\right)\right\} = 0, \quad \text{for all } \lambda_1, \lambda_2$$

and

$$E\left\{d\xi\left(\lambda\right)\right\}^2 = E\left\{d\eta\left(\lambda\right)\right\}^2 = \left\{ \begin{array}{ll} 1/2 f\left(\lambda\right) d\lambda \quad ; & 0 < \lambda < \pi \\ f\left(\lambda\right) d\lambda \quad ; & \lambda = 0, \pi. \end{array} \right.$$

Thus the individual sinusoidal components make an infinitesimal contribution to the variance, namely

$$\gamma\left(0\right) = \int_{-\pi}^{\pi} f\left(\lambda\right) d\lambda,$$

and are orthogonal to each other. Returning to (2.9), and considering the vector $x_T = \begin{bmatrix} x(0) & x(1) & \cdots & x(T-1) \end{bmatrix}'$, we may form

$$U_T x_T = \begin{bmatrix} \cdots & T^{-1/2} \sum_{k=0}^{T-1} x(k) e^{-ik\omega_j} & \cdots \end{bmatrix}'$$

so that the quantities

$$w(\omega_j) = T^{-1/2} \sum_{k=0}^{T-1} x(k) e^{-ik\omega_j}$$

satisfy

$$E\left\{ w(\omega_j) \overline{w(\omega_k)} \right\} = \delta_{j,k} 2\pi f(\omega_j) + o(1).$$

In addition,

$$x(t) = T^{-1/2} \sum_{j=0}^{T-1} w(\omega_j) e^{it\omega_j}, \tag{2.12}$$

which mimics (2.10) with $T^{-1/2} w(\omega_j)$, the mean square of which is

$$(2\pi/T) f(\omega_j) + o\left(T^{-1}\right),$$

occupying the place of $d\zeta(\lambda)$. Thus these quantities $w(\omega_j)$ constitute the statistics (i.e. quantities computed from the data) that, in a sense, 'estimate' $\zeta(\lambda)$ and thus will necessarily play a substantial part in our theory.

We shall be using the $w(\omega_j)$ or twice their squared moduli,

$$I(\omega_j) = 2 |w(\omega_j)|^2,$$

when the data is a sequence $\{y(0), y(1), \ldots, y(T-1)\}$ with

$$y(t) = \mu + \sum_{k=1}^{r} A_k \cos(\lambda_k t + \phi_k) + x(t), \ 0 < \lambda_k < \pi \tag{2.13}$$

and $\{x(t)\}$ is as above. Then $I(\omega_j)$ will display very different behaviour near one or other of the λ_k. Indeed

$$w(\omega_j) = \sum_{k=1}^{r} T^{-1/2} \sum_{t=0}^{T-1} e^{-it\omega_j} A_k \cos(\lambda_k t + \phi_k) + T^{-1/2} \sum_{t=0}^{T-1} x(t) e^{-it\omega_j},$$
$$\tag{2.14}$$

the first term of which is

$$\sum_{k=1}^{r} \frac{A_k e^{i\phi_k}}{2} T^{-1/2} \sum_{t=0}^{T-1} e^{it(\lambda_k - \omega_j)} + \sum_{k=1}^{r} \frac{A_k e^{-i\phi_k}}{2} T^{-1/2} \sum_{t=0}^{T-1} e^{-it(\omega_j + \lambda_k)}. \tag{2.15}$$

For $\omega_j > 0, 0 < \lambda_k < \pi$, the second term is always $O\left(T^{-1/2}\right)$. The kth summand in the first term has squared modulus

$$\frac{A_k^2 \sin^2\left\{T\left(\omega_j - \lambda_k\right)/2\right\}}{4\ T\sin^2\left\{\left(\omega_j - \lambda_k\right)/2\right\}}, \tag{2.16}$$

which is $O\left(T^{-1}\right)$ unless ω_j approaches λ_k when it becomes $O\left(T\right)$. If $\omega_j = \lambda_k$, which will not usually be possible for fixed T, then the expression is $TA_k^2/4$. Thus $I\left(\omega_j\right)$ will, for T large, show pronounced peaks near the λ_k and the locations of these can be used to estimate these quantities. The last term in (2.14) (and the second term in (2.15)) have to be taken into account and the magnitude of that will be important. Its importance relative to the first term in (2.15) will depend, *inter alia*, on the SNR at the λ_k namely, according to a conventional definition,

$$SNR(\lambda_k) = 10\log_{10}\left\{\frac{A_k^2}{4\pi f\left(\lambda_k\right)}\right\}.$$

However what will also be relevant is

$$10\log_{10}\left[\frac{A_k^2}{\max_\lambda\left\{4\pi f\left(\lambda\right)\right\}}\right]$$

since this will indicate how easily a location of a peak in $f\left(\lambda\right)$ could be mistaken for a λ_k value. The full understanding of this requires an understanding of the statistical properties of $w\left(\omega\right)$ calculated from the $x\left(t\right)$ and that is one of the topics treated in the next chapter.

Two points can now be dealt with that were mentioned above. The first of these relates to the quantities $V\left(T\right)$ and V_0 in (2.6). We may easily represent $V\left(T\right)$ in terms of the spectrum, since

$$\begin{aligned}
E\left\{T^{-1}\sum_0^{T-1} x\left(t\right) - \mu\right\}^2 &= T^{-2}\sum_{s,t=0}^{T-1}\gamma\left(s-t\right) \\
&= T^{-2}\sum_{s,t=0}^{T-1}\int_{-\pi}^{\pi} e^{i(s-t)\lambda}f(\lambda)d\lambda \\
&= T^{-1}\int_{-\pi}^{\pi}\frac{\sin^2\left(T\lambda/2\right)}{T\sin^2\left(\lambda/2\right)}f(\lambda)d\lambda.
\end{aligned} \tag{2.17}$$

The quantity $L_T(\lambda) = \sin^2\left(T\lambda/2\right)/\left\{T\sin^2\left(\lambda/2\right)\right\}$, which occurs in (2.16), is called *Féjer's kernel*. It integrates to 2π and concentrates rapidly at $\lambda = 0$.

Thus, if $f(\lambda)$ is continuous at $\lambda = 0$, then (2.17) becomes $\frac{2\pi f(0)}{T}\{1 + o(1)\}$ and V_0, following (2.6), is $2\pi f(0)$.

We introduced the concept of a filter at the beginning of Section 2.2. A particular case is a linear filter, where $f(\cdot)$ is a linear function of the realisation. Thus $z(t)$ in Theorem 1 is then

$$z(t) = \sum_{j=-\infty}^{\infty} \alpha_j x(t-j)$$

$$= \int_{-\pi}^{\pi}\left(\sum_{j=-\infty}^{\infty} \alpha_j e^{-ij\lambda}\right) e^{it\lambda} d\zeta(\lambda) \qquad (2.18)$$

and $d\zeta(\lambda)$ in (2.10) is replaced by $\sum_{j=-\infty}^{\infty} \alpha_j e^{-ij\lambda} d\zeta(\lambda)$. The functions $\sum_{j=-\infty}^{\infty} \alpha_j e^{-ij\lambda}$ and $\sum_{j=-\infty}^{\infty} \alpha_j z^j$ are called the *transfer function* and the *z-transform* of the (linear) filter. We may put

$$\sum_{j=-\infty}^{\infty} \alpha_j e^{ij\lambda} = \rho(\lambda) e^{i\theta(\lambda)} \qquad (2.19)$$

where $\rho(\lambda) = \left|\sum_{j=-\infty}^{\infty} \alpha_j e^{ij\lambda}\right|$ is called the *gain* and $\theta(\lambda)$ the *phase* of the filter. Then $\{z(t)\}$ has spectral density

$$f_z(\lambda) = \left|\sum_{j=-\infty}^{\infty} \alpha_j e^{ij\lambda}\right|^2 f(\lambda). \qquad (2.20)$$

If, say, two processes $\{x_1(t)\}$ and $\{x_2(t)\}$ are simultaneously observable and are *jointly* stationary, then each can be represented in the form (2.10). The covariance properties of the two *spectral measures*, $\zeta_1(\lambda)$ and $\zeta_2(\lambda)$, are clearly important, since these encapsulate the covariance properties of $\{x_1(t)\}$ and $\{x_2(t)\}$. In fact

$$E\left\{d\zeta_1(\lambda_1)\overline{d\zeta_2(\lambda_2)}\right\} = 0, \quad \lambda_1 \neq \lambda_2.$$

Put

$$E\left\{d\zeta_1(\lambda)\overline{d\zeta_2(\lambda)}\right\} = f_{12}(\lambda)d\lambda.$$

Then

$$\gamma_{12}(t) \overset{\triangle}{=} E\{x_1(s)x_2(s+t)\} = \int_{-\pi}^{\pi} e^{it\lambda} f_{12}(\lambda)d\lambda.$$

We call $f_{12}(\lambda)$ the cross spectral density, or cross spectrum, and its argument, $\theta_{12}(\lambda)$, the relative phase. Of course if $x_2(t)$ is $z(t)$, as in (2.18), then

$\theta_{12}(\lambda)$ is just $\theta(\lambda)$ from (2.19). The quantity

$$\rho(\lambda) = \frac{|f_{12}(\lambda)|}{\sqrt{f_{11}(\lambda) f_{22}(\lambda)}}, \qquad (2.21)$$

where the two spectral densities are denoted by $f_{11}(\lambda)$ and $f_{22}(\lambda)$, is called the *coherence*. This is unity when $x_2(t)$ is $z(t)$ as in (2.18). Thus a linear filter affects only the relative phasing of the series. If $x_2(t) = x_1(t-l)$ then, from (2.19), $\theta(\lambda) = \theta_{12}(\lambda) = l\lambda$. The phase and coherence can be given the following meaning. Imagine filtering $\{x_1(t)\}$ and $\{x_2(t)\}$ by a very narrow band pass filter, passing a narrow band of frequencies around λ and eliminating other frequencies. Now we might seek to lag one of the filtered series relative to the other so as to bring them into closest agreement as measured by the correlation between the two. In the limit, as the pass bands become indefinitely narrow, this correlation becomes the coherence and the rephasing needed is $\theta(\lambda)/\lambda$.

A major topic in time series analysis is prediction but we shall not deal with it here except to define and state a result for the variance of the linear predictor. By this predictor we mean that quantity, $x(t; t-1)$ let us say, which is both a linear function of the history of the process to time $t-1$, and which achieves the minimimum mean square prediction error, i.e.

$$\sigma^2 = E\{x(t) - x(t; t-1)\}^2$$

over these linear functions. For a Gaussian process this is also the best predictor, in the least squares sense, whether linear or nonlinear, but that is not so in general. There is a remarkable formula, due to G. Szegö, that connects σ^2 with $f(\lambda)$; namely,

$$\log \sigma^2 = (2\pi)^{-1} \int_{-\pi}^{\pi} \log\{2\pi f(\lambda)\} d\lambda. \qquad (2.22)$$

If $\log f(\lambda)$ is not integrable that can only be because the integral diverges to $-\infty$ and then the formula remains true with $\sigma^2 = 0$.

It may be, and usually is so, that the underlying phenomenon is observable, in principle, continuously through time, though in fact $\{x(t)\}$ is observed at equal intervals of time and this common time interval has been chosen as the time unit. Then generalisations of (2.8) and (2.10) hold, namely,

$$\gamma(t) = \int_{-\infty}^{\infty} e^{it\lambda} f_c(\lambda) d\lambda, \quad x(t) = \int_{-\infty}^{\infty} e^{it\lambda} d\zeta_c(\lambda),$$

where now $f_c(\lambda)$ is the spectral density of the continuous time process and

$\zeta_c(\lambda)$ has the same properties as before but for $-\infty < \lambda < \infty$. The connection between f_c and f is via the sampling formula

$$f_c(\lambda) = \sum_{k=-\infty}^{\infty} f(\lambda + 2k\pi), \quad -\pi < \lambda \leq \pi \tag{2.23}$$

which is fairly easily obtained from (2.8) and (2.10) since for integral k, $e^{it(\lambda + 2k\pi)} = e^{it\lambda}$.

2.4 Maximum likelihood and the Cramér–Rao Theorem

We shall now call the data y (instead of x). This is because the applications later (see Chapter 3) will be in situations where we observe $y(t) = s(t) + x(t)$ and $s(t)$ is a deterministic 'signal' term, made up of sinusoids. Of course the probability law of $y(t)$ comes from that of $x(t)$ and $s(t)$ affects only the time dependent mean value of $y(t)$. However the data, y, need not have come from observations on a time series. We assume however that it has a continuous rather than discrete probability distribution, so that we can work in terms of a probability density function. Thus the distribution function F, which defines the probability that

$$y = \begin{bmatrix} y_1 & y_2 & \cdots & y_n \end{bmatrix}'$$

should lie in any orthant

$$\{y_1 \leq a_1, y_2 \leq a_2, \ldots, y_n \leq a_n\},$$

can be replaced by its derivative, i.e. $f = \frac{\partial F}{\partial y}$. We also assume that the distribution depends on a set, θ, of parameters. For example with $y(t) = s(t) + x(t)$ as above, this vector θ would specify the amplitudes, phases and frequencies of the components of $s(t)$ as well as any parameters needed to specify the probability law of $\{x(t)\}$. An example of the latter, for $\{x(t)\}$ scalar, would be to prescribe $\{x(t)\}$ as an autoregressive (AR) process so that

$$\sum_{j=0}^{h} \alpha_j x(t-j) = \varepsilon(t), \quad \alpha_0 = 1; \quad \sum_{j=0}^{h} \alpha_j z^j \neq 0, \quad |z| \leq 1 \tag{2.24}$$

where, say, the $\varepsilon(t)$ are independent and identically distributed normally, with means zero and variances σ^2. The last part of (2.24) ensures that σ^2 is, in fact, the linear prediction variance of $x(t)$, and that $\{x(t)\}$ is stationary. Then $\alpha_1, \alpha_2, \ldots, \alpha_h, \sigma^2$ would also occur in θ, and perhaps h might also since it would not be known. Thus part of θ might consist of integer parameters.

Another example would be the number, r say, of sinusoidal parameters in $s(t)$. We shall discuss that later but for the moment will deal only with the case where θ varies continuously. Considered as a function of y, $f(y|\theta)$ is, of course, a probability density function. However this may be also be considered as a function of the unknown θ given the random vector y, and is then known as the 'likelihood' and written as a different function, say, $L(\theta; y)$. The 'method of maximum likelihood', ML, chooses as estimator that value of θ which maximises $L(\theta; y)$. The maximiser $\widehat{\theta}$ is called the *maximum likelihood estimator*, or MLE.

Let us begin by considering the case where θ is a scalar quantity. We always assume, minimally, that $f(y|\theta)$, for *almost* all y, is differentiable in θ, so that we may form, where $l(\theta; y) = \log L(\theta; y)$,

$$\frac{dl(\theta; y)}{d\theta}. \tag{2.25}$$

Other conditions will be required later, but we shall not detail them here. We shall merely require that $f(y|\theta)$ be 'regular'.

The reason for the use of $l(\theta; y)$ can be seen from the importance of the following quantity in the Cramér–Rao (CR) bound in Theorem 2 below. Put

$$i(\theta) = E\left\{\frac{dl(\theta; y)}{d\theta}\right\}^2.$$

The quantity $\frac{dl(\theta; y)}{d\theta}$ is called the *score* or the *efficient score*. Its mean square, $i(\theta)$, is called the *information* about θ from y. Unless $L(\theta; y)$ achieves its maximimum at a boundary point of its domain, then $\frac{dl(\widehat{\theta}; y)}{d\theta} = 0$ at the MLE $\widehat{\theta}$ i.e. for our data y the efficient score is zero at the MLE.

If θ^* is an estimator of θ, i.e. a function of y used to estimate θ, then let

$$b(\theta) = E(\theta^*) - \theta \tag{2.26}$$

where again we are considering θ^* as a random variable i.e. a function of the random variable y. Then $b(\theta)$ is called the *bias* in θ^* as an estimator of θ.

Theorem 2 *In a regular estimation problem*

$$E\left\{(\theta^* - \theta)^2\right\} \geq \frac{\{1 + db(\theta)/d\theta\}^2}{i(\theta)}. \tag{2.27}$$

Equality holds if and only if $\theta^* - \theta$ is a linear function of the efficient score.

The proof is so simple that we include it.

Proof

$$E\left(\theta^*|\theta\right) = \theta + b\left(\theta\right) = \int \theta^* f\left(y|\theta\right) dy$$

Thus

$$
\begin{aligned}
1 + \frac{db\left(\theta\right)}{d\theta} &= \int \theta^* \frac{d\log f\left(y|\theta\right)}{d\theta} f\left(y|\theta\right) dy \\
&= \int \left(\theta^* - \theta\right) \frac{d\log f\left(y|\theta\right)}{d\theta} f\left(y|\theta\right) dy \\
&= E\left\{\left(\theta^* - \theta\right) \frac{dl\left(\theta; y\right)}{d\theta}\right\}
\end{aligned}
$$

since

$$\int f\left(y|\theta\right) dy = 1$$

and

$$\int \frac{d\log f\left(y|\theta\right)}{d\theta} f\left(y|\theta\right) dy = \frac{d}{d\theta} \int f\left(y|\theta\right) dy = 0.$$

Thus from Schwarz's inequality

$$\left\{1 + \frac{db\left(\theta\right)}{d\theta}\right\}^2 \le E\left\{\left(\theta^* - \theta\right)^2\right\} i(\theta)$$

and equality can hold, from the necessity part of that result, if and only if

$$\theta^* - \theta = k \frac{dl\left(\theta; y\right)}{d\theta}.$$

Here k is independent of θ^* but will, in general, depend on θ.   □

A more careful examination (see Cramér 1946, p. 480) shows that equality will hold when and only when θ^* *in addition*, is a *sufficient* statistic, i.e. when the conditional distribution of the data given θ^* is independent of θ. What is meant is that if the probability law of the assembly of all data sets that give the value θ^*, which would normally depend on θ^*, does not, then θ^* is said to be sufficient. The terminology comes from the fact that once we know θ^*, if it is sufficient, then the data can tell us nothing more about θ.

If $b(\theta) = 0$ then the right side of (2.27) is $\{i\left(\theta\right)\}^{-1}$ and it is this that is often referred to as the Cramér–Rao bound. This bound is the reciprocal of the information so that as the information goes up the bound becomes smaller, which makes the use of the term seem reasonable. If the lower bound is attained the method is said to be *efficient*. This is rarely so but we give an example of when it is. Let $y(1), \ldots, y(N)$ be normal and independent with mean μ and variance 1. Then $\theta = \mu$ and

$$-2l(\theta; y) = N \log(2\pi) + \sum_{j=1}^{N} \{y(j) - \mu\}^2.$$

The MLE is thus $\hat{\mu} = \bar{y} = N^{-1} \sum_{j=1}^{N} y(j)$. Also $\bar{y}$ is Gaussian with mean μ and variance N^{-1}. It is unbiased and the CR bound is found from

$$\frac{dl(\theta; y)}{d\theta} = -\sum_{j=1}^{N} y(j) + N\mu$$

so that $i(\theta)$ is the variance of $\sum_{j=1}^{N} y(j)$, which is just N times the variance of $y(1)$, or N. Thus the lower bound is attained and $\bar{y}$ is sufficient and efficient. The situation is not a realistic one and in fact $\bar{y}$ is often a bad statistic to use.

It is more often true that the lower bound is obtained asymptotically. Following Cox and Hinkley (1974, p. 304) consider $v(T)(\theta^* - \theta)$ for a suitable $v(T)$. Usually $v(T) = T^{1/2}$ but in connection with frequency estimation $v(T)$ will be $T^{3/2}$. Assume that $v(T)(\theta^* - \theta)$ has a distribution which converges as T increases to a Gaussian distribution with mean zero, and that $v^2(T)/i(\theta)$ converges to a quantity $\sigma^2(\theta)$. *If the variance in the limiting distribution for $v(T)(\theta^* - \theta)$ is $\sigma^2(\theta)$ we say that θ^* is asymptotically efficient.* One might expect to phrase such a definition in terms of $i(\theta) E\left\{(\theta^* - \theta)^2\right\}$ but that would be much harder to handle because often θ^* is such a nonlinear function of the data that it may be difficult even to prove that the expectation is finite, much less to evaluate it accurately.

If an efficient estimator exists it will be a sufficient statistic. The method of Maximum Likelihood will find an efficient estimator if there is one. It will also find an asymptotically efficient statistic.

It is difficult to prescribe a suitable stochastic process for generating the data and usually we are forced to use a Gaussian model in order to calculate a maximum likelihood estimator. It may then be possible to establish that $v(T)\left(\hat{\theta} - \theta\right)$ is asymptotically normal with mean zero and variance $\lim v^2(T)/i(\theta)$, with $i(\theta)$ obtained from the Gaussian theory, and indeed to prove this under more general conditions, without requiring Gaussianity. However it then becomes possible, under these more general conditions, that some more efficient procedure exists. It may also be true that, even under Gaussian assumptions, the MLE, $\hat{\theta}$, is difficult to compute and some approximation, θ^*, may be used. Of course then it is of interest to prove the

CLT for $v(T)(\theta^* - \theta)$ with variance $\lim v^2(T)/i(\theta)$. We shall illustrate all of this later.

We have so far dealt with a single scalar parameter θ but now wish to consider a vector of parameters, θ, with components θ_i, $i = 1, \ldots, m$. Now $i(\theta)$ is replaced by an $m \times m$ matrix, $I(\theta)$, with entries

$$I_{uv}(\theta) = E\left\{ \frac{\partial l(\theta; y)}{\partial \theta_u} \frac{\partial l(\theta; y)}{\partial \theta_v} \right\}. \tag{2.28}$$

In this connection notice that

$$I_{uv}(\theta) = -E\left\{ \frac{\partial^2 l(\theta; y)}{\partial \theta_u \partial \theta_v} \right\} \tag{2.29}$$

because

$$0 = \int \frac{\partial f(y|\theta)}{\partial \theta_u} dy = \int \frac{\partial \log f(y|\theta)}{\partial \theta_u} f(y|\theta)\, dy \tag{2.30}$$

so that

$$0 = \int \frac{\partial \log f(y|\theta)}{\partial \theta_u} \frac{\partial \log f(y|\theta)}{\partial \theta_v} f(y|\theta)\, dy + \int \frac{\partial^2 \log f(y|\theta)}{\partial \theta_u \partial \theta_v} f(y|\theta)\, dy$$

as can be seen by differentiating (2.30) under the integral sign with respect to θ_v. This establishes (2.29). It is sometimes easier to evaluate this than (2.28).

An individual component, θ_u, can be considered in its own right, but that is not what is usually required. Consider a scalar statistic $\alpha^*(y)$ with expectation $\alpha(\theta) + b(\theta)$, where $\alpha(\theta)$ is the number $\alpha^*(y)$ is designed to estimate. Put

$$a_u(\theta) = \frac{\partial}{\partial \theta_u} \{\alpha(\theta) + b(\theta)\},$$

and denote simply by a, the vector, $a(\theta)$, constructed from the $a_u(\theta)$.

Theorem 3 *With the above notation, for the regular situation,*

$$E\left\{ (\alpha^* - \alpha)^2 \right\} \geq a'\{I(\theta)\}^{-1} a.$$

In particular, let $\alpha_u(\theta) = \theta_u$ and assume that $b(\theta) = 0$. Then it follows that

$$E\left\{ (\theta_u^* - \theta_u)^2 \right\} \geq I^{uu}(\theta) \tag{2.31}$$

where $I^{uu}(\theta)$ is the element in row u, column u of $\{I(\theta)\}^{-1}$. It is always true that $I^{uu}(\theta) \geq \{I_{uu}(\theta)\}^{-1}$ so that (2.31) provides a better (sharper) bound. Indeed $\{I_{uu}(\theta)\}^{-1}$ is the residual mean square for the regression of the uth element of a set of random variables with covariance matrix $I\left(\theta\right)$ on the remainder, which is smaller than the variance $I_{uu}\left(\theta\right)$ of the uth element unless $I_{uv}\left(\theta\right) = 0$, $v \neq u$, when the two are the same.

2.5 Central limit theorem and law of the iterated logarithm

This section will be brief, for an extended discussion would take us much too far. The quantities with which we are most concerned below are of the form

$$w_x\left(\omega_j\right) = T^{-1/2} \sum_{t=0}^{T-1} x\left(t\right) e^{-it\omega_j}, \quad \omega_j = 2\pi j/T. \tag{2.32}$$

In fact results will also be needed for quantities of the form

$$T^{-(2k+1)/2} \sum_{t=0}^{T-1} t^k x\left(t\right) e^{-it\omega_j}, \tag{2.33}$$

at least for $k = 0, 1, 2$. The theorems may be formulated in terms of the real and imaginary parts of a set of quantities of the form (2.32) and (2.33).

Some conditions are required, needless to say. To begin with, we cannot establish a reasonable central limit theorem (CLT) for *all* of the quantities (2.32) since this (see (2.12)) would imply something close to a Gaussian distribution for $\{x\left(t\right)\}$ itself. (If $\{x\left(t\right)\}$ is Gaussian then so are all of the quantities (2.32) and (2.33), but here we wish to be more general.) Thus our CLT will relate to a fixed frequency, λ_0 say, and a fixed number, say m, of the ω_j nearest to λ. We shall require

$$f\left(\lambda_0\right) > 0, \tag{2.34}$$

where f is the spectral density of $\{x\left(t\right)\}$, since otherwise the distribution of (2.32) for $|\omega_j - \lambda_0| = O\left(T^{-1}\right)$ will become degenerate (see above (2.12)).

We further need some condition that relates to the decreasing relationship between $x\left(s\right)$ and $x\left(t\right)$ as $|s - t|$ increases. One very general way of formulating this is as follows. Consider the collection of all functions of $\{x\left(s\right); s \leq t\}$ which have finite mean square. Call this H_t. Let $H_{-\infty}$ be the intersection of all of these, i.e. the set of such functions determined by the infinitely far past. Let P_t project, perpendicularly, onto H_t so that $P_t f$ is the part of f that is a linear combination of the functions in H_t got, as it were, by regressing f on the elements of H_t. Then $P_{-\infty}$ projects onto $H_{-\infty}$. We require

$$P_{-\infty} x\left(t\right) = 0 \quad \text{a.s.} \tag{2.35}$$

so that $x(t)$ is not influenced by the infinitely far past. In particular it must have zero mean. Put $\xi(t, s) = P_s x(t) - P_{s-1} x(t)$. Then

$$x(t) = \sum_{j=0}^{\infty} \xi(t, t - j), \qquad (2.36)$$

and

$$\sum_{j=0}^{\infty} E\left\{\xi^2(t, t - j)\right\} < \infty.$$

However, a stronger condition is needed, namely,

$$\sum_{j=0}^{\infty} \left[E\left\{\xi^2(t, t - j)\right\}\right]^{1/2} < \infty. \qquad (2.37)$$

This is the condition, combined with (2.35), on the decay of dependence on the past that is used. None of the conditions so far seems onerous. Something extra is needed (see Hannan 1973, 1979-) which amounts to a slight strengthening of the ergodicity requirement to 'weak mixing', but we shall not discuss that here.

Theorem 4 *Let $k = 0, 1, \ldots, K - 1$ and consider the $2Km$ quantities that are the real and imaginary parts of (2.33), for m values of ω_j nearest to λ_0, $0 < \lambda_0 < \pi$. Then the distribution of these $2Km$ random variables converges to the distribution of a vector of Gaussian random variables with zero means. For different values of j the limiting random variables are independent and the imaginary terms are independent of the real. For fixed j, and for each of the real and imaginary components, the K limiting random variables have covariance matrix with (k, l)th entry $\pi f(\lambda_0)/(k + l + 1)$, $k, l = 0, \ldots, K - 1$.*

Thus for $k = l = 0$ we get $\pi f(\lambda_0)$, or half the limiting mean square of (2.32). Suppose now that T is odd, so that $\omega_j = \pi$ does not arise. Letting $T = 2N + 1$, we might consider writing down as the likelihood of the $2N$ quantities that are the real and imaginary parts of the quantities in (2.32) for $j = 1, \ldots, N$, the function

$$\prod_{j=1}^{N} \left[\left\{2\pi^2 f(\omega_j)\right\}^{-1} e^{-I(\omega_j)/\{4\pi f(\omega_j)\}}\right]. \qquad (2.38)$$

However, we are not entitled, *even asymptotically*, to do this, since Theorem 4 refers to only a fixed number, m, of these quantities, for ω_j nearest to a fixed frequency. Nevertheless (2.38) *is* a likelihood, the $w(\omega_j)$ have very small correlation for T large and mean square near to $2\pi f\left(\omega_j\right)$ and Theorem 4 shows that the CLT does hold in that certain sense. Moreover we do not have a 'true' likelihood. The data is just data and is neither Gaussian (certainly) nor stationary. It is not unreasonable to begin from (2.38) as a likelihood. This is a principle that seems to be initially due to Whittle (1951). We shall use it below. Of course having used (2.38) to suggest a method (or something equivalent to (2.38)) and having derived our estimator we may choose to go back to discuss the properties of this using a general formulation such as that given before Theorem 4. *We are using* (2.38) *only as a likelihood in order to find a good procedure.*

Some support for this procedure, based on (2.38), is found in the following. Consider the quantities

$$\frac{w_x\left(\omega_j\right)}{\left\{\pi f\left(\omega_j\right)\right\}^{1/2}} = u_{2j-1} - i\,u_{2j}, \quad j = 1, \ldots, N, \tag{2.39}$$

where N is the integer part of $(T-1)/2$, which we henceforth write as $\lfloor(T-1)/2\rfloor$. This gives us $2N$ quantities $u_1, \ldots, u_{2N}$, which we are taking as normally distributed and independent with zero means and unit variances in (2.38). Let us form the 'empirical distribution function' of these, i.e. the function of u that is constant save at the values u_k where it jumps by $(2N)^{-1}$. Call this empirical distribution function $F_T\left(u\right)$. Thus $F_T\left(u\right)$ is the proportion of the u_k that are at most u (taking $F_T\left(u\right)$ as continuous from the right). If the quantities in (2.32) were all Gaussian with zero mean and unit variance, as we imply in forming (2.38), albeit falsely but on the basis of a rough piece of reasoning, then the probability that u_k is less than or equal to u would be given by $\Phi\left(u\right)$, which is obtained from (2.4) with $\mu = 0$, $\sigma^2 = 1$, and $x = u$. One might then hope that $F_T\left(u\right)$ would approximate $\Phi\left(u\right)$. Indeed under some (rather strict) conditions

$$\lim_{T\to\infty} \sup_{u} \left|F_T\left(u\right) - \Phi\left(u\right)\right| = 0, \quad \text{a.s.}$$

so that $F_T\left(u\right)$ converges, *uniformly* in u, to $\Phi\left(u\right)$. This provides only vague support for the use of (2.38).

Finally let us briefly discuss the law of the iterated logarithm (LIL). We know that $w_x\left(\lambda\right)$ converges in distribution. Thus if we divide $w_x\left(\lambda\right)$ by a function of T, say $g\left(T\right)$, that increases to ∞ as $T \to \infty$, then

$$g\left(T\right)^{-1} w_x\left(\lambda\right) \to 0,$$

at least in probability. How slowly can $g\left(T\right)$ increase and this fail, in the almost sure sense? The answer is as follows; namely that, for fixed λ,

$$\limsup_{T \to \infty} \frac{|w_x(\lambda)|}{\{2\pi f(\lambda) \, 2 \log \log T\}^{1/2}} = 1, \quad \text{a.s.} \qquad (2.40)$$

Indeed the sequence of quantities in (2.40) has limit points that precisely make up the interval $[0, 1]$ so that there will only be finitely many of them outside $[0, 1 + \varepsilon]$, for any $\varepsilon > 0$. In fact, if we take an interval $\mathcal{I}$ of width $T^{-1}(\log T)^a$ about λ, where $a > 0$, and consider the $w_x(\omega_j)$ in that interval then

$$\limsup_{T \to \infty} \max_{\omega_j \in \mathcal{I}} \frac{|w_x(\omega_j)|}{\{2\pi f(\omega_j) \, 2 \log \log T\}^{1/2}} < \infty, \quad \text{a.s.} \qquad (2.41)$$

so that the order of magnitude of the maximum of the $|w_x(\omega_j)|$ over this small interval $\mathcal{I}$, which will include $O\{(\log T)^a\}$ of the ω_j, is $(\log \log T)^{1/2}$.

This result is important in theoretical considerations, as may also be (2.40), but it is not directly useful and is vastly inferior in that regard to the kind of result (CLT) in Theorem 4. This is not only because of the factor $(\log \log T)^{1/2}$, which increases to infinity, though very slowly, but also because it may require a very large value of T indeed before (2.41) becomes relevant.

3

The Estimation of a Fixed Frequency

3.1 Introduction

In this chapter, we apply the theory of Chapter 2 to sinusoidal models with fixed frequencies. In Section 3.2, the likelihood function under Gaussian noise assumptions is derived, for both white and coloured noise cases, and the relationships between the resulting maximum likelihood estimators and local maximisers of the periodogram is explored. The problem of estimating the fundamental frequency of a periodic signal in additive noise is also discussed. The asymptotic properties of these estimators are derived in Section 3.3. The results of a number of simulations are then used to judge the accuracy of the asymptotic theory in 'small samples'.

The exact CRB for the single sinusoid case is computed in Section 3.4 and this is used in Section 3.5 to obtain accurate asymptotic theory for two special cases. In the first case, we assume that there are two sinusoids, with frequencies very close together. In fact, we assume that they are so close together that we expect sidelobe interference, and that the periodogram will not resolve the frequencies accurately. Although the difference between the frequencies is taken to be of the form $\frac{a}{T}$, where T is the sample size, we show that the maximum likelihood estimators of the two frequencies still have the usual orders of accuracy. It should be noted that so-called 'high resolution' techniques based on autocovariances cannot possibly estimate the frequencies with this accuracy, as the inherent asymptotic standard deviations of these procedures are of order $O\left(T^{-\frac{1}{2}}\right)$, which is larger in order than the actual difference between the true frequencies. The second case we consider is that of a single, very low, frequency, also taken to be of the form $\frac{a}{T}$. Rather than being simpler than the other case, it is more complex, since it is really equivalent to a problem with *three* close frequencies, namely $-\frac{a}{T}$, 0 and $\frac{a}{T}$. Such a model is appropriate if the time series is known *a priori* to contain only a small number of cycles of a sinusoid. The two special cases are illustrated by a number of simulations. The final part of the section is concerned with the general problem of *resolvability*.

The final two sections of the chapter are devoted to two related questions. An important problem in fitting any regression model is that of determining which regressors should be included, and how many. The difference in the context of sinusoidal regression is that the regressors are not exogenous, but parametrised by the single frequency parameter. We thus need to determine *how many* sinusoids should be included in the model. In Section 3.6, we consider how to *estimate* the number of sinusoids, using information theoretic criteria, while in Section 3.7, we look at the question from the more classical approach of hypothesis testing, using the likelihood ratio procedure.

3.2 The maximum likelihood method

The more important problems of frequency estimation are those where the frequency is changing through time. Situations arise, however, where a frequency is essentially fixed. Moreover, one procedure for handling the time varying case is to treat the changing frequency as constant over intervals where it is barely changing and to estimate the frequency for each interval.

The model considered here is, therefore,

$$y(t) = \mu + \sum_{1}^{r} \{\alpha_j \cos(\lambda_j t) + \beta_j \sin(\lambda_j t)\} + x(t). \tag{3.1}$$

The parameters to be estimated are thus μ, and $\alpha_j, \beta_j, \lambda_j, \quad j = 1, \ldots, r$. We shall first take r as known and later discuss its estimation. The residual $x(t)$ will be taken to be generated by a stationary process with zero mean and spectral density $f_x(\omega) > 0$, as in Chapter 2. We put Γ_T for the $T \times T$ matrix with $\gamma_x(t - s)$ in row s, column t. Let

$$y_T' = \begin{bmatrix} y(0) & \cdots & y(T-1) \end{bmatrix}$$

and let X_T be the matrix whose tth row, $t = 0, \ldots, T-1$, is

$$\begin{bmatrix} 1 & \cos(\lambda_1 t) & \sin(\lambda_1 t) & \cdots & \cos(\lambda_r t) & \sin(\lambda_r t) \end{bmatrix}. \tag{3.2}$$

We shall, *when necessary*, indicate true values by a zero subscript, i.e. r_0, λ_{0j} etc., to distinguish these from trial values, but for the most part shall omit this subscript.

The log-likelihood is, for $\{x(t)\}$ Gaussian,

$$\text{constant} - \frac{1}{2} \log |\Gamma_T| - \frac{1}{2} (y_T - X_T B)' \Gamma_T^{-1} (y_T - X_T B) \tag{3.3}$$

where $B' = [\mu\ \alpha_1\ \beta_1\ \cdots\ \alpha_r\ \beta_r]$. For the moment we assume that Γ_T is known and maximise (3.3) as though the λ_j were known, to obtain a criterion depending only on the λ_j. Since

$$\widehat{B}\left(\lambda_1,\ldots,\lambda_r\right) = \left(X_T'\Gamma_T^{-1}X_T\right)^{-1}X_T'\Gamma_T^{-1}y_T,$$

the 'reduced' likelihood, i.e. the likelihood with the matrix B replaced by its estimator, is

$$\text{constant} - \frac{1}{2}\log|\Gamma_T| - \frac{1}{2}y_T'\Gamma_T^{-1}y_T + \widetilde{Q}_T\left(\lambda_1,\ldots,\lambda_r\right)$$

where

$$\widetilde{Q}_T\left(\lambda_1,\ldots,\lambda_r\right) = y_T'\Gamma_T^{-1}X_T\left(X_T'\Gamma_T^{-1}X_T\right)^{-1}X_T'\Gamma_T^{-1}y_T, \qquad (3.4)$$

which we call the 'regression sum of squares'. It is evident that $\widetilde{Q}_T$ has to be maximised with respect to the λ_j in order to find the MLEs of the λ_j.

To proceed further let us observe (Hannan and Wahlberg 1989) that, defining U_T as above (2.9),

$$U_T\Gamma_T^{-1}U_T^* = \text{diag}\left[\{2\pi f_x\left(\omega_j\right)\}^{-1}\right] + O\left(T^{-a}\right), \quad \omega_j = 2\pi j/T$$

where the term that is $O\left(T^{-a}\right)$ is a $T \times T$ matrix whose elements converge, uniformly, at the rate T^{-a}. If $f_x\left(\omega\right)$ is continuously differentiable with a reasonably smooth derivative, then we may take $a > 1/2$. Then, remembering that U_T is unitary, we find, after some manipulation, that

$$\widetilde{Q}_T\left(\lambda_1,\ldots,\lambda_r\right) = \frac{1}{4\pi}\sum_{k=0}^{r}\left[\frac{\sum_j \frac{I_y(\omega_j)}{f_x^2(\omega_j)}\frac{\sin^2\{(\lambda_k-\omega_j)T/2\}}{T^2\sin^2\{(\lambda_k-\omega_j)/2\}}}{\sum_j \frac{1}{f_x(\omega_j)}\frac{\sin^2\{(\lambda_k-\omega_j)T/2\}}{T^2\sin^2\{(\lambda_k-\omega_j)/2\}}}\right] + O\left(T^{-a}\right). \quad (3.5)$$

Here the $k = 0$ term corresponds to the frequency $\lambda_0 = 0$, which accounts for the parameter μ. In deriving (3.5) we have used the fact that

$$\sum_j \frac{1}{f_x\left(\omega_j\right)}\frac{\sin\{(\lambda_k-\omega_j)T/2\}}{T\sin\{(\lambda_k-\omega_j)/2\}}\frac{\sin\{(\lambda_l-\omega_j)T/2\}}{T\sin\{(\lambda_l-\omega_j)/2\}} = O\left(T^{-1}\right), \quad k \neq l$$

since

$$\frac{\sin\{(\lambda_k-\omega_j)T/2\}}{T\sin\{(\lambda_k-\omega_j)/2\}} = O(1)$$

only for $(\lambda_k - \omega_j) = o\left(1\right)$ and otherwise will be $O\left(T^{-1}\right)$, and the λ_k, λ_l are fixed, distinct frequencies. Thus we are led to estimate the λ_j by maximising the first term in (3.5), which we rewrite as

$$\frac{1}{4\pi} \sum_{k=0}^{r} \frac{\sum_j \left\{ \frac{I_y(\omega_j)}{f_x^2(\omega_j)} L_T(\lambda_k - \omega_j) \right\}}{\sum_j \left\{ \frac{1}{f_x(\omega_j)} L_T(\lambda_k - \omega_j) \right\}} \qquad (3.6)$$

where

$$L_T(\omega) = \frac{\sin^2(\omega T/2)}{T^2 \sin^2(\omega/2)}.$$

The neglect of the quantity that is $O\left(T^{-a}\right)$, for $a > 1/2$ will make no difference to the *asymptotic properties* of the estimators of the λ_j.

Put $\lambda_k = 2\pi \left(m_k + \delta_k\right)/T$, where m_k is an integer and $-1/2 < \delta_k < 1/2$. Then, when $\delta_k \neq 0$ and $m_k - j = o\left(T\right)$,

$$L_T(\lambda_k - \omega_j) = \frac{\sin^2(\pi\delta_k)}{\pi^2(m_k - j + \delta_k)} + o(1), \qquad (3.7)$$

while when $\delta_k = 0$, then $L_T(\lambda_k - \omega_j)$ is 1 for $\lambda_k = \omega_j$ and 0 otherwise. Thus $L_T(\lambda_k - \omega_j)$ will be very small except over a range of j values corresponding to a band of values of width $o(1)$. The differentiability condition on $f_x(\omega)$ means that it is nearly constant over this last range of ω values and leads to (3.6) being approximated by

$$\frac{1}{4\pi} \sum_{k=0}^{r} \frac{\sum_j \left\{ I_y(\omega_j) L_T(\lambda_k - \omega_j) \right\}}{f_x(\lambda_k) \sum_j L_T(\lambda_k - \omega_j)} = \frac{1}{4\pi} \sum_{k=0}^{r} \frac{I_y(\lambda_k)}{f_x(\lambda_k)}, \qquad (3.8)$$

since $\sum_j L_T(\lambda_k - \omega_j) = 1$ and $\sum_j I_y(\omega_j) L_T(\lambda_k - \omega_j) = I_y(\lambda_k)$, which we leave the reader to check. Thus, to this last approximation, we may find the estimators of the λ_k by choosing these as the locations of the r greatest relative maxima of $I_y(\omega)$, ignoring local maxima which are so close to others that they may be assumed to be due to 'sidelobes'. If $f_x(\omega)$ were known, we could use (3.6) or even (3.8). However that information is unlikely to be available. If $\{x(t)\}$ is Gaussian 'white noise', (3.4) becomes

$$\sigma^{-2} y_T' X_T \left(X_T' X_T\right)^{-1} X_T' y_T,$$

where σ^2 is the variance of $x(t)$. Ignoring σ^2 we put

$$Q_T(\lambda_1, \ldots, \lambda_r) = y_{T'} X_T \left(X_T' X_T\right)^{-1} X_T' y_T \qquad (3.9)$$

which we again call the regression sum of squares. Now

$$
T^{-1}X_T'X_T = \begin{bmatrix} 1 & 0 & \cdots & 0 \\ 0 & 1/2 & & 0 \\ \vdots & & \ddots & \vdots \\ 0 & 0 & \cdots & 1/2 \end{bmatrix} + O\left(T^{-1}\right) \qquad (3.10)
$$

the $O\left(T^{-1}\right)$ term being composed of terms which are of the forms

$$
(cT)^{-1}\sum \cos\left(at\right)
$$

and

$$
(cT)^{-1}\sum \sin\left(at\right),
$$

where c is 1 or ± 2 and a is one of λ_k, $2\lambda_k$ or $\lambda_k \pm \lambda_l$. All of these terms are clearly $O\left(T^{-1}\right)$, provided λ_k is bounded away from 0 and π and $(\lambda_k - \lambda_l)$ and $(\lambda_k + \lambda_l)$ from 0 and 2π. Since π is (in a sense) an arbitrary frequency determined by the sampling frequency, this means, in practice, that (3.10) *fails if some of the λ_k approach 0, or two or more of the λ_k approach each other.*

If we take only the first term in (3.10), then (3.9) becomes

$$
T\bar{y}^2 + \sum_{k=1}^{r} I_y\left(\lambda_k\right)
$$

and we thus once more arrive at the conclusion that the estimators may be obtained by choosing the locations of the $\hat{r}$ greatest maxima of $I_y\left(\lambda\right)$, again ignoring local maxima so close to others that they may be assumed to be due to 'sidelobes'.

The estimators obtained by the methods cited above (i.e. via maximisation of (3.4), (3.6), (3.9) or $I_y\left(\lambda\right)$) will, under more general conditions than those cited above, have the same asymptotic properties. These will be discussed below. *However their small sample properties may be very different if the above-emphasised properties hold.* We shall also deal with this further below. It is somewhat more difficult to use (3.9) than $I_y\left(\lambda\right)$, although not overwhelmingly so, and there is a good argument for using it. Before going on to treat such matters we discuss notation. The various quantities, (3.4), $I_y\left(\lambda\right)$ etc., lead to different estimators. To avoid a profusion of notations we shall use $\widehat{\lambda}_k$ for each of these estimators, relying on the context or some explicit statement to make clear which method is being discussed.

One way of proceeding is as follows. For convenience let us assume that the λ_{0k} are ordered according to decreasing values of $A_{0j} = \left(\alpha_{0j}^2 + \beta_{0j}^2\right)^{1/2}$.

Use $Q_T(\lambda)$, with $r = 1$, or $I_y(\lambda)$, computed from $y(t) - \overline{y}$, to obtain $\widehat{\lambda}_1$ as the maximising value. Compute $\widehat{\alpha}_1$ and $\widehat{\beta}_1$ by regressing $y(t) - \overline{y}$ on $\cos(\widehat{\lambda}_1 t)$ and $\sin(\widehat{\lambda}_1 t)$, and compute $Q_T^{(1)}(\lambda)$ or $I_y^{(1)}(\lambda)$ from the residuals $y(t) - \overline{y} - \widehat{\alpha}_1 \cos(\widehat{\lambda}_1 t) - \widehat{\beta}_1 \sin(\widehat{\lambda}_1 t)$. Determine $\widehat{\lambda}_2$ as the maximising value of one or the other of these, and so on. Having found $\widehat{\lambda}_1, \ldots, \widehat{\lambda}_r$, recompute $\widehat{\alpha}_j$ and $\widehat{\beta}_j$, $j = 1, \ldots, r$ by computing

$$\widehat{B}(\lambda_j; j = 1, \ldots, r) = \left(X_T' X_T\right)^{-1} X_T' y_T.$$

If necessary one could perform a further iteration, by beginning from the residuals from the regression on $\cos(\widehat{\lambda}_j t)$, and $\sin(\widehat{\lambda}_j t)$, $j = 2, \ldots, r$ to re-estimate $\widehat{\lambda}_1$, then doing the same but omitting $\widehat{\lambda}_2$ from the regression procedure, and replacing $\widehat{\lambda}_1$ by the new estimate of λ_{01}, to get a new estimate of λ_{02} and so on. For details see Bloomfield (1976). Again estimators obtained by this procedure will have the same asymptotic properties.

One thing that is not explained is how the maximisers of the $I_y^{(j)}(\lambda)$ or $Q_T^{(j)}(\lambda)$ are to be obtained. As we shall see, $\left(\widehat{\lambda}_j - \lambda_{0j}\right)$ is of order $O\left(T^{-3/2}\right)$, in probability. This extraordinary accuracy, which is one of the reasons for the importance of frequency estimation, indicates that $I_y(\lambda)$ and $Q_T(\lambda)$ must be very irregular functions of λ. Indeed this has been indicated already by the discussion in Section 2.5, leading to (2.38) as a likelihood, since this acts as though the $I_y(\omega_j)$, separated in frequency by only $2\pi/T$, are independent random variables. Again this matter will be dealt with later. An obvious procedure is to examine $I_y^{(k)}(\omega_j)$, which may be cheaply computed by a fast Fourier transform algorithm, and to use ω_{j0} which maximises $I_y^{(k)}(\omega_j)$ as an initiating value. This ω_{j0} will give an estimate accurate to $O\left(T^{-1}\right)$, one iteration of the procedure discussed in Chapter 5 improves the accuracy to $O_P\left(T^{-3/2}\right)$ and one more yields an estimator as accurate as the maximiser of $I_y^{(k)}(\omega_j)$. It should be noted here that an iterative procedure such as that which uses Newton's method to find zeros of the first derivative of $I_y(\lambda)$ is not guaranteed to converge to the local maximiser. For details, see Rice and Rosenblatt (1988) and Chapter 4. Further detail about such an iteration will be given later, as well as a discussion of the use of ω_{j0}.

To finish this section we deal with the case where $\lambda_{0j} = j\lambda_0$ and λ_0 is unknown. Thus all frequencies are harmonics of the fundamental frequency λ_0. We are then led to maximise

$$\widetilde{Q}_H(\lambda) = \widetilde{Q}_T(\lambda, 2\lambda, \ldots, r\lambda),$$

$$Q_H(\lambda) = Q_T(\lambda, 2\lambda, \ldots, r\lambda)$$

or

$$\sum_{j=1}^{r} \frac{I_y\left(j\lambda\right)}{f_x\left(j\lambda\right)}. \tag{3.11}$$

If the spectral density of $\{x\left(t\right)\}$ were known, this would be fairly easy. However, the spectral density is usually *not* known and we must thus estimate the values or relative values at the harmonics. For details, see Quinn and Thomson (1991). A more straightforward approach is to maximise

$$\sum_{j=1}^{r} I_y\left(j\lambda\right)$$

which has been constructed under the supposition that $\{x\left(t\right)\}$ is white. Although the estimator produced by such an approach will be less efficient (i.e. will have a larger asymptotic variance) unless $\{x\left(t\right)\}$ is indeed white, the benefits gained from not estimating the spectral density values are obvious. An alternative idea is to model $\{x\left(t\right)\}$ as a long order autoregression, and maximise (3.11) using the estimator of $f_x\left(j\lambda\right)$ obtained from the autoregressive estimators. This is the subject of ongoing investigations by Quinn and Thomson, in the more general context where some of the data are missing.

3.3 Properties of the periodogram and the MLE

The basic results of the present section are established under conditions of the type discussed in Section 2.5.

We begin with $I_x\left(\omega\right), \;\; \omega \in [-\pi, \pi]$. Let $\mathcal{I}$ be an interval about a fixed frequency, λ, of width $\left(\log T\right)^a / T, \; 0 \le a < \infty$. Then

Theorem 5

$$\sup_{\omega \in \mathcal{I}} I_x\left(\omega\right) = O\left(\log\log T\right), \quad \text{a.s.} \tag{3.12}$$

$$\limsup_{T \to \infty} \sup_{\omega} \frac{I_x\left(\omega\right)}{4\pi f_x\left(\omega\right)\log T} \le 1, \quad \text{a.s.} \tag{3.13}$$

These results are mostly due to An, Chen and Hannan (1983). If $f_x\left(\omega\right) > 0$, and requiring conditions a little stronger, the latter show that

$$\limsup_{T \to \infty} \sup_{\omega} \frac{I_x\left(\omega\right)}{\log T} = \sup_{\omega} 4\pi f_x\left(\omega\right), \quad \text{a.s.} \tag{3.14}$$

The first part of Theorem 5 is relevant because it gives a strong bound on $I_x\left(\omega\right)$ over a narrow band near λ. If λ is the frequency of a term in (3.1) we may be concerned only with such a narrow band since ω_{j0} (see above) will be within π/T of λ if T is large. The second part is important for the following kind of reason. Suppose we wish to compare

$$\frac{1}{T} \sum_{t=0}^{T-1} \left[y(t) - \bar{y} - \sum_{j=1}^{r-1} \left\{ \widehat{\alpha}_j \cos\left(\widehat{\lambda}_j t\right) + \widehat{\beta}_j \sin\left(\widehat{\lambda}_j t\right) \right\} \right]^2 \qquad (3.15)$$

with

$$\frac{1}{T} \sum_{t=0}^{T-1} \left[y(t) - \bar{y} - \sum_{j=1}^{r} \left\{ \widehat{\alpha}_j \cos\left(\widehat{\lambda}_j t\right) + \widehat{\beta}_j \sin\left(\widehat{\lambda}_j t\right) \right\} \right]^2. \qquad (3.16)$$

(Note that the $\widehat{\alpha}_j$ and $\widehat{\beta}_j$ would be different in the two expressions because different variables will have been used in the two regressions, i.e. $\cos\left(\widehat{\lambda}_r t\right)$ and $\sin\left(\widehat{\lambda}_r t\right)$ are added in the second case.) The reduction in error mean square from the first case to the second will converge, as $T \to \infty$ to $A_{0r}^2/2$, since

$$\frac{1}{T} \sum_{t=0}^{T-1} \left\{ \alpha_{0r} \cos\left(\lambda_{0r} t\right) + \beta_{0r} \sin\left(\lambda_{0r} t\right) \right\}^2 \to \frac{1}{2}\left(\alpha_{0r}^2 + \beta_{0r}^2\right) = \frac{1}{2}A_{0r}^2.$$

This is not a proof but the proof is not difficult. If however there are only $(r-1)$ frequencies present, i.e. only $(r-1)$ terms in (3.1), then the frequency $\widehat{\lambda}_r$ will be located at the maximiser of $I_y^{(r-1)}(\omega)$ which will be close to the maximiser of $I_x(\omega)$ and then

$$\frac{1}{T} \sum_{t=0}^{T-1} \left\{ \widehat{\alpha}_r \cos\left(\widehat{\lambda}_r t\right) + \widehat{\beta}_r \sin\left(\widehat{\lambda}_r t\right) \right\}^2$$

is near to

$$\frac{1}{2}\left(\widehat{\alpha}_r^2 + \widehat{\beta}_r^2\right) = \frac{1}{T} \sup_{\omega} I_y^{(r-1)}(\omega) \sim \frac{1}{T} \sup_{\omega} I_x(\omega) = O\left(\frac{\log T}{T}\right), \quad \text{a.s.}$$

Thus $\widehat{\lambda}_r$ might be mistakenly treated as a true frequency if T were small and $f_x(\omega)$ had a marked peak. Of course, the above shows that as $T \to \infty$ the reduction in error mean square becomes insignificant.

There is a more precise result than (3.13) and (3.14) due to Turkman and Walker (1984). The result has been established for processes of the form

$$x(t) = \sum_{j=0}^{\infty} \alpha(j)\varepsilon(t-j),$$

where

$$\sum_{j=0}^{\infty} j^{\delta} \left| \alpha\left(j\right) \right| < \infty, \; \delta > 0, \; f_x\left(\omega\right) > 0$$

and the $\varepsilon\left(t\right)$ are normally distributed and independent and identically distributed. The result relates to weak convergence and is contained in the following Theorem.

Theorem 6 *Under the above conditions*

$$\Pr\left\{ \max_{\omega} \frac{I_x\left(\omega\right)}{4\pi f_x\left(\omega\right)} \leq \log T + \frac{1}{2}\log\log T + \frac{1}{2}\log\frac{\pi}{3} + x \right\} \to e^{-e^{-x}}. \quad (3.17)$$

Each of (3.13), (3.14) and (3.17) must be used with care for they are extreme value results and, as is well known, the sample size, T, may need to be very large before they are relevant. It is well known that such large values of the periodogram can occur for a stationary series with a continuous spectrum. Of course in practice the distinction between a very peaked spectral density and a periodic component fades. We shall further illustrate these phenomena later in this section but first consider the asymptotic distribution of the estimates, $\widehat{\alpha}_j, \widehat{\beta}_j, \widehat{\lambda}_j$ when r is known.

Theorem 7 *Under the conditions indicated at the beginning of this section, the distribution of*

$$T^{1/2}\left[\; \left(\widehat{\mu} - \mu_0\right) \quad \widehat{\theta}_1' \quad \cdots \quad \widehat{\theta}_r' \; \right]',$$

where $\widehat{\theta}_j' = \left[\; \left(\widehat{\alpha}_j - \alpha_{0j}\right) \quad \left(\widehat{\beta}_j - \beta_{0j}\right) \quad T\left(\widehat{\lambda}_j - \lambda_{0j}\right) \;\right]$, converges to the normal with zero mean and covariance matrix whose only non-zero entries are the $(1,1)$ entry, which is $2\pi f_x\left(0\right)$, and the remaining r 3×3 diagonal blocks, the jth of which is

$$2\pi f_x\left(\lambda_{0j}\right) \begin{bmatrix} 1/2 & 0 & \beta_{0j}/4 \\ 0 & 1/2 & -\alpha_{0j}/4 \\ \beta_{0j}/4 & -\alpha_{0j}/4 & \left(\alpha_{0j}^2 + \beta_{0j}^2\right)/6 \end{bmatrix}^{-1} \quad (3.18)$$

Thus the asymptotic variance of $T^{3/2}\left(\widehat{\lambda}_j - \lambda_{0j}\right)$ is

$$\frac{48\pi f_x\left(\lambda_{0j}\right)}{A_{0j}^2}.$$

This theorem, in the generality given here, is due to Hannan (1973), but for the case where $\{x(t)\}$ is white noise was proved earlier by Walker (1971). In fact, the result, and especially the rate of convergence, were given by Whittle (1952). It may be better to express the central limit theorem in terms of $T^{1/2}\left(\widehat{A}_j - A_{0j}\right)$ and $T^{1/2}\left(\widehat{\phi}_j - \phi_{0j}\right)$, where $\alpha_{0j} = A_{0j}\cos\phi_{0j}$ and $\beta_{0j} = -A_{0j}\sin\phi_{0j}$. The asymptotic covariance matrix is then

$$2\pi f_x(\lambda_{0j})\begin{bmatrix} 2 & 0 & 0 \\ 0 & \frac{8}{A_{0j}^2} & -\frac{12}{A_{0j}^2} \\ 0 & -\frac{12}{A_{0j}^2} & \frac{24}{A_{0j}^2} \end{bmatrix}. \tag{3.19}$$

Thus the estimator of amplitude is asymptotically independent of the estimators of frequency and phase, but $\widehat{\phi}_j$ and $\widehat{\lambda}_j$ have asymptotic correlation $-\sqrt{3}/2 \sim -0.866$. The LIL also holds so that

$$\limsup_{T\to\infty} \frac{T^{3/2}\left|\widehat{\lambda}_j - \lambda_{0j}\right|}{\left\{2\log\log T\, 48\pi\, f_x(\lambda_{0j})/A_{0j}^2\right\}^{1/2}} \le 1, \text{a.s.} \tag{3.20}$$

a result due to Hannan and Mackisack (1986). The same holds for the ϕ_{0j} etc. so that, for example,

$$\limsup_{T\to\infty} \frac{T^{1/2}\left|\widehat{\phi}_j - \phi_{0j}\right|}{\left\{2\log\log T\, 16\pi\, f_x(\lambda_{0j})/A_{0j}^2\right\}^{1/2}} \le 1, \text{a.s.} \tag{3.21}$$

These LIL results are much less useful than the CLT results in Theorem 7 and in (3.19), partly because of the factor $(2\log\log T)^{1/2}$ that occurs in (3.20) and (3.21). This is only 1.75 for $T = 100$ and 1.97 for $T = 1000$, however, and the basic reason is that the bound, unity, may be considerably exceeded when T is not large even though only finitely many values of the left sides of (3.20) and (3.21) will lie outside $[0, 1 + \varepsilon]$, $\varepsilon > 0$, for all T. The LIL is mainly useful in theoretical developments.

For the case of an harmonic sequence, where $\lambda_{0j} = j\lambda_{01}$ the following result, due to Quinn and Thomson (1991), holds. Put

$$\widehat{\theta}_T = T^{1/2}\left[(\widehat{\mu} - \mu_0)\ \left(\widehat{A} - A_0\right)'\ \left(\widehat{\phi} - \phi_0\right)'\ T\left(\widehat{\lambda}_1 - \lambda_{01}\right)\right]'$$

where

$$\widehat{A} - A_0 = \begin{bmatrix} \widehat{A}_1 - A_{01} & \cdots & \widehat{A}_r - A_{0r} \end{bmatrix}'$$

and

$$\widehat{\phi} - \phi_0 = \left[\begin{array}{ccc} \widehat{\phi}_1 - \phi_{01} & \cdots & \widehat{\phi}_r - \phi_{0r} \end{array} \right]'.$$

Then

Theorem 8 *Under the same conditions as Theorem 7, in the harmonic case, $\widehat{\theta}_T$ has a distribution which converges to the normal with zero mean and covariance matrix*

$$\left[\begin{array}{cccc} 2\pi f_x(0) & 0 & 0 & 0 \\ 0 & R & 0 & 0 \\ 0 & 0 & P & C \\ 0 & 0 & C' & \sigma_\lambda^2 \end{array} \right]$$

where

$$\sigma_\lambda^2 = \frac{48\pi}{\sum_{j=1}^r \frac{j^2 A_{0j}^2}{f_x(j\lambda_{01})}},$$

R *is diagonal with jth diagonal entry $4\pi f_x(j\lambda_{01})$, P has (j,k)th entry*

$$\delta_{j,k} \frac{4\pi f_x(j\lambda_{01})}{A_{0j}^2} + \frac{1}{4} jk\sigma_\lambda^2$$

and C has jth element $j\sigma_\lambda^2$.

It is true that the LIL will hold for the quantities in Theorem 8, but we omit the details.

To illustrate, we show some results of a simulation. Two cases are shown, one for the case where $\{x(t)\}$ is white noise and the other when it has spectral density

$$f_x(\omega) = \frac{1}{2\pi \left|1 + 0.81 e^{i2\omega}\right|^2}$$

which we refer to as the coloured noise case. The maximum of $2\pi f_x(\omega)$ is at $\pi/2$ where it is 27.7. The case $r = 2$ with $\lambda_{01} = 1$ and $\lambda_{02} = 2$ was studied with $A_{01}^2 = 2$ and $A_{02}^2 = 1/2$. There were 100 replications. It should be obvious that there will be problems associated with the coloured noise simulations. For the two sample sizes 128 and 1024, Figures 4.5 and 4.6 depict two replications and their periodograms. The results of the simulations are summarised in Table 3.1. The first entry in each set of three for each T is the mean of the $\widehat{\lambda}_j$ for these replications and the second is the standard deviation. The third is the standard deviation from the asymptotic theory. $\{x(t)\}$ was pseudo-Gaussian. As the order of estimating the two frequencies is irrelevant, the smaller of the two estimates was deemed $\widehat{\lambda}_1$ and the larger $\widehat{\lambda}_2$.

Table 3.1. Estimates of frequency

	White Noise		Coloured Noise	
T	$\widehat{\lambda}_1$	$\widehat{\lambda}_2$	$\widehat{\lambda}_1$	$\widehat{\lambda}_2$
128	0.99815	1.99800	1.07772	1.59537
	0.01648	0.10895	0.18550	0.10975
	0.00239	0.00478	0.00241	0.00619
256	0.99558	1.99002	1.00554	1.58371
	0.00423	0.10002	0.05206	0.05896
	0.00085	0.00169	0.00085	0.00219
512	1.00002	1.99996	1.00013	1.73606
	0.00031	0.00060	0.00033	0.21409
	0.00030	0.00060	0.00030	0.00077
1024	1.00000	1.99999	1.00002	1.94490
	0.00010	0.00020	0.00010	0.14337
	0.00011	0.00021	0.00011	0.00027

Although the results are in good agreement with the asymptotic theory in the white noise case, that is not so for coloured noise except for λ_{01} when $T = 512$ and 1024. The cause of the latter is the large value of $2\pi f_x(\omega)$ at $\omega = \pi/2$, namely 27.7. The method of finding $\widehat{\lambda}$ is to begin from $2\pi j_0/T$ which maximises $I_y^{(r)}(\omega)$, $r = 0, 1$, and then to use an iterative method to locate $\widehat{\lambda}$. The method used was that due to Quinn and Fernandes (1991) to be discussed in Chapter 4. However if $2\pi j_0/T$ is near to $\pi/2$ rather than λ_{02}, for $r = 1$, for example, then the method or, any other standard technique, will not locate $\widehat{\lambda}_2$. The SNR would usually be defined for a frequency λ_{0j} as

$$\text{SNR} = 10 \log_{10} \left\{ \frac{A_{0j}^2}{4\pi f_x(\lambda_{0j})} \right\} \tag{3.22}$$

but, as the simulations show, an equally relevant figure is

$$10 \log_{10} \left\{ \frac{A_{0j}^2}{4\pi \max_\omega f_x(\omega)} \right\}.$$

In the case of $j = 2$ this is -20. For $j = 1$ it is -14. Of course T also matters. The peak in $I_y^{(r)}(\omega)$ near λ_{0j} will have a main component of $TA_{0j}^2/4$ due to the signal. We show this in Table 3.2 in relation to $\max 2\pi f_x(\omega)\{\log T + 1/2\log(\pi/3) + 2.25\}$. This comes from (3.17) with $e^{-e^{-x}} = 0.9$.

Table 3.2. Comparison of peak value of $I_y^{(r)}(\omega)$ with
$2\pi \max_\omega f_x(\omega)\left\{\log T + \frac{1}{2}\log\log T + \frac{1}{2}\log(\pi/3) + 2.25\right\}$

T	$TA_{01}^2/4$	$TA_{02}^2/4$	Coloured	White
128	64	16	219	7.9
256	128	32	240	8.7
512	256	64	261	9.4
1024	512	128	282	10.2

Of course the value of $I_y^{(r)}(\omega)$ at λ_{0j} may differ from $TA_{0j}^2/4$, mainly because of the cross-product term

$$T^{-1}\sum_{t=0}^{T-1} x(t)\, e^{it\omega} \sum_{u=0}^{T-1} s(u)\, e^{-iu\omega},$$

where $s(u)$ is the 'signal' $\sum_{j=1}^{r} A_{0j}\cos(\lambda_{0j}u + \phi_{0j})$.

Table 3.1 suggests what has been found by simulation, namely that for the coloured noise case only for $T = 1024$ can good results be found for $j = 1$. However there is some disagreement with Table 3.2, namely where $j = 2$ and $T = 128$ for the white noise case and, for example, where $j = 2$ and $T = 1024$ for the coloured noise case. In the first case the low SNR, namely -6, combined with the small sample size means that the asymptotic theory is not relevant. In the second case Table 3.2 suggests that the maximiser of $I_y^{(1)}(\omega_k)$ should be found near $\pi/2$ much of the time. Indeed formula (3.17) suggests that $\pi/2$ will be found nearly all of the time. In fact only in 13 cases in 100 does that occur. Since $(13\pi/2 + 87 \times 2)/100 \doteq 1.944$ we can see why the mean of the $\widehat{\lambda}_2$ when $T = 1024$ is too low. The reason is as follows. In the first place $I_y^{(1)}(\omega)$ is maximised only over $\omega = \omega_k = 2\pi k/T$ and not over all frequencies. More important is the following. The asymptotic theory springs from the fact that $I_x(\omega)/\{2\pi f_x(\omega)\}$ can be treated as $I_\varepsilon(\omega)$ where $\{\varepsilon(t)\}$ is white noise. Now the maximiser of $I_\varepsilon(\omega)$ is equally likely to be found in any part of $(0,\pi)$. However, the multiplication of $I_\varepsilon(\omega)$ by $2\pi f_x(\omega)$ to obtain $I_x(\omega)$, approximately, means that the maximiser of that quantity is fairly sure to be near $\pi/2$. It is unlikely that the maximising value of $I_\varepsilon(\omega)$ will be near $\pi/2$, or for that matter a very large value. Thus the largest value, near $\pi/2$, of $I_y^{(1)}(\omega_k)$ will not be as large as (3.17) might suggest and the maximising value of $I_y^{(1)}(\omega_k)$ will, most of the time, be near $\lambda_{02} = 2$. The SNR, in the sense of (3.17), was too small in this case, especially when $j = 2$, for results to agree with the asymptotic theory at these sample sizes in the coloured noise case.

Some further simulations are given by Rice and Rosenblatt (1988). There are 100 replications with $T = 100$. The noise is white with unit variance and the frequency is 0.5 radians. The SNR values (S) are $-9, -3, 3, 9$, and 15. At -9 with $T = 100$ we can expect poor agreement with the asymptotic theory since Table 3.1 shows such poor agreement with the asymptotic theory when $T = 128$ and for the frequency λ_{02}. However this is not really relevant to the Rice and Rosenblatt simulations since they chose an initial value for $\widehat{\lambda}$ by examining the criterion over 50 values in an interval of width $2\pi/T$ around λ_{01}. Table 3.3 shows, for each of A, λ and ϕ, the mean of the estimates less the true value, divided by the asymptotic standard deviation (in the first column) and the ratios of the standard deviations of the 100 estimates to the asymptotic standard deviations. For completeness, we have repeated the simulations, but using the Quinn and Fernandes approach commencing at the nearest Fourier frequency, which is $2\pi \times 8/T$. The results are reported in Table 3.4. As both procedures commence near the same frequency, they should be directly comparable. Note that for a result to appear unusual, either the first column for each SNR, multiplied by 10, i.e. the square root of the number of replications, should depart from values of a standard normal variate, (indicating bias, or outliers), or else the second column entries should be substantially different from 1. Of note, therefore, are the Rice and Rosenblatt values corresponding to the SNR of -9, all of their amplitude values $(\widehat{A})$ (this is probably caused by using the *same* pseudo-Gaussian sequence for each SNR), and the frequency and phase estimates $(\widehat{\lambda}$ and $\widehat{\phi})$ at high SNR using the Quinn and Fernandes procedure. The latter is obviously due to a bias in the procedure which is of lower order but effectively higher at high SNR.

3.4 The Cramér–Rao Bound

Because we are, in this book, concerned with the estimation of frequency and because the frequency can be so accurately estimated, so that the CRBs for μ, ϕ and A are effectively independent of frequency, we shall deal mainly

Table 3.3. Estimating a Single Frequency in White Noise – Rice and Rosenblatt

S	−9		−3		3		9		15	
$\widehat{A}$	0.94	1.03	0.45	1.30	0.31	1.33	0.24	1.34	0.20	1.34
$\widehat{\lambda}$	−0.71	6.98	0.02	1.26	0.05	1.25	0.06	1.25	0.07	1.25
$\widehat{\phi}$	0.05	1.34	−0.05	1.31	−0.08	1.28	−0.10	1.28	−0.10	1.28

Table 3.4. Estimating a Single Frequency in White Noise – Quinn and Fernandes

S	−9		−3		3		9		15	
$\widehat{A}$	0.23	0.93	0.04	0.90	0.18	1.03	−0.11	1.08	−0.02	1.01
$\widehat{\lambda}$	−0.08	1.28	0.05	0.94	0.12	1.04	0.40	0.99	0.58	1.07
$\widehat{\phi}$	0.04	1.15	−0.07	0.90	−0.12	0.90	−0.39	0.96	−0.57	1.02

with the CRB for frequency. However we shall see that this does, in some cases (e.g. for very low frequencies) depend also on the phase.

In principle it would be possible to consider the case where $\{x\,(t)\}$ is not a sequence of independent random variables. However the considerations in Section 3.2 show how difficult that would be. We therefore consider only the CRB for the estimator of frequency for the Gaussian white noise case. Of course Theorem 7 shows that the asymptotic distribution of $T^{3/2}\big(\widehat{\lambda}_j - \lambda_{0j}\big)$ is normal under rather general conditions, and it is more or less obvious, from the discussion in Section 2.4, that the asymptotic CRB, for which we shall use the acronym ACRB, will be $48\pi\, f_x\,(\lambda_{0j})/\,\big(T^3 A_{0j}^2\big)$, for the Gaussian independent and identically distributed case. Indeed this could be shown to be so for the Gaussian case in general, i.e. for a reasonably general Gaussian stationary $\{x\,(t)\}$, as can be seen from the development leading to (3.5). Of course more work would need to be done since it needs to be shown that the term that is negligible remains so after differentiation. We shall state the result as a theorem without proof.

Theorem 9 *For the Gaussian regular case the ACRB for the variance of unbiased estimators of frequency is*

$$48\pi\, f_x\,(\lambda_{0j})/\,\big(T^3 A_{0j}^2\big)\,.$$

We remark that the ACRB used here is that corresponding to the bound given in the discussion below Theorem 3.

We shall give a somewhat more precise discussion only for the case of one frequency. If the frequencies are well-separated, the results for a single frequency will represent each individual frequency well, and, if the spectral density $f_x(\omega)$ is fairly smooth, then the substitution $\sigma^2 \to 2\pi f_x(\lambda_{0j})$ will lead to a good approximation to the general stationary Gaussian case, at least for T not too small. It is convenient to reparametrise, replacing ϕ by $\psi = \phi + \frac{T-1}{2}\lambda$ and to put

$$D_T(\lambda) = \frac{\sin(T\lambda)}{T \sin \lambda}. \tag{3.23}$$

Theorem 10 *For the case of Gaussian white noise with variance σ^2, the CRB, ignoring the factor due to the bias, is, for a single frequency, with μ also estimated and $\psi_0 = \phi_0 + \frac{T-1}{2}\lambda_0$,*

$$\frac{\sigma_0^2}{A_0^2}\left(a\cos^2\psi_0 + b\sin^2\psi_0\right)^{-1} \tag{3.24}$$

where

$$\begin{aligned}
a &= \frac{T(T^2-1)}{24} + \frac{T}{8}D_T''(\lambda_0) - \frac{T}{4}\left\{D_T'\left(\frac{\lambda_0}{2}\right)\right\}^2 \\
&\quad - \frac{T}{8}\frac{\left\{D_T'(\lambda_0) - 2D_T'\left(\frac{\lambda_0}{2}\right)D_T\left(\frac{\lambda_0}{2}\right)\right\}^2}{1 + D_T(\lambda_0) - 2\left\{D_T\left(\frac{\lambda_0}{2}\right)\right\}^2}
\end{aligned} \tag{3.25}$$

and

$$b = \frac{T(T^2-1)}{24} - \frac{T}{8}D''(\lambda_0) - \frac{T}{8}\frac{\{D'(\lambda_0)\}^2}{1 - D(\lambda_0)}. \tag{3.26}$$

We prove this theorem in an appendix to this chapter as the proof is easy but not readily available. It follows that the minimum of the CRB is

$$\frac{\sigma_0^2}{A_0^2}\frac{1}{\max(a,b)},$$

the minimum occurring when $\psi_0 \equiv 0 \bmod \pi$, if $a > b$ and when $\psi_0 \equiv \pi/2 \bmod \pi$, if $a < b$.

As $T \to \infty$, the CRB becomes $24T^{-3}\sigma^2 A_0^{-2}\{1 + o(1)\}$ in accordance with the preceding theorem.

Yau and Bresler (1992), in an extensive survey of the subject, evaluate

CRBs, on the assumption of Gaussian white noise, under general circumstances including many different frequency components and many receivers as well as sampling times not necessarily at a fixed distance apart. They assume that the phases and amplitudes are known and thus do not estimate a constant term or mean. Thus their results do not quite include the above theorem. Of course for a very low frequency (see Section 3.5) mean correction is important. Yau and Bresler point out the virtue of 'centering' the phase, which in the case of Theorem 10 amounts to taking ψ and not ϕ as the phase parameter. This avoids the somewhat confusing appearance of the frequency λ_0 (or, in results to be given in Section 3.5, of the frequency difference, for two adjacent frequencies) in the criterion for the best and worst cases for the CRB for $\widehat{\lambda}$.

3.5 Very low and closely adjacent frequencies

It was seen in Section 3.2 (see (3.9) and the discussion after), that if some frequencies are near zero or are close together, then $Q_T(\lambda_1, \ldots, \lambda_r)$ should be used to evaluate the maximum likelihood estimators of frequency. In other words, the full likelihood procedure for the Gaussian white noise case should be used, and not that based on $I_y(\lambda)$. Of course if it were known that the frequencies were well separated from each other and far from zero, except for pairs of close frequencies, we might use $Q_T(\lambda_1, \lambda_2)$ for the pairs. Similarly, if we knew that there was only one frequency close to zero, we would use $Q_T(\lambda)$ for that frequency. In either of these cases, we would use $I_y(\lambda)$ for the remainder.

Let us first consider a single frequency approaching zero. We put $\lambda_{0,T} = a_0/T$ and thus allow $\lambda_{0,T}$ to vary as $T \to \infty$. We do this in order to construct an asymptotic theory, realising that T and λ_0 may be fixed in practice, but hoping that the actual T and λ_0 values will be such that the asymptotic theory will be relevant. Of course the exact distribution of the estimators will be very complicated.

It is preferable to introduce a new parametrisation. Put $\tau = \frac{T-1}{2}$ and note that

$$D_T\left(\frac{a}{T}\right) \to \frac{\sin a}{a} = \operatorname{sinc}\left(\frac{a}{\pi}\right),$$

which we call $s(a)$. Put

$$y(t) = \nu + \alpha\left[\cos\{\lambda(t-\tau)\} - D_T\left(\frac{\lambda}{2}\right)\right] + \beta\sin\{\lambda(t-\tau)\} + x(t).$$

In terms of our original notation,

$$\nu = \mu + \alpha D_T\left(\frac{\lambda}{2}\right), \quad \alpha = A\cos\psi, \quad \beta = -A\sin\psi, \quad \psi = \phi + \tau\lambda. \quad (3.27)$$

Note that the definitions of α and β have changed. We do this because it ensures that, putting

$$S_T(\lambda) = \begin{bmatrix} 1 & \cos(\lambda\tau) - D_T\left(\frac{\lambda}{2}\right) & -\sin(\lambda\tau) \\ \vdots & \vdots & \vdots \\ 1 & \cos\{\lambda(t-\tau)\} - D_T\left(\frac{\lambda}{2}\right) & \sin\{\lambda(t-\tau)\} \\ \vdots & \vdots & \vdots \\ 1 & \cos(\lambda\tau) - D_T\left(\frac{\lambda}{2}\right) & \sin(\lambda\tau) \end{bmatrix} \quad (3.28)$$

then the matrices $S_T(\lambda)'S_T(\lambda), S_T^{(1)}(\lambda)'S_T^{(1)}(\lambda), S_T^{(1)}(\lambda)'S_T(\lambda)$ are diagonal, where $S_T^{(1)}(\lambda) = \frac{d}{d\lambda}S_T(\lambda)$. We estimate $\nu, \alpha, \beta, \lambda$ and σ^2. It may be shown that

$$T^{-1}S_T(\lambda)'S_T(\lambda) \rightarrow \begin{bmatrix} 1 & 0 & 0 \\ 0 & \frac{1}{2}\{1 + s(a) - 2s^2\left(\frac{a}{2}\right)\} & 0 \\ 0 & 0 & \frac{1}{2}\{1 - s(a)\} \end{bmatrix}$$

$$T^{-2}S_T(\lambda)'S_T^{(1)}(\lambda) \rightarrow \begin{bmatrix} 0 & 0 & 0 \\ 0 & \frac{1}{4}s'(a) - \frac{1}{2}s\left(\frac{a}{2}\right)s'\left(\frac{a}{2}\right) & 0 \\ 0 & 0 & -\frac{1}{4}s'(a) \end{bmatrix}$$

$$T^{-3}S_T^{(1)}(\lambda)'S_T^{(1)}(\lambda) \rightarrow \begin{bmatrix} 0 & 0 & 0 \\ 0 & \frac{1}{24} + \frac{1}{8}s''(a) - \frac{1}{4}\{s'\left(\frac{a}{2}\right)\}^2 & 0 \\ 0 & 0 & \frac{1}{24} - \frac{1}{8}s''(a) \end{bmatrix}.$$

$$(3.29)$$

Put $B_0 = \begin{bmatrix} \nu_0 & \alpha_0 & \beta_0 \end{bmatrix}'$ and, for brevity, call the three matrices on the right in (3.29) $S_a, S_{1,a}$ and $S_{2,a}$. Then we have the following theorems, noting that ψ_0, now called $\psi_{0,T}$, depends on T since it is $\phi_0 + \frac{(T-1)a_0}{2T} = \phi_0 + \frac{a_0}{2} - \frac{a_0}{2T}$.

Theorem 11 *Let $\theta = [\nu\ \alpha\ \beta\ a]'$. Under the same conditions as Theorem 7, $\widehat{\theta} \rightarrow \theta_0$, a.s., and the distribution of $T^{1/2}(\widehat{\theta} - \theta_0)$ converges to the normal with mean zero and covariance matrix*

$$2\pi f_x(0)\begin{bmatrix} S_{a_0} & S'_{1,a_0}B_0 \\ B'_0 S_{1,a_0} & B'_0 S_{2,a_0}B_0 \end{bmatrix}^{-1}.$$

In particular the asymptotic variance of $T^{1/2}\left(\widehat{a}-a_0\right)$ is

$$\frac{48\pi f_x\left(0\right)}{A_0^2}\left\{\xi\left(a_0\right)\cos^2\psi_{0,T}+\zeta\left(a_0\right)\sin^2\psi_{0,T}\right\}^{-1} \tag{3.30}$$

where

$$\xi\left(a_0\right)=1+3s''\left(a_0\right)-6\left\{s'\left(\frac{a_0}{2}\right)\right\}^2-3\frac{\left\{s'\left(a_0\right)-2s\left(\frac{a_0}{2}\right)s'\left(\frac{a_0}{2}\right)\right\}^2}{1+s\left(a_0\right)-2s^2\left(\frac{a_0}{2}\right)}$$

and

$$\zeta\left(a_0\right)=1-3s''\left(a_0\right)-\frac{3\left\{s'\left(a_0\right)\right\}^2}{1-s\left(a_0\right)}.$$

These results are originally due to Hannan and Quinn (1989) and were refined by Huang (2000). Of course (3.30) can be obtained directly from (3.24), (3.25) and (3.26) by evaluating $D\left(\lambda_0\right)$, $D'\left(\lambda_0\right)$ and $D''\left(\lambda_0\right)$ at $\lambda_0=a_0/T$. In the Gaussian white noise case for $\{x\left(t\right)\}$, (3.30) shows that the ACRB is attained. One consequence of (3.30) is that the accuracy of $\widehat{\lambda}_0$, for very low frequencies, depends on $\psi_{0,T}=\phi_0+\frac{T-1}{2T}a_0\sim\phi_0+a_0/2$. The two special cases are $\psi_{0,T}\equiv0\bmod\pi$, and $\psi_{0,T}\equiv\frac{\pi}{2}\bmod\pi$. One of these gives a maximum and the other a minimum. In Figure 3.1, we show how the ratio of the asymptotic variance of $T^{1/2}\left(\widehat{a}-a_0\right)$ to $48\pi f_x\left(\lambda_0\right)/A^2$, the value predicted by the asymptotic fixed frequency theory, varies as a function of a and $\psi_{0,T}$.

The variance above may be written in the form

$$\left\{B+3\kappa\left(a_0\right)\cos^2\psi_{0,T}\right\}^{-1}$$

where $B>0$ and

$$\kappa\left(a_0\right)=2s''\left(a_0\right)-2\left\{s'\left(\frac{a_0}{2}\right)\right\}^2-\frac{\left\{s'\left(a_0\right)-2s\left(\frac{a_0}{2}\right)s'\left(\frac{a_0}{2}\right)\right\}^2}{1+s\left(a_0\right)-s^2\left(\frac{a_0}{2}\right)}+\frac{\left\{s'\left(a_0\right)\right\}^2}{1-s\left(a_0\right)}. \tag{3.31}$$

The sign of this determines whether $\psi_{0,T}\equiv0$ or $\psi_{0,T}\equiv\frac{\pi}{2}\bmod\pi$, gives the minimum variance, the other giving the maximum, with $\psi_{0,T}\equiv0\bmod\pi$ corresponding to the minimum when $\kappa\left(a_0\right)>0$. The situation is reversed when $\kappa\left(a_0\right)<0$. Figure 3.2 depicts the function κ for a_0 between 0 and 10π.

In Table 3.5 we present the results of 100 simulations for $\{x(t)\}$ pseudo-Gaussian white noise with variance σ^2. We took $\lambda_{0,T}=\pi/T$, $\mu_0=0$, $A_0=1$ and ϕ_0 either 0 (i.e. $\alpha_0=1$, $\beta_0=0$) or $\pi/2$ (i.e. $\alpha_0=0$, $\beta_0=-1$). The SNR is shown in the table. Thus $\psi_{0,T}=\phi_0+\frac{T-1}{2}\lambda_{0,T}$ is either $\pi/2-\pi/\left(2T\right)$ or $\pi-\pi/\left(2T\right)\equiv-\pi/\left(2T\right)\bmod\pi$. Also $\kappa\left(a_0\right)<0$ so that $\phi_0=0$ gives

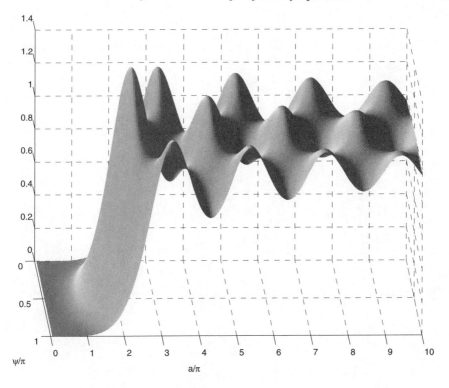

Fig. 3.1. 3D plot of asymptotic variance vs. CRB for fixed frequency

$\psi_{0,T} = \pi/2 - \pi/(2T)$, which is near to the value of $\psi_{0,T}$ that minimises the asymptotic variance. The agreement is good at both SNRs and sample sizes. The agreement is much better than with the result from Theorem 9 which gives a variance for $\widehat{a}$ that is far too small. For example, when $T = 256$, and $\phi_0 = \pi/2$ with SNR = 22, the formula from Theorem 9 gives 2.96×10^{-4} for the asymptotic variance of $\widehat{a}$ whereas Theorem 11 gives 8.93×10^{-2}, which is near to the observed value of 8.66×10^{-2}. Thus Theorem 11 gives much more accurate results for λ_0 very small.

For each estimate the first entry in each set of three is the mean of the 100 values, the second is their variance and the third is the variance from the asymptotic theory.

Theorem 11 brings to mind another question, that of resolvability. In the low frequency context, this is the question of how near to zero can λ_0 be and still be resolved, i.e. distinguished from the zero frequency component μ_0. Of course this needs to be stated more precisely and in particular the sample size and SNR need to be taken into account. One way to formulate

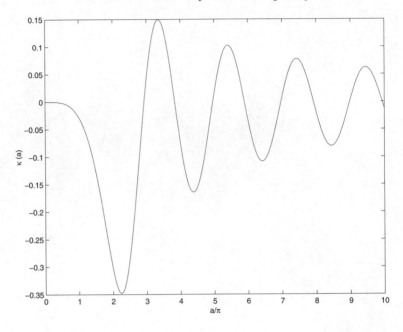

Fig. 3.2. $\kappa\,(a)$, for $a \in [0, 10\pi]$

Table 3.5. One Unknown Low Frequency, π/T

SNR	ϕ_0	T	$\widehat{\mu}$	$\widehat{\alpha}$	$\widehat{\beta}$	$\widehat{a}$
22	0	64	-6.79×10^{-4}	1.00	-2.99×10^{-3}	3.14
			2.54×10^{-4}	4.93×10^{-4}	3.62×10^{-3}	1.41×10^{-2}
			2.62×10^{-4}	3.88×10^{-4}	3.71×10^{-3}	1.34×10^{-2}
22	$\pi/2$	64	3.51×10^{-2}	-3.58×10^{-2}	-1.01	3.16
			5.40×10^{-2}	4.73×10^{-2}	4.93×10^{2}	1.69×10^{-1}
			1.01×10^{-1}	8.94×10^{-2}	1.01×10^{-1}	3.52×10^{-1}
22	0	256	-8.17×10^{-4}	1.00	-1.52×10^{-3}	3.14
			7.74×10^{-5}	9.08×10^{-5}	7.53×10^{-4}	2.50×10^{-3}
			6.51×10^{-5}	1.06×10^{-4}	9.59×10^{-4}	3.36×10^{-3}
22	$\pi/2$	256	3.85×10^{-2}	-3.79×10^{-2}	-1.03	3.11
			2.69×10^{-2}	2.43×10^{-2}	2.44×10^{-2}	8.66×10^{-2}
			2.56×10^{-2}	2.24×10^{-2}	2.47×10^{-2}	8.93×10^{-2}
7	0	64	-7.23×10^{-3}	1.01	-2.54×10^{-2}	3.10
			1.04×10^{-2}	7.04×10^{-3}	5.42×10^{-2}	1.60×10^{-1}
			8.30×10^{-3}	1.23×10^{-2}	1.17×10^{-1}	4.25×10^{-1}
7	0	256	-4.09×10^{-3}	1.01	-1.64×10^{-2}	3.12
			2.11×10^{-3}	2.99×10^{-3}	2.46×10^{-2}	8.55×10^{-2}
			2.06×10^{-3}	3.37×10^{-3}	3.03×10^{-2}	1.06×10^{-1}

the problem is to treat it asymptotically and to require of an estimator, $\widehat{\lambda}_T$, that, for resolvability,

$$\Pr\left\{\left|\widehat{\lambda}_T - \lambda_{0,T}\right| \geq c\lambda_{0,T}\right\} \to 0, \quad \text{for all } c > 0. \tag{3.32}$$

Assuming that the obvious central limit theorem holds, the left side behaves asymptotically as

$$2Q\left(\frac{c\lambda_{0,T}}{\sqrt{CRB}}\right)$$

where

$$Q(x) = \frac{1}{\sqrt{2\pi}} \int_x^\infty e^{-\frac{1}{2}u^2} du.$$

We now ask how fast $\lambda_{0,T}$ can approach zero as $T \to \infty$ and (3.32) hold, that is, how fast can $\lambda_{0,T}$ approach zero and $r_T = \text{CRB}/\lambda_{0,T}^2$ still converge to 0. We are led to consider (3.30), which is of course the ACRB for $T^{1/2}(\widehat{a} - a_0)$ or, equivalently for $T^{3/2}\left(\widehat{\lambda} - \lambda_{0,T}\right)$. Thus we put $\lambda_{0,T} = hT^{-\delta}$, $\delta > 1$, so that $a_0 = hT^{1-\delta}$. We may take $\delta > 1$ since we know that $\widehat{a}$ is consistent when $\delta = 1$. Thus $r_T = h^{-2}T^{2\delta-3}$ times (3.30) with $a_0 = hT^{1-\delta}$, and we must find the upper bound to the values of δ for which this quantity converges to zero. There is a lack of precision in the derivation of the results we present since the CRB has been evaluated only for Gaussian white noise $\{x(t)\}$. We present them in a somewhat more general context so that resolvability is being interpreted in terms of estimation over methods derived from the optimisation of a Gaussian likelihood. A very precise analysis of the Gaussian white noise case has been presented in Molchan (1989), which suggested, to a degree, the results presented here.

It is easily shown that

$$r_T = \frac{48\pi f_x(0)}{A_0^2} h^{-2}T^{2\delta-3}\left(\frac{h^2 T^{6-6\delta}}{6350400}\cos^2\psi_{0,T} + \frac{h^4 T^{4-4\delta}}{25200}\sin^2\psi_{0,T}\right)^{-1}$$
$$\times \{1 + o(1)\}. \tag{3.33}$$

Now, as $\psi_{0,T} = \phi_0 + \frac{1}{2}hT^{1-\delta} + O(T^{-\delta})$, it follows that if $\phi_0 \neq 0 \mod \pi$, then (3.33) converges to zero if $6\delta - 7 < 0$, i.e. if $\delta < 7/6$. However, if $\phi_0 \equiv 0 \mod \pi$, then $\sin^2\psi_{0,T} = O(T^{2-2\delta})$, both terms are of the same order and we need $8\delta - 9 < 0$, i.e. $\delta < 9/8$. As the definition of phase is essentially arbitrary, and in practice the phase is unknown, it is best to use the stronger sufficient condition. We must emphasise that the above argument depends on the central limit theorem holding for $\delta < 7/6$. That this is the case is easily shown using arguments similar to those used to prove the theorem for fixed λ_0.

Theorem 12 *Let $\lambda_{0T} = hT^{-\delta}$, where $\delta < 3/2$. Then condition (3.32) will be satisfied and $\lambda_{0,T}$ will be resolved from zero if $\delta < 7/6$, unless $\phi_0 = 0$ when the stricter condition, $\delta < 9/8$ applies.*

We next turn to the case of two frequencies, which is more important. We now take $\lambda_{0,1}$ fixed, $0 < \lambda_{0,1} < \pi$, and $\lambda_{0,2} = \lambda_{0,1} + a_0/T$. We consider the vector of parameters

$$\theta_{0,T} = \begin{bmatrix} \mu_0 & \alpha_{01} & \alpha_{02} & \beta_{01} & \beta_{02} & a_0 & T\lambda_{0,1} \end{bmatrix}'$$

and the asymptotic distribution of $T^{1/2}(\widehat{\theta}_T - \theta_{0,T})$. The model is

$$
\begin{aligned}
y(t) &= \mu + \alpha_1 \cos\{\lambda_1(t-\tau)\} + \beta_1 \sin\{\lambda_1(t-\tau)\} \\
&\quad + \alpha_2 \cos\{\lambda_2(t-\tau)\} + \beta_2 \sin\{\lambda_2(t-\tau)\} + x(t)
\end{aligned}
$$

where $\tau = (T-1)/2$ and $0 \le t \le T-1$. The covariance matrix in the limiting distribution of $T^{1/2}(\widehat{\theta}_T - \theta_{0,T})$ becomes

$$4\pi \begin{bmatrix} f_x(0) & 0 \\ 0 & f_x(\lambda_0)\mathcal{I}^{-1} \end{bmatrix} \tag{3.34}$$

where

$$\mathcal{I} = \begin{bmatrix} S & 0 & A \\ 0 & S & B \\ A' & B' & C \end{bmatrix},$$

$$S = \begin{bmatrix} 1 & s(a_0/2) \\ s(a_0/2) & 1 \end{bmatrix},$$

$$\begin{bmatrix} A' & B' \end{bmatrix} = s(a_0/2) \begin{bmatrix} \alpha_{02} & 0 & \beta_{02} & 0 \\ \alpha_{02} & -\alpha_{01} & \beta_{02} & -\beta_{01} \end{bmatrix}$$

and C is given by

$$\begin{bmatrix} A_{02}^2 & A_{02}^2 - 3\eta s''(a_0/2) \\ A_{02}^2 - 3\eta s''(a_0/2) & A_{01}^2 + A_{02}^2 - 6\eta s''(a_0/2) \end{bmatrix},$$

where $\eta = \alpha_{01}\alpha_{02} + \beta_{01}\beta_{02}$.

In the proof in Hannan and Quinn (1989) it is required that $\widehat{\lambda}_1$ and $\widehat{\lambda}_1 + \widehat{a}/T$ be bounded away from 0 and π by $O\{(\log T)^{-1}\}$. This is a fairly unimportant requirement but, in any case, Huang (1996), suggests that it is not needed. Hannan and Quinn also require that $\widehat{a} \ge (C_T T^{-1} \log T)^{1/2}$ where $C_T \uparrow \infty$ and $C_T T^{-1} \log T \downarrow 0$. Since, as we shall see, the resolution of the two frequencies requires a_0 to be larger in order than $O\left(T^{1-\delta}\right)$, where $\delta < 5/4$, at most, this condition on $\widehat{a}$ also seems innocuous.

Theorem 13 *Under the same conditions on $\{x(t)\}$ and with the above requirements and definitions, $T^{1/2}\left(\widehat{\theta}_T - \theta_{0,T}\right)$ has a distribution which converges to the normal with zero mean and covariance matrix (3.34)*

Table 3.6 shows the results of 100 simulations, corroborating the theorem. Here $\{x(t)\}$ is a pseudo-normal independent sequence with means zero and variances 1, and $\begin{bmatrix} \mu_0 & \alpha_{01} & \alpha_{02} & \beta_{01} & \beta_{02} \end{bmatrix} = \begin{bmatrix} 0 & 1 & 0 & 0 & 1 \end{bmatrix}$. Also $\lambda_{0,1} = \pi/10$ and $a_0 = \pi$. Three SNRs are used, namely 22, 12 and 7. The T values chosen were rather small, namely 64 and 128, and close agreement cannot then be expected save at the highest SNR. There is little point in taking T large since we know that the agreement will then be good. Also, as has been seen, the accuracy of the estimators of $\lambda_{0,1}$ and a_0 depend on the relative phasing of the two components. The asymptotic variances are largest when $\phi_{01} - \phi_{02} \equiv a_0/2 \bmod \pi$, and smallest when $\phi_{01} - \phi_{02} \equiv (a_0 + \pi)/2 \bmod \pi$. In our case ϕ_{01} is $0 \bmod \pi$ and ϕ_{02} is $\pi/2 \bmod \pi$. Since $a_0 = \pi$, it follows that $\phi_{01} - \phi_{02} \equiv a_0/2 \bmod \pi$ and the case chosen is the worst possible. Thus, apart from fairly high SNR, the simulation is a severe test of the theory.

The agreement with the asymptotic theory is worst for $\widehat{a}$ at the lowest SNR and $T = 64$, and best at the highest SNR and $T = 128$, as is to be expected. The theoretical variances are much nearer the observed values than those from the standard theory for well-separated frequencies. For example $48\pi f_x\left(\lambda_{01}\right)/\left(A_{01}^2 T^3\right)$ is 3.45×10^{-8} at $T = 128$ and SNR $= 22$, which is much too small.

We can now proceed as for Theorem 12 to study the resolvability of the two frequencies. We require that the errors in the estimators of the two frequencies converge to zero, in probability, faster than the difference between the two frequencies. Since

$$\Pr\left\{\left|\widehat{\lambda}_1 - \lambda_{01}\right| \geq cT^{-\delta}\right\} \to 0, \text{ for all } c > 0, \ \delta < 3/2, \tag{3.35}$$

follows from Theorem 13, the essential question is whether

$$\Pr\left\{\left|\widehat{a} - a_0\right| \geq cT^{1-\delta}\right\} \to 0. \tag{3.36}$$

This can be determined by studying $\mathrm{ACRB}(\widehat{a})/T^{2-2\delta}$ and establishing those values of δ for which this quantity converges to zero. The following theorem states the result.

Table 3.6. Simulation Results: Two Unknown Close Frequencies

SNR	T	$\widehat{\mu}$	$\widehat{\alpha}_1$	$\widehat{\alpha}_2$	$\widehat{\beta}_1$	$\widehat{\beta}_2$	$\widehat{\lambda}_1$	$\widehat{a}$
22	64	5.4×10^{-4}	0.993	0.012	0.016	0.988	0.314	3.17
		5.8×10^{-5}	0.023	0.019	0.019	0.023	2.0×10^{-5}	0.011
		4.7×10^{-5}	0.031	0.025	0.025	0.031	2.4×10^{-5}	0.007
	128	9.0×10^{-4}	1.006	-0.003	0.019	0.980	0.314	3.18
		2.0×10^{-5}	0.017	0.013	0.015	0.018	3.6×10^{-6}	0.007
		2.3×10^{-5}	0.015	0.013	0.013	0.015	3.0×10^{-6}	0.003
12	64	1.6×10^{-3}	0.933	0.089	0.110	0.909	0.311	3.56
		5.9×10^{-4}	0.174	0.137	0.136	0.175	2.6×10^{-4}	0.488
		4.7×10^{-4}	0.309	0.253	0.253	0.309	2.5×10^{-4}	0.067
	128	2.8×10^{-3}	0.974	0.034	0.076	0.925	0.313	3.42
		2.0×10^{-4}	0.113	0.088	0.098	0.119	3.3×10^{-5}	0.150
		2.4×10^{-4}	0.154	0.126	0.126	0.154	3.0×10^{-5}	0.034
7	64	1.2×10^{-3}	0.846	0.208	0.280	0.778	0.304	4.73
		1.9×10^{-3}	0.368	0.286	0.292	0.361	1.2×10^{-3}	3.80
		1.6×10^{-3}	1.020	0.837	0.837	1.020	8.1×10^{-4}	0.222
	128	3.8×10^{-3}	0.917	0.099	0.146	0.859	0.312	3.90
		6.8×10^{-4}	0.235	0.179	0.213	0.254	9.4×10^{-5}	0.560
		7.8×10^{-4}	0.511	0.418	0.511	0.418	1.0×10^{-4}	0.111

Theorem 14 *The two frequencies λ_{01} and $\lambda_{01} + hT^{-\delta}$ will be resolved in the sense of* (3.35) *and* (3.36) *in the following cases:*

   (i) $\phi_{01} - \phi_{02} \not\equiv 0 \mod \pi$, *if $\delta < 5/4$;*

   (ii) $\phi_{01} - \phi_{02} \equiv 0 \mod (2\pi)$ *and $A_{01} = A_{02}$, if $\delta < 5/4$;*

   (iii) $\phi_{01} - \phi_{02} \equiv \pi \mod (2\pi)$ *or $\{\phi_{01} - \phi_{02} \equiv 0 \mod (2\pi)$ and $A_{01} \neq A_{02}\}$, if $\delta < 7/6$.*

In the central case, where $\delta < 5/4$, T may need to be very large before resolution will, in practice, be attained when δ is near its critical value. We demonstrate the first part of the theorem only: the other two parts are slightly more complicated but straightforward.

The result of Theorem 14 follows from evaluating the ACRB, i.e. the variance in its limiting distribution of $\widehat{a}$ from Theorem 13. In fact we are merely evaluating the CRB for the Gaussian white noise case. The ACRB is

$$T^{-1}48\pi f_x\left(\lambda_{01}\right)\frac{\left(A_{01}^2 + A_{02}^2\right)U - 2A_{01}A_{02}V\cos\left(\phi_{01} - \phi_{02} - a_0/2\right)}{A_{01}^2 A_{02}^2\left\{U^2 - V^2\cos^2\left(\phi_{01} - \phi_{02} - a_0/2\right)\right\}}, \quad (3.37)$$

where

$$U = 1 - 3\frac{s'(a_0/2)\, s(a_0/2)}{1 - s^2(a_0/2)}$$

and

$$V = 3\left[s''(a_0/2) + \frac{s'(a_0/2)\, s(a_0/2)}{1 - s^2(a_0/2)} \right].$$

Putting $a_0 = hT^{1-\delta}$, U and V become, respectively,

$$\frac{1}{60}h^2 T^{2-2\delta} + \frac{1}{12600}h^4 T^{4-4\delta} + O\left(T^{6-6\delta}\right)$$

and

$$\frac{1}{60}h^2 T^{2-2\delta} + \frac{1}{50400}h^4 T^{4-4\delta} + O\left(T^{6-6\delta}\right).$$

Thus the dominant term in (3.37) is

$$T^{-1} 2880\pi f_x(\lambda_{01})\, T^{2\delta-2} h^{-2} \frac{(A_{01}^2 + A_{02}^2) - 2A_{01}A_{02}\cos(\phi_{01} - \phi_{02} - a_0/2)}{A_{01}^2 A_{02}^2 \sin^2(\phi_{01} - \phi_{02} - a_0/2)}.$$

Assuming that the CLT still holds, condition (3.36) will thus hold if

$$T^{2\delta-2-1+2\delta-2} \to 0$$

with T, i.e., if $\delta < 5/4$.

3.6 The estimation of the number of components

We now consider the determination of r in (3.1). One natural procedure is to use a test of significance for each new term as it is introduced. Such a test could be based on the 'maximum likelihood ratio' i.e. the ratio of the maximised likelihood for r terms to the maximised likelihood for $(r-1)$ terms. The bigger this quantity is the more evidence there is for the rth term being needed. If we examine (3.3) where X_T is as in (3.2) and B as below (3.3) then when $\{x(t)\}$ is Gaussian and white, $\Gamma_T = \gamma_x(0) I_T$ and $\gamma_x(0)$ is estimated by

$$v(r) = \frac{1}{T}\sum_{t=0}^{T-1} \widehat{x}_r^2(t)$$

where $\widehat{x}_r(t)$ is the residual from the regression on X_T with λ_j replaced by $\widehat{\lambda}_j$. Inserting this we see that the maximised log-likelihood is of the form

$$c - \frac{T}{2}\log v(r) - \frac{T}{2}$$

so that the log-likelihood ratio is

$$\frac{T}{2} \log \frac{v\,(r-1)}{v\,(r)}.$$

However

$$v\,(r) \;\approx\; v\,(r-1) - \frac{1}{T} \sum_t \left\{ \widehat{\alpha}_r \cos\left(\widehat{\lambda}_r t\right) + \widehat{\beta}_r \sin\left(\widehat{\lambda}_r t\right) \right\}^2$$

$$=\; v\,(r-1) - \frac{1}{2}\left(\widehat{\alpha}_r^2 + \widehat{\beta}_r^2\right) + O\,(1) \tag{3.38}$$

if $\alpha_{0r} = \beta_{0r} = 0$, i.e. $r_0 = r-1$. This is because $\widehat{\alpha}_r$ and $\widehat{\beta}_r$ will both converge to zero, so that in fact the last term will be $O\left(T^{-1} \log \log T\right)$. Neglecting the last term, the log-likelihood ratio is

$$\frac{T}{2} \log \left\{ 1 + \frac{1}{T} \frac{\max_\omega I_{r-1}\,(\omega)}{v\,(r-1)} \right\}$$

where $I_{r-1}\,(\omega)$ is the periodogram constructed from the $\widehat{x}_{r-1}\,(t)$. The above is nearly

$$\frac{1}{2v\,(r-1)} \max_\omega I_{r-1}\,(\omega). \tag{3.39}$$

Of course $v\,(r-1)$ converges to $\gamma_x\,(0)$ if $r_0 = r-1$. Proceeding more carefully we would get

$$-\frac{T}{2} \log \left\{ \frac{y_T' y_T - \max Q_T\,(\lambda_1, \lambda_2, \ldots, \lambda_r)}{y_T' y_T - \max Q_T\,(\lambda_1, \lambda_2, \ldots, \lambda_{r-1})} \right\}$$

the maxima being over all indicated arguments. This will reduce to (3.39) omitting a term that is relatively small. Turning to the case where $\{x\,(t)\}$ is no longer white, we are led to consider Theorem 6. This relates to the case where $r = 0$ but for $r_0 = r-1$ the same asymptotic theory would hold for $\max_\omega I_{r-1}\,(\omega)$, namely

$$\lim_{T \to \infty} \Pr\left\{ \frac{\max_\omega I_{r-1}\,(\omega)}{4\pi f_x\,(\omega)} \leq \log T + \frac{1}{2} \log \log T + \frac{1}{2} \log \frac{\pi}{3} + x \right\} = e^{-e^{-x}}.$$

As was said in Section 3.3 the results (3.17) and (3.13) may be relevant only for very large T.

If $r-1 < r_0$ then, from (3.38), $v\,(r)$ will decrease, approximately, by $\frac{1}{2} A_{0r}^2$ as the argument increases to r_0. If however $r-1 = r_0$, the decrease will be, from (3.38) again, $\max_\omega I_{r-1}\,(\omega)/T$. This leads us to consider minimising a criterion of the form

$$\varphi_c\,(r) = \log v\,(r) + cr \frac{\log T}{T}$$

which changes for $r_0 > r - 1$ by about

$$\log \left\{ 1 - \frac{A_{01}^2}{2v \left(r - 1 \right)} \right\} + c \frac{\log T}{T}$$

as we go from r to $r - 1$. Quinn (1989) suggested this technique for the case of white noise, and with restrictions on the allowable frequencies. However, when $r_0 = r - 1$, and the noise is coloured, the above changes by approximately

$$-\frac{1}{T} \frac{\max_\omega I_{r_0} \left(\omega \right)}{v \left(r_0 \right)} + c \frac{\log T}{T} = -\frac{4\pi \max_\omega f_x \left(\omega \right) \log T}{T \int_{-\pi}^{\pi} f_x \left(\omega \right) d\omega} \left\{ 1 + o \left(1 \right) \right\} + c \frac{\log T}{T}$$

and so the constant c must depend on the spectral density of the noise. The above follows because $v \left(r_0 \right) \to \gamma_x \left(0 \right)$ and this is evaluated using (2.8). Of course (3.14) does not assert that the maximiser of $I_x \left(\omega \right)$ is found at, or very near, the maximiser of $2\pi f_x \left(\omega \right)$ but that was established by Wang (1993) to whom the following theorem is due.

Theorem 15 *Under the same conditions as Theorem 6, if $\widehat{r}$ is chosen as the lowest value of r for which $\varphi_c \left(r \right) < \varphi_c \left(r + 1 \right)$ then for*

$$c > \frac{4\pi \max_\omega f_x \left(\omega \right)}{\int_{-\pi}^{\pi} f_x \left(\omega \right) d\omega}, \tag{3.40}$$

$\widehat{r}$ converges almost surely to r_0. If the inequality in (3.40) is reversed, convergence does not take place.

The constant on the right in (3.40) is 2 when $\{x \left(t \right)\}$ is white noise but otherwise is unknown unless f is known. It could be estimated by a two-step procedure, first choosing r_1 large and using $\widehat{x}_{r_1} \left(t \right)$ to estimate the constant in (3.40) and then finding $\widehat{r}$. However the criterion $\varphi_c \left(r \right)$ does not seem to be the right one. We go on to discuss the construction of an appropriate criterion.

We use the ideas explained in Rissanen (1989) and due to him. Rissanen considers Statistics to be concerned with the summarisation of data which he interprets as the construction of a binary code for the data, of minimum length. The code is constructed on the basis of a model for the data which will usually be a probabilistic model and will certainly be in the cases considered here. Thus there is data to be summarised, and models coming from the investigator, and no more. Let $y = \left[\begin{array}{ccc} y \left(1 \right) & \cdots & y \left(T \right) \end{array} \right]'$ be the string of data and $P \left(\cdot \right)$ be the probability law coming from the model. In concept the data must be quantised since one cannot finitely encode to arbitrary accuracy. This is no conceptual problem but in using the method one may use any evaluation based on a model for continuous variation of the $y(t)$, for convenience and as an approximation. (Perhaps the bulk of mathematical models for continua has this kind of function.) Thus $P(\cdot)$ is defined on a discrete set. Then it is known, (see, Shannon 1948), that the lowest possible code length, on average, is $-\log_2 P(y)$ i.e.

$$-\sum_y P(y)\log_2 P(y) \leq -\sum_y P(y)\log_2 Q(y) \qquad (3.41)$$

for any Q such that

$$\sum_y Q(y) \leq 1, \quad Q(y) \geq 0, \qquad (3.42)$$

which implies

$$\sum_y 2^{-\{-\log_2 Q(y)\}} \leq 1;$$

this is called the *Kraft inequality* for $-\log_2 Q(y)$. Most coding principles provide code lengths satisfying this Kraft inequality. If (3.41) holds for all $Q(y)$ with $\sum_y Q(y) = 1$ and $Q(y) \geq 0$, it certainly holds for $Q(y)$ satisfying (3.42). But that is equivalent to

$$\sum_y P(y)\log_2\left\{\frac{Q(y)}{P(y)}\right\} \leq \log_2\left\{\sum_y P(y)\frac{Q(y)}{P(y)}\right\} = 0.$$

This last inequality is just that between the geometric mean and the arithmetic mean and is a special case of Jensen's inequality. Equality can hold in (3.41) only when $Q(y) = P(y)$, for all y.

For long strings y, i.e. when T is large, coding systems exist that attain lengths very near to $-\log_2 P(y)$. It is also true for stationary $\{y(t)\}$ for example (see the Shannon, McMillan, Breiman theorem in Billingsley 1995) that $-\log_2 P(y)$ is not merely the best on average but is also 'more or less' the lowest possible value for the particular y we have. Here 'more or less' means for long strings, excluding a set, A, of y values with $P(A)$ very small, and with $-\log_2 P(y)$, $y \notin A$, very near to the smallest possible code length for that y.

If $P(y)$ depends on a parameter vector θ, in which case we write $P(y;\theta)$, then $-\log_2 P(y;\theta)$ is minimised at the MLE, $\widehat{\theta}$. Thus, so far, Rissanen's analysis arrives at Maximum Likelihood. More needs to be said, however, for if all possible models were considered, then the model that allocated all probability to the observed point would win, since for the observed y, say y_0, $-\log_2 P(y_0) = 0$, the smallest possible value, and this is silly since clearly no summarisation is then being effected. The problem is that the code must also be decodable by a decoder knowing, *a priori*, only some general information, such as the totality of models considered. The decoder then has to be told which model has been used for the encoding received so that decoding can take place and the data can be recovered. Thus $\widehat{\theta}$ must be communicated. One of the main problems with the implementation of Rissanen's ideas is the evaluation of the cost of encoding the model used. With a finite set of models one could merely number them and send the number. If there are M models then this will cost about $\log_2 M$ bits. Moreover, $\widehat{\theta}$ will also have to be quantised and a reasonable way to do that, for a scalar $\widehat{\theta}$, is to use some multiple (say twice) the (asymptotic) standard deviation of $\widehat{\theta}$ to determine the quantisation level. Of course this will depend on θ but usually will be of the form $c_\theta T^{-a}$ and $-\log_2\left(c_\theta T^{-a}\right) = -\log_2 c_\theta + a \log_2 T$, which is dominated by the last term for large T. The quantisation level determines the number of possible values of $\widehat{\theta}$, assuming that $\widehat{\theta}$ is bounded *a priori* so that this rule follows. If θ is not scalar, then (ignoring the correlations between the components of $\widehat{\theta}$) the cost to the first order will be $\sum_j a_j \log_2 T$ where $c_{\theta_j} T^{-a_j}$ is the quantisation level for $\widehat{\theta}_j$. Thus the overall encoding cost is put at

$$-\log_2 P\left(y;\widehat{\theta}\right) + \sum_j a_j \log_2 T. \tag{3.43}$$

Clearly the second term above depends on asymptotics (otherwise the term $-\log_2 c_{\theta_j}$ could not be ignored). There are other difficulties also (caused, for example, by ignoring associations). However we shall say no more here about that or about alternative procedures (see Rissanen 1989). In contexts in which $\{y(t)\}$ is generated by a stochastic process other than one of independent random variables one must recognise the narrow range of models that are available. In practice $P(y;\theta)$ has, in such circumstances, been assumed to be Gaussian. Of course (3.43) will be useful over a much wider range of models. Under Gaussian assumptions,

$$P(y;\theta) = (2\pi)^{-T/2}\left(\det \Gamma_T\right)^{-1/2} e^{-\frac{1}{2}(y_T - \mu_T)' \Gamma_T^{-1}(y_T - \mu_T)}, \tag{3.44}$$

where μ_T is a vector of deterministic means. Assuming that K parameters are needed to specify μ_T and that h parameters are needed to describe the process generating $y_T - \mu_T$ then

$$-\log P\left(y_T;\theta\right) = \frac{T}{2}\log\left(2\pi\right) + \frac{1}{2}\log\det \Gamma_T + \frac{1}{2}\operatorname{tr}\left\{\Gamma_T^{-1}\left(y_T - \mu_T\right)\left(y_T - \mu_T\right)'\right\}.$$

Recalling (3.4) we are led to put

$$\log \det \widehat{\Gamma}_T = \sum_j \log \left\{ 2\pi \widehat{f}(\omega_j) \right\},$$

and from (2.22) this is just T multiplied by the approximating sum to

$$\frac{1}{2\pi} \int_{-\pi}^{\pi} \log \left\{ 2\pi \widehat{f}(\omega) \right\} d\omega = \log \widehat{\sigma}_{K,h}^2 ,$$

where $\widehat{f}(\omega)$ is the estimator of $f(\omega)$ from the estimated structure. Also, the exponent is

$$\frac{1}{2} \sum_j \frac{|w(\omega_j)|^2}{2\pi f(\omega_j)},$$

which can be regarded as nearly equalling

$$\frac{T}{4\pi} \int_{-\pi}^{\pi} \frac{|w(\omega)|^2}{2\pi f(\omega)} d\omega$$

which will be approximately $T/2$ in a fairly wide range of circumstances. This leads to the approximation

$$- \log_2 P\left(y; \widehat{\theta}\right) = \text{constant} + \frac{T}{2} \log_2 \widehat{\sigma}_{K,h}^2.$$

Thus, multiplying by $2/T$ for convenience, changing to natural logarithms, and ignoring irrelevant constants we arrive at the expression

$$\log \widehat{\sigma}_{K,h}^2 + \frac{\log T}{T} \left(2 \sum_j a_j \right). \tag{3.45}$$

This is to be minimised with respect to K and h. In our case the parameters in μ_T are μ and the r triples (A_k, ϕ_k, λ_k). For these, the values of a_j are $\frac{1}{2}$ except for λ_j where it is $\frac{3}{2}$. Assuming that the h parameters specifying $f_x(\omega)$ have standard deviations that are $O\left(T^{-1/2}\right)$, and subtracting the common $T^{-1} \log T$ term, we obtain the expression

$$\log \widehat{\sigma}_{r,h}^2 + \frac{(5r + h) \log T}{T}, \quad r \le R, \ h \le H. \tag{3.46}$$

It is assumed that some prior bound is put on R and H. One parametrisation is to regard $f(\omega)$ as constant over bands of width $2\pi m/T$ and to use the estimator

$$T^{-1} \log \det \widehat{\Gamma}_T \sim \frac{2m}{T} \sum_{k=1}^{M} \log \left\{ \frac{1}{2m} \sum_{j=(k-1)m+1}^{km} I(\omega_j) \right\}.$$

Here we have used the fact that $f(-\omega) = f(\omega)$ and have put $T/2 \sim Mm$. This may be too approximate. There is also a problem with $\omega = \pi$, where only one degree of freedom is available. A more accurate formula is derived as follows. We put $M = \lfloor \frac{T-1}{2m} \rfloor$, $p = \lfloor \frac{T-1}{2} \rfloor - mM$, $\delta_T = 0$ for T odd and 1 for T even, and

$$
\log \widehat{\sigma}_{r,m}^2 = \frac{2m}{T} \sum_{k=1}^{M} \log \left\{ \frac{1}{2m} \sum_{j=(k-1)m+1}^{km} I_{r-1}(\omega_j) \right\} +
$$
$$
\frac{2p + \delta_T}{T} \log \left[\frac{1}{2p + \delta_T} \left\{ \sum_{j=mM+1}^{mM+p} I_{r-1}(\omega_j) + \frac{1}{2} \delta_T I_{r-1}(\pi) \right\} \right].
$$

$$(3.47)$$

Then (3.45) becomes

$$
\varphi(r, m) = \log \widehat{\sigma}_{r,m}^2 + \frac{5r \log T}{T} + \frac{M \log m + \log(p + \delta_T)}{T}. \qquad (3.48)
$$

The last term is nearly $\log m / (2m)$. The $\log m$ term occurs because only m frequencies are used to estimate $f(\omega)$ over the typical bound. It is assumed that m is large.

An alternative would be to fit an autoregression (AR) to the residuals $\widehat{x}_r(t)$. There is a very large literature on AR fitting (see Hannan and Deistler 1988, for example.) One method is as follows. Let

$$
c_r(k) = T^{-1} \sum_{t=0}^{T-1} \{\widehat{x}_r(t) - \overline{x}_r\} \{\widehat{x}_r(t+k) - \overline{x}_r\}
$$

where $\overline{x}_r$ is the mean of the $\widehat{x}_r(t)$. Using the Durbin–Levinson algorithm, we estimate the autoregressive coefficients α_j in (2.24) and σ^2, the innovation variance, successively for increasing h, by $\widehat{\alpha}_h(j)$, and $\widehat{\sigma}_h^2$, where h is the order of the AR:

$$
\widehat{\alpha}_h(h) = -\frac{\sum_{j=0}^{h-1} \widehat{\alpha}_{h-1}(j) c_r(h-j)}{\widehat{\sigma}_{h-1}^2}
$$
$$
\widehat{\alpha}_h(j) = \widehat{\alpha}_{h-1}(j) + \widehat{\alpha}_h(h) \widehat{\alpha}_{h-1}(h-j); \quad j = 1, \ldots, h-1 \quad (3.49)
$$
$$
\widehat{\sigma}_h^2 = \{1 - \widehat{\alpha}_h^2(h)\} \widehat{\sigma}_{h-1}^2,
$$

where $\widehat{\sigma}_0^2 = c_r(0)$ and $\widehat{\alpha}_{h,0} = 1$. In the present circumstances, where the $\widehat{\alpha}_h(j)$ are not themselves needed but only the $\widehat{\sigma}_h^2$, we might, preferably, proceed as follows. The advantage comes from a better estimation procedure.

We form, for $0 \leq t \leq T - 1$, $h > 0$, with $\widehat{r}_h(-1) = 0$

$$\widehat{\varepsilon}_0(t) = \widehat{r}_0(t) = \widehat{x}_r(t) - \overline{x}_r$$

$$\widehat{\sigma}_0^2 = T^{-1} \sum_{t=0}^{T-1} \{\widehat{x}_r(t) - \overline{x}_r\}^2$$

$$\widehat{\alpha}_h(h) = \frac{T^{-1} \sum_{t=0}^{T-1} \widehat{\varepsilon}_{h-1}(t) \widehat{r}_{h-1}(t-1)}{\widehat{\sigma}_{h-1}^2} \tag{3.50}$$

$$\widehat{\varepsilon}_h(t) = \widehat{\varepsilon}_{h-1}(t) + \widehat{\alpha}_h(h) \widehat{r}_{h-1}(t-1)$$

$$\widehat{r}_h(t) = \widehat{r}_{h-1}(t-1) + \widehat{\alpha}_h(h) \widehat{\varepsilon}_{h-1}(t)$$

$$\widehat{\sigma}_h^2 = \{1 - \widehat{\alpha}_h^2(h)\} \widehat{\sigma}_{h-1}^2.$$

Whether we use (3.49) or (3.50), we may calculate

$$\varphi(r, h) = \log \widehat{\sigma}_h^2 + \frac{(5r + h) \log T}{T}. \tag{3.51}$$

For each r the value of h or m is found, $h \leq H$, $m \leq M$, where H and M are chosen, *a priori*, not too large in relation to T, to minimise $\varphi(r, m)$ or $\varphi(r, h)$. Then $\widehat{r}$ is the first r after which this minimum increases.

These formulae are asymptotic. There may be a tendency for them to underestimate r_0 because the penalty terms, i.e., the last terms on the right in (3.48) and (3.51), increase too quickly with T. We are, calling $b_j T^{-a_j}$ the asymptotic standard deviation of the jth component, neglecting $\log b_j$ in $-\log b_j + a_j \log T$. As h increases, for example, b_j may tend to increase and $-\log b_j$ to fall so that neglecting it may make the penalty increase too quickly with h.

For further details, see Kavalieris and Hannan (1994).

3.7 Likelihood ratio test for the number of frequencies

We include the following results for completeness, as well as to show how complicated hypothesis testing for the number of periodicities is even in the simplest case. We again consider the model

$$y(t) = \mu + \sum_{j=1}^{r} \{\alpha_j \cos(\lambda_j t) + \beta_j \sin(\lambda_j t)\} + x(t),$$

but now restrict the frequencies to be Fourier frequencies, i.e. of the form $2\pi k/T$, $1 \leq k \leq n = \lfloor (T-1)/2 \rfloor$. If $\{x(t)\}$ is Gaussian and white, with means 0 and variances σ^2, then the maximised log-likelihood for fixed σ^2

and $\lambda_1, \ldots, \lambda_r$ given $\{y(0), \ldots, y(T-1)\}$ would be

$$\text{constant} \ - \ \frac{T}{2} \log \left(\sigma^2 \right) - \frac{1}{2\sigma^2} \left[\sum_{t=0}^{T-1} \{y(t) - \bar{y}\}^2 - \sum_{j=1}^{r} I_y(\lambda_j) \right].$$

The maximised log-likelihood for fixed σ^2 in this case is thus

$$\text{constant} - \frac{T}{2} \log \left(\sigma^2 \right) - \frac{1}{2\sigma^2} \left[\sum_{t=0}^{T-1} \{y(t) - \bar{y}\}^2 - \sum_{j=1}^{r} S_j \right],$$

where S_j is the jth largest of $\{I_y(2\pi k/T); 1 \le k \le n\}$. The unconstrained maximised log-likelihood is then

$$l_r = \text{constant} - \frac{T}{2} \log \left(\widehat{\sigma}_r^2 \right) - \frac{T}{2},$$

where

$$\widehat{\sigma}_r^2 = \frac{1}{T} \left[\sum_{t=0}^{T-1} \{y(t) - \bar{y}\}^2 - \sum_{j=1}^{r} S_j \right].$$

The likelihood ratio test of

$$H_0 : r = r_0$$

against

$$H_A : r = r_A,$$

where $r_A > r_0$, thus rejects H_0 on large values of

$$
\begin{aligned}
\Upsilon \ &= \ \frac{\widehat{\sigma}_{r_0}^2}{\widehat{\sigma}_{r_A}^2} \\
&= \ \frac{\sum_{t=0}^{T-1} \{y(t) - \bar{y}\}^2 - \sum_{j=1}^{r_0} S_j}{\sum_{t=0}^{T-1} \{y(t) - \bar{y}\}^2 - \sum_{j=1}^{r_A} S_j}.
\end{aligned}
$$

Our wish to derive the *exact* distribution of this statistic under Gaussian and white assumptions is thwarted by the fact that the first r_0 of the S_j do not not necessarily coincide with the periodogram values at the *true* frequencies. Nevertheless, we can quantify this probability. Assume that H_0 is true, and let $\Delta_T = \{2\pi k/T; 1 \le k \le n\} \setminus \{\lambda_1, \ldots, \lambda_{r_0}\}$. Let B_T be the event given by

$$B_T = \left\{ \max_{\omega \in \Delta_T} I_y(\omega) < \min_{1 \le j \le r_0} I_y(\lambda_j) \right\}.$$

Then, since the $I_y(\omega)/\sigma^2$ for $\omega \in \Delta_T$ are independent and identically distributed as χ_2^2 random variables,

$$
\begin{aligned}
\Pr\{B_T\} &= E\left[\chi\left\{\max_{\omega \in \Delta_T} I_y(\omega) < \min_{1 \le j \le r_0} I_y(\lambda_j)\right\}\right] \\
&= E\left[\left\{1 - e^{-\min_{1 \le j \le r_0} I_y(\lambda_j)/(2\sigma^2)}\right\}^{n-r_0}\right].
\end{aligned}
$$

where $\chi(C)$ denotes the indicator or characteristic function of the set C. Now,

$$
\begin{aligned}
I_y(\lambda_j) &= \frac{2}{T}\left|\frac{T}{2}(\alpha_j - i\beta_j) + \sum_{t=0}^{T-1} x_t e^{-i\lambda_j t}\right|^2 \\
&= \frac{T}{2}A_j^2 + 2\,\mathrm{Re}\left\{(\alpha_j + i\beta_j)\sum_{t=0}^{T-1} x_t e^{-i\lambda_j t}\right\} + I_x(\lambda_j),
\end{aligned}
$$

and the three terms above are $O(T)$, $O_P(T^{1/2})$ and $O_P(1)$ respectively. Let δ be any positive real number less than $\min_{1 \le j \le r_0} A_j^2/(4\sigma^2)$. Then

$$
\begin{aligned}
&\Pr\{B_T\} \\
&> E\left(\left\{1 - e^{-\min_{1 \le j \le r_0} I_y(\lambda_j)/(2\sigma^2)}\right\}^{n-r_0}\chi\left[\bigcap_{j=1}^{r_0}\{I_y(\lambda_j) > 2T\delta\sigma^2\}\right]\right) \\
&> \left(1 - e^{-T\delta}\right)^{n-r_0} E\left(\chi\left[\bigcap_{j=1}^{r_0}\{I_y(\lambda_j) > 2T\delta\sigma^2\}\right]\right) \\
&= \left(1 - e^{-T\delta}\right)^{n-r_0}\Pr\left[\bigcap_{j=1}^{r_0}\{I_y(\lambda_j) > 2T\delta\sigma^2\}\right].
\end{aligned}
$$

But

$$
\begin{aligned}
&\Pr\{I_y(\lambda_j) > 2T\delta\sigma^2\} \\
&= \Pr\left[2T^{-1/2}\,\mathrm{Re}\left\{(\alpha_j + i\beta_j)\sum_{t=0}^{T-1} x_t e^{-i\lambda_j t}\right\} > 2T^{1/2}\delta\sigma^2 \right.\\
&\qquad\qquad\qquad\qquad\qquad\qquad \left. -T^{1/2}A_j^2/2 - T^{-1/2}I_x(\lambda_j)\right] \\
&> \Pr\left[2T^{-1/2}\,\mathrm{Re}\left\{(\alpha_j + i\beta_j)\sum_{t=0}^{T-1} x_t e^{-i\lambda_j t}\right\} > 2T^{1/2}\delta\sigma^2 - T^{1/2}A_j^2/2\right]
\end{aligned}
$$

$$= 1 - \Phi \left(\frac{2T^{1/2} \delta \sigma^2 - T^{1/2} A_j^2 / 2}{2^{1/2} A_j \sigma} \right)$$

$$= \Phi \left(T^{1/2} \varepsilon_j \right)$$

$$= 1 - \frac{1}{\varepsilon_j \sqrt{2\pi T}} e^{-T \varepsilon_j^2 / 2} + O \left(T^{-3/2} e^{-T \varepsilon_j^2 / 2} \right),$$

where

$$\varepsilon_j = \frac{A_j^2 - 4\delta \sigma^2}{2^{3/2} A_j \sigma} > 0.$$

Hence

$$1 - \Pr \{B_T\}$$

$$< 1 - \left(1 - e^{-T\delta} \right)^{n-r_0} \prod_{j=1}^{r_0} \left\{ 1 - \frac{1}{\varepsilon_j \sqrt{2\pi T}} e^{-T \varepsilon_j^2 / 2} + O \left(T^{-3/2} e^{-T \varepsilon_j^2 / 2} \right) \right\}$$

$$= 1 - \left\{ 1 - n e^{-T\delta} + O \left(n^2 e^{-2T\delta} \right) \right\} \left[1 - \frac{c}{\varepsilon \sqrt{2\pi T}} e^{-T \varepsilon^2 / 2} \{ 1 + o(1) \} \right],$$

where $\varepsilon = \min_{1 \leq j \leq r_0} \varepsilon_j$ and c is the multiplicity of this smallest ε_j. Consequently, for large enough T, and some positive constants K and γ,

$$1 - \Pr \{B_T\} < KT e^{-T\gamma}.$$

Thus when H_0 is true,

$$\Pr \{\Upsilon \leq x\} = \Pr \{\Upsilon \leq x \mid B_T\} \Pr \{B_T\} + E_T,$$

where

$$E_T \leq 1 - \Pr \{B_T\} < KT e^{-T\gamma}.$$

Now

$$\sum_{t=0}^{T-1} \{y(t) - \overline{y}\}^2 = \begin{cases} \sum_{j=1}^{n} I_y \left(\frac{2\pi j}{T} \right) + I_y (\pi) & ; \quad T \text{ even} \\ \sum_{j=1}^{n} I \left(\frac{2\pi j}{T} \right) & ; \quad T \text{ odd}. \end{cases}$$

We first consider the case where T is odd. From the above, and noting that

$$I_y \left(\frac{2\pi k}{T} \right) = I_x \left(\frac{2\pi k}{T} \right)$$

when the frequency $2\pi k/T$ is *not* one of the λ_j, we have

$$\Pr\left\{\Upsilon \leq x | B_T\right\} = \Pr\left\{\frac{\sum_{t=0}^{T-1}\left\{y\left(t\right)-\overline{y}\right\}^2 - \sum_{j=1}^{r_0} S_j}{\sum_{t=0}^{T-1}\left\{y\left(t\right)-\overline{y}\right\}^2 - \sum_{j=1}^{r_A} S_j} \leq x\right\}$$

$$= \Pr\left\{\frac{\sum_{\lambda\in\Delta_T} I_x\left(\lambda\right)}{\sum_{\lambda\in\Delta_T} I_x\left(\lambda\right) - \sum_{j=1}^{r} T_j} \leq x\right\}$$

$$= \Pr\left\{1 - \frac{\sum_{j=1}^{r} T_j}{\sum_{\lambda\in\Delta_T} I_x\left(\lambda\right)} \geq \frac{1}{x}\right\}$$

$$= \Pr\left\{\frac{\sum_{j=1}^{r} T_j}{\sum_{\lambda\in\Delta_T} I_x\left(\lambda\right)} \leq 1 - \frac{1}{x}\right\},$$

where T_j is the jth largest of $\{I_x\left(\lambda\right); \lambda \in \Delta_T\}$ and $r = r_A - r_0$. Of course, when $r_0 = 0$ and $r_A = 1$, under H_0

$$\Upsilon = \frac{\sum_{t=0}^{T-1}\left\{y\left(t\right)-\overline{y}\right\}^2}{\sum_{t=0}^{T-1}\left\{y\left(t\right)-\overline{y}\right\}^2 - S_1} = \frac{\sum_{t=0}^{T-1}\left\{x\left(t\right)-\overline{x}\right\}^2}{\sum_{t=0}^{T-1}\left\{x\left(t\right)-\overline{x}\right\}^2 - S_1} = \frac{1}{1-G},$$

where

$$G = \frac{S_1}{\sum_{t=0}^{T-1}\left\{y\left(t\right)-\overline{y}\right\}^2}$$

is Fisher's g-statistic and $S_1 = \max_k I_y\left(2\pi k/T\right)$. We shall find the exact distribution of

$$\frac{\sum_{j=1}^{r} T_j}{\sum_{\lambda\in\Delta_T} I_x\left(\lambda\right)}$$

under H_0 and will thus be able to find critical regions from the equation

$$\Pr\left\{\Upsilon \leq x\right\} \sim \Pr\left\{\frac{\sum_{j=1}^{r} T_j}{\sum_{\lambda\in\Delta_T} I_x\left(\lambda\right)} \leq 1 - \frac{1}{x}\right\},$$

with no error when $r_0 = 0$, and with relative error of order at most $Te^{-T\gamma}$ when $r_0 > 0$. But

$$\sum_{\lambda\in\Delta_T} I_x\left(\lambda\right) = \sum_{j=1}^{n-r_0} T_j.$$

The distribution of

$$\frac{\sum_{j=1}^{r} T_j}{\sum_{\lambda\in\Delta_T} I_x\left(\lambda\right)}$$

is thus the same as the distribution of

$$\frac{\sum_{j=1}^{r} X_{(j)}}{\sum_{j=1}^{m} X_j},$$

where the X_j are independent and identically distributed exponentially with mean 1, $X_{(j)}$ is the jth largest of the X_k and $m = n - r$. Let

$$U = \sum_{j=1}^{r} X_{(j)}, \quad V = \sum_{j=r+1}^{m} X_{(j)}.$$

We are interested in the distribution of $Z = U/(U + V)$. The joint moment generating function of U and V is then

$$\phi(s,t) = E\left\{e^{-(sU+tV)}\right\}$$

$$= m\binom{m-1}{r-1}\int_0^\infty e^{-(s+1)x}\left\{\int_x^\infty e^{-(s+1)z}dz\right\}^{r-1}$$

$$\times \left\{\int_0^x e^{-(t+1)z}dz\right\}^{m-r} dx$$

$$= m\binom{m-1}{r-1}\int_0^\infty e^{-(s+1)x}\left(\frac{1}{s+1}\right)^{r-1}e^{-(s+1)(r-1)x}$$

$$\times \left(\frac{1}{t+1}\right)^{m-r}\left\{1 - e^{-(t+1)x}\right\}^{m-r} dx$$

$$= \frac{m\binom{m-1}{r-1}}{(s+1)^{r-1}(t+1)^{m-r}}\int_0^\infty e^{-(s+1)rx}\left\{1 - e^{-(t+1)x}\right\}^{m-r} dx.$$

Substituting $z = e^{-(t+1)x}$, we obtain

$$\phi(s,t)$$

$$= \frac{m!}{(s+1)^{r-1}(t+1)^{m-r+1}(m-r)!(r-1)!}\int_0^1 z^{\frac{(s+1)r}{t+1}-1}(1-z)^{m-r} dz$$

$$= \frac{m!}{(s+1)^{r-1}(t+1)^{m-r+1}(r-1)!\prod_{j=0}^{m-r}\left\{\frac{(s+1)r}{t+1}+j\right\}}$$

$$= \frac{m!}{(s+1)^{r}(t+1)^{m-r}r!\prod_{j=1}^{m-r}\left\{\frac{(s+1)r}{t+1}+j\right\}}$$

$$= \binom{m}{r}\frac{(s+1)^{-r}}{\prod_{j=1}^{m-r}\left\{t+1+\frac{(s+1)r}{j}\right\}}. \tag{3.52}$$

To obtain the joint probability density function, we must invert the above, first with respect to s and secondly with respect to t, or vice versa. We choose the latter, since the poles with respect to t are simple. The inverse transform with respect to t is then

$$\binom{m}{r}(s+1)^{-r}\sum_{k=1}^{m-r}\frac{e^{-\left\{1+\frac{(s+1)r}{k}\right\}v}}{\prod_{j=1,j\neq k}^{m-r}\left\{\frac{(s+1)r}{j}-\frac{(s+1)r}{k}\right\}}$$

$$=\frac{\binom{m}{r}}{(s+1)^{m-1}}\sum_{k=1}^{m-r}\frac{e^{-\left\{1+\frac{(s+1)r}{k}\right\}v}k^{m-r-2}(m-r)!}{r^{m-r-1}\prod_{j=1,j\neq k}^{m-r}(k-j)}$$

$$=\frac{m!}{r!r^{m-r-1}(s+1)^{m-1}}\sum_{k=1}^{m-r}(-1)^{m-r-k}\frac{k^{m-r-2}e^{-\left\{1+\frac{(s+1)r}{k}\right\}v}}{(k-1)!(m-r-k)!}$$

$$=\frac{\binom{m}{r}}{r^{m-r-1}(s+1)^{m-1}}\sum_{k=1}^{m-r}(-1)^{m-r-k}\binom{m-r}{k}k^{m-r-1}e^{-\left\{1+\frac{(s+1)r}{k}\right\}v},$$

and the inverse transform of this with respect to s is the joint probability density function of U and V

$$f_{U,V}(u,v)$$

$$=\frac{\binom{m}{r}e^{-(u+v)}}{(m-2)!r^{m-r-1}}\sum_{k=1}^{m-r}(-1)^{m-r-k}\binom{m-r}{k}k^{m-r-1}\left(u-\frac{rv}{k}\right)_+^{m-2}$$

$$=\frac{\binom{m}{r}e^{-(u+v)}}{(m-2)!r^{m-r-1}}\sum_{k=1}^{m-r}(-1)^{m-r-k}\binom{m-r}{k}k^{-r+1}(ku-rv)_+^{m-2},$$

where

$$x_+=\begin{cases}x & ; & x>0\\0 & ; & x\leq 0.\end{cases}$$

The probability density function of Z is then, for $z\in(0,1)$,

$$f_Z(z)$$

$$=(1-z)^{-2}\int_0^\infty xf_{U,V}\left(\frac{zx}{1-z},x\right)dx$$

$$=\frac{\binom{m}{r}(1-z)^{-2}}{(m-2)!r^{m-r-1}}\sum_{k=1}^{m-r}(-1)^{m-r-k}\frac{\binom{m-r}{k}}{k^{r-1}}$$

$$\times\int_0^\infty x\left(\frac{kzx}{1-z}-rx\right)_+^{m-2}e^{-\left(\frac{zx}{1-z}+x\right)}dx$$

$$= \frac{\binom{m}{r}(1-z)^{-m}}{(m-2)!r^{m-r-1}} \sum_{k=1}^{m-r} (-1)^{m-r-k} \frac{\binom{m-r}{k}}{k^{r-1}} (kz - r + rz)_+^{m-2}$$

$$\times \int_0^\infty x^{m-1} e^{-\frac{x}{1-z}} dx$$

$$= \frac{(m-1)\binom{m}{r}}{r^{m-r-1}} \sum_{k=1}^{m-r} (-1)^{m-r-k} \frac{\binom{m-r}{k}}{k^{r-1}} \{(k+r)z - r\}_+^{m-2},$$

and the distribution function $F_Z(z)$ is given on $(0,1)$ by

$$F_Z(z) = \frac{(m-1)\binom{m}{r}}{r^{m-r-1}} \sum_{k=1}^{m-r} (-1)^{m-r-k} \frac{\binom{m-r}{k}}{k^{r-1}} \int_0^z \{(k+r)u - r\}_+^{m-2} du$$

$$= \frac{\binom{m}{r}}{r^{m-r-1}} \sum_{k=1}^{m-r} (-1)^{m-r-k} \frac{\binom{m-r}{k}}{k^{r-1}(k+r)} \{(k+r)z - r\}_+^{m-1}.$$

The complete description is

$$F_Z(z) = \begin{cases} 0 & ; \quad z \le \frac{r}{m} \\ \frac{\binom{m}{r}}{r^{m-r-1}} \sum_{k=1}^{m-r} (-1)^{m-r-k} \frac{\binom{m-r}{k}}{k^{r-1}(k+r)} & \\ \quad \times \{(k+r)z - r\}_+^{m-1} & ; \quad \frac{r}{m} < z < 1 \\ 1 & ; \quad z \ge 1. \end{cases} \tag{3.53}$$

Note that in order that $(k+r)z - r$ be positive for some k, we must have $mz - r > 0$, i.e. $z > r/m$. This makes sense because we must have $U/r > V/(m-r)$.

In the special case where $r = 1$, we obtain for $z \in (0,1)$ the equations

$$f_Z(z) = m(m-1) \sum_{k=1}^{m-1} (-1)^{m-1-k} \binom{m-1}{k} \{(k+1)z - 1\}_+^{m-2}$$

and

$$F_Z(z) = m \sum_{k=1}^{m-1} (-1)^{m-1-k} \frac{1}{k+1} \binom{m-1}{k} \{(k+1)z - 1\}_+^{m-1}$$

$$= \sum_{k=1}^{m-1} (-1)^{m-1-k} \binom{m}{k+1} \{(k+1)z - 1\}_+^{m-1}$$

$$= \sum_{k=2}^{m} (-1)^{m-k} \binom{m}{k} (kz - 1)_+^{m-1}. \tag{3.54}$$

The usual description is, however,

$$
\begin{aligned}
F_Z(z) &= 1 - \sum_{k=1}^{\lfloor 1/z \rfloor} (-1)^{k-1} \binom{m}{k} (1-kz)^{m-1} \\
&= \sum_{k=0}^{m} (-1)^k \binom{m}{k} (1-kz)_+^{m-1} .
\end{aligned}
\tag{3.55}
$$

These *must* be equivalent, and in fact the usual expression is obtained by carrying out the transform inversion in the opposite order. We shall do these computations for completeness. From (3.52), we have, using partial fractions,

$$
\begin{aligned}
&\phi(s,t) \\
&= \binom{m}{r} \frac{(m-r)!\,(s+1)^{-r}}{r^{m-r}\,\prod_{j=1}^{m-r}\left\{(s+1) + \frac{j(t+1)}{r}\right\}} \\
&= \frac{m!}{rr!} \sum_{j=1}^{m-r} (-1)^{j-1} \frac{1}{(t+1)^{m-r-1}\,(j-1)!\,(m-r-j)!} \frac{(s+1)^{-r}}{s+1+\frac{j(t+1)}{r}} .
\end{aligned}
$$

For fixed t, this has simple poles at $s = -1 - j(t+1)/r$, $1 \le j \le m - r$ and a pole of order r at $s = -1$. Inversion with respect to s is complicated by the $(s+1)^{-r}$ term. Now, since the inverse transform of $s^{-r}/(s+a)$ is

$$
\begin{aligned}
&\frac{1}{2\pi i} \oint \frac{s^{-r}}{s+a} e^{su}\,ds \\
&= \frac{1}{(r-1)!} \frac{d^{r-1}}{ds^{r-1}} \left.\frac{e^{su}}{s+a}\right|_{s=0} + (-a)^{-r} e^{-au} \\
&= \frac{1}{(r-1)!} \sum_{k=0}^{r-1} \binom{r-1}{k} \frac{d^k}{ds^k}(s+a)^{-1} \left.\frac{d^{r-1-k}}{ds^{r-1-k}} e^{su}\right|_{s=0} + (-a)^{-r} e^{-au} \\
&= \frac{1}{(r-1)!} \sum_{k=0}^{r-1} (-1)^k \binom{r-1}{k} k! a^{-k-1} u^{r-1-k} + (-a)^{-r} e^{-au} \\
&= \sum_{k=0}^{r-1} (-1)^k \frac{a^{-k-1} u^{r-1-k}}{(r-1-k)!} + (-a)^{-r} e^{-au} \\
&= \sum_{k=0}^{r-1} (-1)^{r-1-k} \frac{a^{k-r} u^k}{k!} + (-a)^{-r} e^{-au} ,
\end{aligned}
$$

it follows that the inverse transform with respect to s of $\phi\left(s,t\right)$ is

$$
\frac{m!}{rr!}e^{-u}\sum_{j=1}^{m-r}(-1)^{j-1}\frac{1}{(t+1)^{m-r-1}\left(j-1\right)!\left(m-r-j\right)!}
$$

$$
\times\left\{\sum_{k=0}^{r-1}(-1)^{r-1-k}\frac{j^{k-r}\left(t+1\right)^{k-r}u^{k}}{k!r^{k-r}}+(-1)^{r}r^{r}j^{-r}\left(t+1\right)^{-r}e^{-uj(t+1)/r}\right\}
$$

$$
=\frac{m!}{r!}e^{-u}\sum_{j=1}^{m-r}(-1)^{j-1}\frac{1}{\left(j-1\right)!\left(m-r-j\right)!}
$$

$$
\times\left\{\sum_{k=0}^{r-1}(-1)^{r-1-k}\frac{r^{r-k-1}u^{k}}{k!j^{r-k}\left(t+1\right)^{m-1-k}}+(-1)^{r}\frac{r^{r-1}e^{-uj(t+1)/r}}{j^{r}\left(t+1\right)^{m-1}}\right\}.
$$

The joint probability density function of U and V is thus

$$
f_{U,V}\left(u,v\right)
$$

$$
=\frac{m!}{r!}e^{-u-v}\sum_{j=1}^{m-r}(-1)^{j-1}\frac{1}{\left(j-1\right)!\left(m-r-j\right)!}
$$

$$
\times\left\{\sum_{k=0}^{r-1}(-1)^{r-1-k}\frac{r^{r-k-1}u^{k}v^{m-2-k}}{k!j^{r-k}\left(m-2-k\right)!}+(-1)^{r}\frac{r^{r-1}\left(v-uj/r\right)_{+}^{m-2}}{j^{r}\left(m-2\right)!}\right\}.
$$

Thus, on $\left(0,1\right)$,

$$
f_{Z}\left(z\right)=\left(1-z\right)^{-2}\int_{0}^{\infty}xf_{U,V}\left(\frac{zx}{1-z},x\right)dx
$$

$$
=\frac{m!}{r!}\left(1-z\right)^{-2}\sum_{j=1}^{m-r}(-1)^{j-1}\frac{1}{\left(j-1\right)!\left(m-r-j\right)!}
$$

$$
\times\left[\sum_{k=0}^{r-1}(-1)^{r-1-k}\frac{r^{r-k-1}}{k!j^{r-k}\left(m-2-k\right)!}\int_{0}^{\infty}x^{m-1-k}\left(\frac{zx}{1-z}\right)^{k}e^{-\frac{x}{1-z}}dx\right.
$$

$$
\left.+(-1)^{r}\frac{r^{r-1}}{j^{r}\left(m-2\right)!}\int_{0}^{\infty}x\left\{x-\frac{jzx}{r\left(1-z\right)}\right\}_{+}^{m-2}e^{-\frac{x}{1-z}}dx\right]
$$

$$= \frac{m!}{r!} \sum_{j=1}^{m-r} (-1)^{j-1} \frac{1}{(j-1)!\,(m-r-j)!}$$

$$\times \left[\sum_{k=0}^{r-1} (-1)^{r-1-k} \frac{r^{r-k-1}}{k!\,j^{r-k}\,(m-2-k)!} \left(\frac{z}{1-z}\right)^k (1-z)^{m-2}\,(m-1)! \right.$$

$$\left. + (-1)^r \frac{r^{r+1-m}}{j^r\,(m-2)!} \{r-(j+r)\,z\}_+^{m-2}\,(m-1)! \right]$$

$$= \frac{\binom{m}{r}}{r^{m-1-r}} \sum_{j=1}^{m-r} (-1)^{j+r-1} \frac{\binom{m-r}{j}}{j^{r-1}} \left[(m-1)\,\{r-(j+r)\,z\}_+^{m-2} \right.$$

$$\left. - \sum_{k=0}^{r-1} (-1)^k\,(m-1)\binom{m-2}{k} j^k r^{m-2-k} z^k (1-z)^{m-2-k} \right].$$

The distribution function of Z on $(0,1)$ is thus obtained from

$$1 - F_Z(z) = \int_z^1 f_Z(z)\,dz$$

$$= \frac{\binom{m}{r}}{r^{m-1-r}} \sum_{j=1}^{m-r} (-1)^{j+r-1} \frac{\binom{m-r}{j}}{j^{r-1}} \left[\frac{\{r-(j+r)\,z\}_+^{m-1}}{j+r} \right.$$

$$\left. - \sum_{k=0}^{r-1} (-1)^k\,(m-1)\binom{m-2}{k} j^k r^{m-2-k} \int_z^1 x^k (1-x)^{m-2-k}\,dx \right]$$

$$= \frac{\binom{m}{r}}{r^{m-1-r}} \sum_{j=1}^{m-r} (-1)^{j+r-1} \frac{\binom{m-r}{j}}{j^{r-1}} \left[\frac{\{r-(j+r)\,z\}_+^{m-1}}{j+r} \right.$$

$$\left. - \sum_{k=0}^{r-1} (-1)^k\, j^k r^{m-2-k} \sum_{l=0}^{k} \binom{m-1}{l} z^l (1-z)^{m-1-l} \right]$$

$$= \frac{\binom{m}{r}}{r^{m-1-r}} \sum_{j=1}^{m-r} (-1)^{j+r-1} \frac{\binom{m-r}{j}}{j^{r-1}(j+r)} \{r-(j+r)\,z\}_+^{m-1}$$

$$+ \binom{m}{r} \sum_{k=0}^{r-1} (-1)^{r+k}\, r^{r-1-k} C_{m,r,k} \sum_{l=0}^{k} \binom{m-1}{l} z^l (1-z)^{m-1-l},$$

$$\tag{3.56}$$

where

$$C_{m,r,k} = \sum_{j=1}^{m-r} (-1)^j \frac{\binom{m-r}{j}}{j^{r-1-k}}.$$

The expression is obviously more complicated than (3.53). Nevertheless, the last term is considerably simplified computationally by noting that

$$\sum_{j=1}^{n} (-1)^j \binom{n}{j} = (1-1)^n - 1 = -1,$$

that

$$\frac{d}{dz} \sum_{j=1}^{n} \frac{\binom{n}{j}}{j} z^j = \sum_{j=1}^{n} \binom{n}{j} z^{j-1} = \frac{(1+z)^n - 1}{z} = \frac{(1+z)^n - 1}{(1+z) - 1}$$

$$= \sum_{j=0}^{n-1} (1+z)^j,$$

and consequently that

$$\sum_{j=1}^{n} \frac{\binom{n}{j}}{j} z^j = \int_0^z \sum_{j=0}^{n-1} (1+x)^j \, dx = \sum_{j=1}^{n} \frac{(1+z)^j - 1}{j}$$

and, in particular,

$$\sum_{j=1}^{n} (-1)^j \frac{\binom{n}{j}}{j} = -\sum_{j=1}^{n} \frac{1}{j}.$$

Similarly,

$$\frac{d}{dz} \sum_{j=1}^{n} \frac{\binom{n}{j}}{j^2} z^j = \sum_{j=1}^{n} \frac{\binom{n}{j}}{j} z^{j-1} = \sum_{j=1}^{n} \frac{(1+z)^j - 1}{j\{(1+z) - 1\}}$$

$$= \sum_{j=1}^{n} \frac{1}{j} \sum_{k=0}^{j-1} (1+z)^k,$$

and so

$$\sum_{j=1}^{n} \frac{\binom{n}{j}}{j^2} z^j = \sum_{j=1}^{n} \frac{1}{j} \sum_{k=1}^{j} \frac{(1+z)^k - 1}{k}$$

and

$$\sum_{j=1}^{n} (-1)^j \frac{\binom{n}{j}}{j^2} = \sum_{j=1}^{n} \frac{1}{j} \sum_{k=1}^{j} \frac{1}{k} = \frac{1}{2} \left\{ \left(\sum_{j=1}^{n} \frac{1}{j} \right)^2 - \sum_{j=1}^{n} \frac{1}{j^2} \right\}.$$

The remaining terms can be similarly simplified (see Table 10 of David and

Kendall 1949). For example,

$$\sum_{j=1}^{n} (-1)^j \frac{\binom{n}{j}}{j^3} = \frac{1}{6} \left\{ \left(\sum_{j=1}^{n} \frac{1}{j} \right)^3 - 3 \left(\sum_{j=1}^{n} \frac{1}{j^2} \right) \left(\sum_{j=1}^{n} \frac{1}{j} \right) + 2 \left(\sum_{j=1}^{n} \frac{1}{j^3} \right) \right\}$$

and

$$\sum_{j=1}^{n} (-1)^j \frac{\binom{n}{j}}{j^4} = \frac{1}{24} \left\{ \left(\sum_{j=1}^{n} \frac{1}{j} \right)^4 - 6 \left(\sum_{j=1}^{n} \frac{1}{j^2} \right) \left(\sum_{j=1}^{n} \frac{1}{j} \right)^2 + 3 \left(\sum_{j=1}^{n} \frac{1}{j^2} \right)^2 \right.$$
$$\left. + 8 \left(\sum_{j=1}^{n} \frac{1}{j^3} \right) \left(\sum_{j=1}^{n} \frac{1}{j} \right) - 6 \left(\sum_{j=1}^{n} \frac{1}{j^4} \right) \right\}.$$

When $r = 1$, we obtain

$$\begin{aligned}
F_Z(z) &= 1 - m \sum_{j=1}^{m-1} (-1)^j \frac{\binom{m-1}{j}}{j+1} \{1 - (j+1) z\}_+^{m-1} \\
&\quad + m \sum_{j=1}^{m-1} (-1)^j \binom{m-1}{j} (1-z)^{m-1} \\
&= 1 - \sum_{j=2}^{m} (-1)^{j-1} \binom{m}{j} (1 - jz)_+^{m-1} - m(1-z)^{m-1} \\
&= \sum_{j=0}^{m} (-1)^j \binom{m}{j} (1 - jz)_+^{m-1},
\end{aligned}$$

in agreement with (3.55). This expression, and (3.54), are difficult to compute directly, because of the alternating signs and the combinatorial coefficients, which can quickly cause overflow problems. In practice, the expressions should be calculated using recursions which are available for the computation of splines (see, for example, de Boor 1978). We now derive an algorithm to compute the above. For $z \in (0, 1)$, let

$$\begin{aligned}
H_m(z) &= \sum_{j=0}^{m} (-1)^{j-1} \binom{m}{j} (1 - jz)_+^{m-1} \\
&= z^{m-1} \sum_{j=0}^{m} (-1)^{j-1} \binom{m}{j} (z^{-1} - j)_+^{m-1} = z^{m-1} G_m(z^{-1}),
\end{aligned}$$

say. Then,

$$G_m(u)$$

$$= \sum_{j=0}^{m} (-1)^j \binom{m}{j} (u-j)_+^{m-2} (u-j)$$

$$= -\sum_{j=1}^{m} (-1)^j \frac{m!}{(j-1)!\,(m-j)!} (u-j)_+^{m-2}$$

$$+ u \sum_{j=0}^{m} (-1)^j \binom{m}{j} (u-j)_+^{m-2}$$

$$= -m \sum_{j=1}^{m} (-1)^j \binom{m-1}{j-1} (u-1-j+1)_+^{m-2}$$

$$+ u \sum_{j=0}^{m} (-1)^j \binom{m}{j} (u-j)_+^{m-2}$$

$$= (m-u) \sum_{j=0}^{m-1} (-1)^j \binom{m-1}{j} (u-1-j)_+^{m-2}$$

$$- u \sum_{j=1}^{m} (-1)^j \binom{m-1}{j-1} (u-j)_+^{m-2} + u_+^{m-1} + u\,(-1)^m (u-m)_+^{m-2}$$

$$+ u \sum_{j=1}^{m-1} (-1)^j \left\{ \binom{m-1}{j} + \binom{m-1}{j-1} \right\} (u-j)_+^{m-2}$$

$$= (m-u) G_{m-1}(u-1) + u_+^{m-1} + u \sum_{j=1}^{m-1} (-1)^j \binom{m-1}{j} (u-j)_+^{m-2}$$

$$= (m-u) G_{m-1}(u-1) + u G_{m-1}(u).$$

Hence, for $z \in (0,1)$,

$$H_m(z) = z^{m-1} \left\{ (m-z^{-1}) G_{m-1}(z^{-1}-1) + z^{-1} G_{m-1}(z^{-1}) \right\}$$

$$= z^{m-1} \left\{ (mz-1) \frac{(1-z)^{m-2}}{z^{m-1}} H_{m-1}\left(\frac{z}{1-z}\right) + \frac{z^{-1}}{z^{m-2}} H_{m-1}(z) \right\}$$

$$= (mz-1)(1-z)^{m-2} H_{m-1}\left(\frac{z}{1-z}\right) + H_{m-1}(z).$$

Such recursions may be readily programmed using languages or packages which allow many levels of recursion. For example, Mathematica^{TM} code to compute the above is

```
H[m_, z\_] := 0 /; z <= 0
H[m_, z\_] := 1 /; z >= 1
H[2, z_] := 0 /; z <= 1/2
H[2, z_] := 2 z - 1 /; z > 1/2 && z < 1
H[m_, z_] := (H[m, z] = H[m - 1, z]
                + (m z - 1) (1 - z)^(m - 2) H[m - 1, z/(1 - z)])
```

The probability $F_Z(0.8)$, for example, can then be calculated when $m = 50$ by evaluating `H[50,0.8]`.

The case where T is even is slightly more complicated, for when T is odd, the total sum of squares is

$$\sum_{t=0}^{T-1} \{y(t) - \overline{y}\}^2 = \sum_{j=1}^{n} I_y\left(\frac{2\pi j}{T}\right) + I_y(\pi).$$

The test statistic thus has the same distribution as

$$Z' = \frac{U}{U + V + W},$$

where $2W$ is distributed as χ^2 with one degree of freedom, independently of U and V, which are defined as before. Thus

$$Z' = \frac{Z}{1 + \frac{W}{U+V}}.$$

Imagine now that $X_1, \ldots, X_m$ are independently and identically distributed exponentially with parameter θ. Then $U + V = \sum_{i=1}^{m} X_i$ is complete and sufficient for θ. Consequently,

$$Z = \frac{U}{U + V},$$

which has a distribution that does not depend on θ, is independent of $U+V$. It is thus also independent of $Y = W/(U + V)$, since W is independent of $X_1, \ldots, X_m$. The distribution function of Z' is thus the same as that of

$$\frac{Z}{1 + Y}.$$

We shall prove a slightly more general result by assuming that $2W$ is distributed as χ^2 with degrees of freedom q. This will also cover the case where it is known *a priori* that the true frequency is confined to a certain range, and consequently that not all eligible frequencies are searched. It follows then that mY/q is then distributed as F with degrees of freedom q and m.

Thus

$$1 - F_{Z'}(z) = \int_0^\infty \Pr\{Z > z(x+1)\} \frac{x^{q/2-1}(1+x)^{-m-q/2}}{B(q/2, m)} dx$$

where $B(a, b) = \Gamma(a)\Gamma(b)/\Gamma(a+b)$. Simplification of this integral does not seem to be possible using (3.53). However, it *is* possible using (3.56), since

$$\int_0^\infty \{r - k(x+1)z\}_+^{m-1} x^{q/2-1}(1+x)^{-m-q/2} dx$$

$$= \begin{cases} 0 & ; \quad r - kz \le 0 \\ \int_0^{\frac{r-kz}{kz}} \{r - k(x+1)z\}^{m-1} x^{q/2-1}(1+x)^{-m-q/2} dx & ; \quad r - kz > 0. \end{cases}$$

Substituting $u = x/(x+1)$ the above integral becomes, since $x = u/(1-u)$,

$$\int_0^{\frac{r-kz}{r}} \left(r - \frac{kz}{1-u}\right)^{m-1} u^{q/2-1}(1-u)^{1-q/2}(1-u)^{m+q/2}(1-u)^{-2} du$$

$$= \int_0^{\frac{r-kz}{r}} (r - kz - ru)^{m-1} u^{q/2-1} du.$$

Finally, putting $u = (r - kz)x/r$, we obtain

$$(r-kz)^{m-1} \int_0^1 (1-x)^{m-1} (r-kz)^{q/2} r^{-q/2} x^{q/2-1} dx$$

$$= (r-kz)^{m+q/2-1} r^{-q/2} B(q/2, m).$$

Similarly, again substituting $u = x/(1+x)$, the other terms may be simplified by noting that

$$\int_0^{\frac{1-z}{z}} \{z(x+1)\}^l \{1 - z(x+1)\}^{m-1-l} x^{q/2-1}(1+x)^{-m-q/2} dx$$

$$= z^l \int_0^{1-z} (1-u)^{-l} \left(1 - \frac{z}{1-u}\right)^{m-1-l} \left(\frac{u}{1-u}\right)^{q/2-1} (1-u)^{m+q/2-2} du$$

$$= z^l \int_0^{1-z} (1 - z - u)^{m-1-l} u^{q/2-1} du$$

$$= z^l (1-z)^{m+q/2-1} B(q/2, m-1-l).$$

Hence, for $z \in (0, 1)$,

$$1 - F_{Z'}(z)$$

$$= \frac{\binom{m}{r}}{r^{m+q/2-1-r}} \sum_{j=1}^{m-r} (-1)^{j+r-1} \frac{\binom{m-r}{j}}{j^{r-1}(j+r)} \{r - (j+r) z\}_+^{m+q/2-1}$$

$$+ \binom{m}{r} \sum_{k=0}^{r-1} (-1)^{r+k} r^{r-1-k} C_{m,r,k}$$

$$\times \sum_{l=0}^{k} \binom{m+q/2-1}{l} z^l (1-z)^{m+q/2-1-l},$$

where for real x and integer j, we define $\binom{x}{j}$ by

$$\binom{x}{j} = \frac{\Gamma(x+1)}{j!\,\Gamma(x+1-j)},$$

which agrees with the usual definition when x is an integer.

Finally, we derive the probability density and distribution functions which are appropriate when m is large. Assuming that q is fixed, the asymptotic distribution of G is the same as that of U/m, since $(U+V+W)/m$ converges in probability to 1. The moment generating function of U is given by $\phi_m(s) = \phi(s, 0)$, where $\phi(s, t)$ is defined in (3.52). Thus

$$\phi_m(s) = \binom{m}{r} \frac{(s+1)^{-r}}{\prod_{j=1}^{m-r}\left\{1 + \frac{(s+1)r}{j}\right\}}$$

$$= \frac{m!}{r!} (s+1)^{-r} \frac{\Gamma(1+(s+1)r)}{\Gamma(m-r+1+(s+1)r)}.$$

Using Stirling's formula, we have

$$\frac{m!}{\Gamma(m-r+1+(s+1)r)}$$

$$\sim \frac{m^{m+1/2} e^{-m}}{\sqrt{2\pi}} \frac{\sqrt{2\pi}}{\{m-r+(s+1)r\}^{m-r+(s+1)r+1/2} e^{-\{m-r+(s+1)r\}}}$$

$$= m^{-sr} e^{sr} \left\{1 - \frac{sr}{m}\right\}^{m+sr+1/2}$$

$$\sim m^{-sr},$$

with the approximation holding uniformly in a neighbourhood of $s = 0$. Hence $m^{sr} \phi_m(s)$ converges to

$$\phi(s) = \frac{1}{r!} (s+1)^{-r} \Gamma(1+(s+1)r)$$

uniformly in a neighbourhood of 0 and we obtain the asymptotic distribution of $U' = U - r \log m$ by inverting $\phi(s)$. This inverse is just e^{-u} times the inverse of

$$\tilde{\phi}(s) = \frac{1}{r!} s^{-r} \Gamma(1 + sr)$$

which has poles at 0, with multiplicity r, and $-1/r, -2/r, \ldots$, each with multiplicity 1. Thus the asymptotic probability density function of U' is

$$
\begin{aligned}
f(u) \\
&= e^{-u} \frac{1}{2\pi i} \oint \frac{1}{r!} s^{-r} \Gamma(1 + sr) e^{us} ds \\
&= \frac{1}{r!} e^{-u} \left[\sum_{k=1}^{\infty} \left(-\frac{k}{r} \right)^{-r} e^{-ku/r} \lim_{s \to -k/r} (s + k/r) \Gamma(1 + sr) \right. \\
&\qquad \left. + \frac{1}{(r-1)!} \frac{d^{r-1}}{ds^{r-1}} \{ \Gamma(1 + sr) e^{us} \} \bigg|_{s=0} \right] \\
&= \frac{1}{r!} e^{-u} \left\{ \sum_{k=1}^{\infty} \left(-\frac{k}{r} \right)^{-r} e^{-ku/r} \frac{(-1)^{k-1}}{r(k-1)!} \right. \\
&\qquad \left. + \frac{1}{(r-1)!} \sum_{j=0}^{r-1} \binom{r-1}{j} r^j u^{r-1-j} \frac{d^j}{ds^j} \Gamma(s) \bigg|_{s=1} \right\} \\
&= \frac{r^{r-1}}{r!} e^{-u} \left\{ \sum_{k=1}^{\infty} (-1)^{r+k-1} \frac{e^{-ku/r}}{k^{r-1} k!} + \sum_{j=0}^{r-1} \frac{\gamma_j (u/r)^{r-1-j}}{j!(r-1-j)!} \right\}, \quad (3.57)
\end{aligned}
$$

where $\gamma_0 = \Gamma(1) = 1$ and, for $j > 0$,

$$\gamma_j = \frac{d^j}{ds^j} \Gamma(s) \bigg|_{s=1} = \frac{d^{j-1}}{ds^{j-1}} \Gamma'(s) \bigg|_{s=1} = \frac{d^{j-1}}{ds^{j-1}} \{ \psi(s) \Gamma(s) \} \bigg|_{s=1},$$

and $\psi(x)$ is the digamma function, given by

$$\psi(x) = \frac{d}{dx} \log \Gamma(x).$$

The γ_j may be calculated recursively from

$$
\begin{aligned}
\gamma_j &= \sum_{k=0}^{j-1} \binom{j-1}{k} \psi^{(j-1-k)}(s) \frac{d^k}{ds^k} \Gamma(s) \bigg|_{s=1} = \sum_{k=0}^{j-1} \binom{j-1}{k} \psi^{(j-1-k)}(1) \gamma_k \\
&= \sum_{k=0}^{j-1} \binom{j-1}{k} \psi^{(j-1-k)}(1) \gamma_k,
\end{aligned}
$$

where, by definition, $\psi^{(0)}(s) = \psi(s)$. This may be further simplified by using the fact that $\psi^{(0)}(1) = -\gamma$ (Euler's constant), and for $k > 0$,

$$\psi^{(k)}(1) = (-1)^{k+1} k! \zeta(k+1),$$

where ζ is the Riemann zeta function. The asymptotic distribution function is then

$$1 - F(u)$$

$$= \int_u^\infty \frac{r^{r-1}}{r!} e^{-x} \left\{ \sum_{k=1}^\infty (-1)^{r+k-1} \frac{e^{-kx/r}}{k^{r-1}k!} + \sum_{j=0}^{r-1} \frac{\gamma_j (x/r)^{r-1-j}}{j!(r-1-j)!} \right\} dx$$

$$= \frac{r^r}{r!} e^{-u} \left\{ \sum_{k=1}^\infty (-1)^{r+k-1} \frac{e^{-ku/r}}{(k+r)k^{r-1}k!} + \sum_{j=0}^{r-1} \frac{\gamma_j r^{j-r}}{j!} \sum_{k=0}^{r-1-j} \frac{u^k}{k!} \right\}.$$

which is an awkward series to compute. However, when $r = 1$, $1 - F(u)$ is easily shown to be $1 - e^{-e^{-u}}$, which suggests that there may be closed formulae for the distribution function for $r > 1$. The key lies in finding an alternative expression for the series

$$\sum_{k=1}^\infty \frac{z^k}{k^\alpha k!}.$$

For fixed z, let

$$g(\alpha) = \int_0^1 x^{\alpha-1} e^{-zx} dx.$$

Then

$$g^{(j)}(\alpha) = \int_0^1 (\log x)^j x^{\alpha-1} e^{-zx} dx$$

and so

$$g^{(j)}(1) = \int_0^1 (\log x)^j e^{-zx} dx.$$

But

$$g(\alpha) = \int_0^1 x^{\alpha-1} e^{-zx} dx = \sum_{k=0}^\infty \frac{(-z)^k}{k!(\alpha+k)}.$$

Hence

$$g^{(j)}(1) = j! \sum_{k=0}^\infty \frac{(-1)^{k+j} z^k}{k!(1+k)^{j+1}} = j! \sum_{k=1}^\infty \frac{(-1)^{k+j-1} z^{k-1}}{k!k^j}.$$

Thus, letting $z = e^{-u/r}$, we have

$$f(u)$$

$$= \frac{r^{r-1}}{r!} e^{-u} \left\{ \sum_{k=1}^{\infty} (-1)^{r+k-1} \frac{z^k}{k^{r-1}k!} + \sum_{j=0}^{r-1} \frac{\gamma_j (u/r)^{r-1-j}}{j! (r-1-j)!} \right\}$$

$$= \frac{r^{r-1}}{r!} e^{-u} \left\{ \sum_{j=0}^{r-1} \frac{\gamma_j (u/r)^{r-1-j}}{j! (r-1-j)!} - \frac{1}{(r-1)!} e^{-u/r} g^{(r-1)} (1) \right\}$$

$$= \frac{r^{r-1}}{(r-1)!r!} e^{-u} \left\{ \sum_{j=0}^{r-1} \binom{r-1}{j} \gamma_j (u/r)^{r-1-j} - z \int_0^1 (\log x)^{r-1} e^{-zx} dx \right\}.$$

Now

$$\int_0^{\infty} (\log x)^{r-1} e^{-zx} dx = \frac{d^{r-1}}{d\alpha^{r-1}} \int_0^{\infty} x^{\alpha-1} e^{-zx} dx \bigg|_{\alpha=1}$$

$$= \frac{d^{r-1}}{d\alpha^{r-1}} \left\{ z^{-\alpha} \Gamma(\alpha) \right\} \bigg|_{\alpha=1}$$

$$= \sum_{j=0}^{r-1} \binom{r-1}{j} \gamma_{r-1-j} (-\log z)^j z^{-1}.$$

Consequently,

$$f(u)$$

$$= \frac{r^{r-1}}{(r-1)!r!} e^{-u} \left\{ \sum_{j=0}^{r-1} \binom{r-1}{j} \gamma_j (u/r)^{r-1-j} - z \int_0^1 (\log x)^{r-1} e^{-zx} dx \right\}$$

$$= \frac{r^{r-1}}{(r-1)!r!} e^{-u} \left\{ z \int_0^{\infty} (\log x)^{r-1} e^{-zx} dx - z \int_0^1 (\log x)^{r-1} e^{-zx} dx \right\}$$

$$= \frac{r^{r-1}}{(r-1)!r!} e^{-u(1+1/r)} \int_1^{\infty} (\log x)^{r-1} e^{-xe^{-u/r}} dx ,$$

which exists for all u. A similar formula for the distribution function is

$$F(u) = \frac{r^{r-1}}{(r-1)!} \sum_{j=0}^{r} \frac{e^{-ju/r}}{j!} \int_1^{\infty} x^{j-r-1} (\log x)^{r-1} e^{-xe^{-u/r}} dx,$$

which reduces, when $r = 2$, to

$$\left(1 + e^{-u/2} \right) e^{-e^{-u/2}} - e^{-u} E_1 \left(e^{-u/2} \right),$$

where

$$E_1(x) = \int_x^{\infty} \frac{e^{-u}}{u} du.$$

Appendix: The CR Bound

The usual model for a single sinusoid in Gaussian white noise is

$$y(t) = \mu + A\cos(\lambda t + \phi) + x(t), \; 0 < \lambda < \pi, \; 0 \le t \le T - 1.$$

However the calculations can be greatly simplified by reparametrisation as

$$y(t) = \nu + \alpha\left[\cos\{\lambda(t - \tau)\} - D_T(\lambda/2)\right] + \beta\sin\{\lambda(t - \tau)\} + x(t)$$

where

$$\begin{aligned}
D_T(\lambda) &= \frac{\sin(T\lambda)}{T\lambda}, \; \tau = \frac{T-1}{2}, \; \nu = \mu + \alpha D_T(\lambda/2), \\
\alpha &= A\cos\psi, \; \beta = -A\sin\psi, \; \psi = \phi + \tau\lambda.
\end{aligned}$$

We now estimate the vector $\begin{bmatrix} \sigma^2 & \lambda & \nu & \alpha & \beta \end{bmatrix}'$. Let

$$Y_T' = \begin{bmatrix} y(0) & y(1) & \cdots & y(T-1) \end{bmatrix}$$

and

$$S(\lambda) = \begin{bmatrix}
1 & \cos(\lambda\tau) - D_T\left(\frac{\lambda}{2}\right) & -\sin(\lambda\tau) \\
\vdots & \vdots & \vdots \\
1 & \cos\{\lambda(t-\tau)\} - D_T\left(\frac{\lambda}{2}\right) & \sin\{\lambda(t-\tau)\} \\
\vdots & \vdots & \vdots \\
1 & \cos(\lambda\tau) - D_T\left(\frac{\lambda}{2}\right) & \sin(\lambda\tau)
\end{bmatrix}.$$

The log-likelihood is then

$$l(\sigma^2, \lambda, \nu, \alpha, \beta; Y_T) = -\frac{T}{2}\log(2\pi\sigma^2) - \frac{1}{2\sigma^2}\{Y_T - S(\lambda)C\}'\{Y_T - S(\lambda)C\}$$

where

$$C = \begin{bmatrix} \nu \\ B \end{bmatrix}, B = \begin{bmatrix} \alpha \\ \beta \end{bmatrix}.$$

The information matrix is easily seen to be, letting $S^{(1)}(\lambda) = \frac{d}{d\lambda}S(\lambda)$,

$$\frac{1}{\sigma^2}\begin{bmatrix}
\frac{T}{2\sigma^2} & 0 & 0 \\
0 & C'S^{(1)}(\lambda)'S^{(1)}(\lambda)C & C'S^{(1)}(\lambda)'S(\lambda) \\
0 & S(\lambda)'S^{(1)}(\lambda)C & S(\lambda)'S(\lambda)
\end{bmatrix}$$

and $S^{(1)}(\lambda)'S^{(1)}(\lambda)$, $S^{(1)}(\lambda)'S(\lambda)$ and $S(\lambda)'S(\lambda)$ are all diagonal. Moreover all elements in the first row and column of the first two of these are zero. Thus it is easy to evaluate the element of the inverse of this matrix in the row and column corresponding to λ. It is just

$$\frac{\sigma^2}{B'S^{(1)}(\lambda)'S^{(1)}(\lambda)B - B'S_1^{(1)}(\lambda)'S_1(\lambda)\left\{S_1(\lambda)'S_1(\lambda)\right\}^{-1}S_1(\lambda)'S_1^{(1)}(\lambda)B}$$

where the subscripted matrices are the originals with first columns removed. The elements in these matrices are simple to compute. Using

$$\sum_{t=0}^{T-1}\sin^2\left\{\lambda(t-\tau)\right\} = \frac{1}{2}\sum_{t=0}^{T-1}\left[1 - \cos\left\{2\lambda(t-\tau)\right\}\right],$$

$$\sum_{t=0}^{T-1}\sin\left\{\lambda(t-\tau)\right\}\cos\left\{\lambda(t-\tau)\right\} = \frac{1}{2}\sum_{t=0}^{T-1}\sin\left\{2\lambda(t-\tau)\right\}$$

and

$$\sum_{t=0}^{T-1}e^{i\lambda(t-\tau)} = e^{-i\lambda\tau}\frac{e^{i\lambda T}-1}{e^{i\lambda}-1} = D_T(\lambda/2),$$

we obtain

$$S_1(\lambda)'S_1(\lambda) = \frac{T}{2}\left[\begin{array}{cc} 1 + D_T(\lambda) - 2D_T^2(\lambda/2) & 0 \\ 0 & 1 - D_T(\lambda) \end{array}\right].$$

Similarly, it may be shown that

$$S_1^{(1)}(\lambda)'S_1(\lambda) = \frac{T}{4}\left[\begin{array}{cc} D_T'(\lambda) - 2D_T(\lambda/2)D_T'(\lambda/2) & 0 \\ 0 & -D_T'(\lambda) \end{array}\right]$$

and

$$S_1^{(1)}(\lambda)'S_1^{(1)}(\lambda)$$
$$= \frac{T}{24}\left[\begin{array}{cc} T^2 - 1 + 3D_T''(\lambda) - 6\left[D_T'(\lambda/2)\right]^2 & 0 \\ 0 & T^2 - 1 - 3D_T''(\lambda) \end{array}\right].$$

Thus (3.24), (3.25) and (3.26) follow immediately.

4

Techniques Derived from ARMA Modelling

4.1 ARMA representation

In Chapter 3, maximum likelihood techniques for the estimation of frequency were discussed. These techniques tend to be computationally intensive, as the functions to be maximised are nonlinear and vary rapidly with frequency. Thus good initial estimates are needed. However, they are not not generally provided by maximisation over the ω_j, as these frequencies provide a mesh of size $2\pi/T$, while initial estimates are needed to an accuracy of order $T^{-1-\delta}$, where $\delta > 0$, for example, for most iterative techniques to work. There is thus a need for a technique which is more robust to these relatively poor initial estimates. Several inefficient estimators of frequency have been motivated by the fact that sinusoids are the solutions to second order difference equations whose auxiliary polynomials have all of their zeros on the unit circle. We shall explore several of these techniques in Chapter 5 but consider this difference equation property now. There is a second order filter which annihilates a (discrete-time) sinusoid at a given frequency: if $\{y(t)\}$ satisfies

$$y(t) = A \cos(\lambda t + \phi) + x(t), \tag{4.1}$$

then it also satisfies

$$y(t) - 2\cos\lambda y(t-1) + y(t-2) = x(t) - 2\cos\lambda x(t-1) + x(t-2). \tag{4.2}$$

Hence $\{y(t)\}$ satisfies an ARMA(2,2) equation, which does not have a stationary or invertible solution. This representation nevertheless suggests that λ may be estimated by some ARMA-based technique. We could, for example, ignore the second and third terms on the right side of (4.2), and fit autoregressive models of order two, either unconstrained or with the second coefficient constrained to be 1. Such techniques are still commonly used, but the estimator of λ using such approaches will be shown in Section 5.2 to have very poor properties. Instead, we take advantage of the nature of the right side. In Section 4.2 we shall construct such an estimator. In the remainder of the chapter, we analyse, interpret and generalise this technique.

4.2 An iterative ARMA technique

In Chapter 5 we shall show that frequency estimation techniques based on autoregressive estimation are inconsistent, i.e., asymptotically biased. It might therefore be expected that ARMA fitting will also yield inconsistent estimators. Even if an ARMA-based technique produced a consistent estimator, such an estimator ought not to be statistically efficient, since ARMA estimators generally have asymptotic variances which are of order $O\left(T^{-1}\right)$, while we know that the MLE of frequency has an asymptotic variance of order $O\left(T^{-3}\right)$. Any ARMA-based technique, therefore, will have to be of a non-standard type.

Suppose we wish to estimate α and β in

$$y\left(t\right) - \beta y\left(t-1\right) + y\left(t-2\right) = x\left(t\right) - \alpha x\left(t-1\right) + x\left(t-2\right)$$

while preserving the constraint $\alpha = \beta$. If α is known, and the $x\left(t\right)$ are independent and identically distributed, then β can be estimated by Gaussian maximum likelihood, that is, by minimising

$$\sum_{t=0}^{T-1} x_{\alpha,\beta}^{2}\left(t\right) = \sum_{t=0}^{T-1} \left\{\xi\left(t\right) - \beta\xi\left(t-1\right) + \xi\left(t-2\right)\right\}^{2}$$

with respect to β, where $\xi\left(t\right) = y\left(t\right) + \alpha\xi\left(t-1\right) - \xi\left(t-2\right)$ and $\xi\left(t\right) = 0$, $t < 0$. As this is quadratic in β, the minimising value is the regression coefficient of $\xi\left(t\right) + \xi\left(t-2\right)$ on $\xi\left(t-1\right)$, viz.

$$\frac{\sum_{t=0}^{T-1}\left\{\xi\left(t\right) + \xi\left(t-2\right)\right\}\xi\left(t-1\right)}{\sum_{t=0}^{T-1}\xi^{2}\left(t-1\right)} = \alpha + \frac{\sum_{t=0}^{T-1}y\left(t\right)\xi\left(t-1\right)}{\sum_{t=0}^{T-1}\xi^{2}\left(t-1\right)}$$

$$= \alpha + h_{T}\left(\alpha\right), \qquad (4.3)$$

say. We can then put α equal to this value and re-estimate β using the above, continuing until α and β are sufficiently close, estimating λ from the equation $\alpha = 2\cos\lambda$. We are not saying that this approach is valid *a priori*. We are merely proposing an estimator based on the observation that a sinusoid may be filtered out using a second order filter. Once the estimator is obtained, the principle which generated the estimator is essentially irrelevant. In fact, there is no guarantee that the procedure even minimises $\sum_{t=0}^{T-1} x_{\alpha,\beta}^{2}\left(t\right)$.

For reasons which will become apparent in the proofs which follow, we shall replace the correction term $h_T(\alpha)$ above by twice that value, accelerating the convergence which would otherwise ensue. The factor 2 was, in fact, suggested after reviewing simulations of the above technique. Initial attempts to prove anything about that technique had not led anywhere, but the theory fell into place once the very concept of an accelerating factor was introduced. The complete algorithm is as follows:

Algorithm 1 *Quinn & Fernandes (1991)*

   (i) *Put $\alpha_1 = 2\cos\widehat{\lambda}_1$, where $\widehat{\lambda}_1$ is an initial estimator of the true value λ_0.*

   (ii) *For $j \geq 1$, put $\xi(t) = y(t) + \alpha_j\xi(t-1) - \xi(t-2)$, $t = 0,\ldots,T-1$ where $\xi(t) = 0$, $t < 0$.*

   (iii) *Put*

$$\beta_j = \alpha_j + 2\frac{\sum_{t=0}^{T-1} y(t)\xi(t-1)}{\sum_{t=0}^{T-1} \xi^2(t-1)}.$$

   (iv) *If $|\beta_j - \alpha_j|$ is suitably small, put $\widehat{\lambda} = \cos^{-1}(\beta_j/2)$. Otherwise, let $\alpha_{j+1} = \beta_j$ and go to Step (ii).*

If the accuracy of the initial estimator is known, the number of iterations may be calculated in advance, as will be shown in Theorem 17. For example, if $\widehat{\lambda}_1$ is the periodogram maximiser, then we may put $\widehat{\lambda} = \cos^{-1}(\beta_2/2)$.

4.3 Interpretation of the technique

As the above technique is iterative, there are two important convergence questions to ask. First, does the sequence converge in some sense; second, what are the properties of this limiting value? Both these questions will be answered later, but it is instructive here to interpret the limiting value. Assuming that it is $\widehat{\beta} = 2\cos\widehat{\lambda}$, it is easily seen that $\widehat{\lambda}$ is a zero of

$$\sum_{t=0}^{T-1} y(t)\xi(t-1)$$

where $\xi(t) = y(t) + 2\cos\lambda\xi(t-1) - \xi(t-2)$ and $\xi(t) = 0$, $t < 0$. Since

$$\xi(t) = (\sin\lambda)^{-1}\sum_{j=0}^{T-1}\sin\{\lambda(j+1)\}\,y(t-j)$$

it follows that $\widehat{\lambda}$ is a zero of

$$\nu_T(\lambda) = \sum_{j=1}^{T-1} \sin(j\lambda) \sum_{t=j}^{T-1} y(t)y(t-j) = T \sum_{j=1}^{T-1} \sin(j\lambda) c_j$$

where

$$c_j = T^{-1} \sum_{t=j}^{T-1} y(t)y(t-j)$$

is the jth sample autocovariance of $\{y(t); t = 0, 1, \ldots, T-1\}$. Using the result of Theorem 17, it may be shown that $\nu_T(\lambda)$ is less than 0 when λ is less than but close to $\widehat{\lambda}$ and greater than 0 when λ is greater than but close to $\widehat{\lambda}$. Thus

$$\int_0^\lambda \nu_T(\omega)\, d\omega = -T \sum_{j=1}^{T-1} j^{-1} \{\cos(j\lambda) - 1\} c_j$$

has a local minimiser at $\widehat{\lambda}$. Let

$$I_y(\lambda) = \frac{2}{T} \left| \sum_{t=0}^{T-1} y(t) e^{-it\lambda} \right|^2$$

be the periodogram of $\{y(t); t = 0, \ldots, T-1\}$. Then

$$I_y(\lambda) = 2 \sum_{j=1-T}^{T-1} e^{-ij\lambda} c_j$$

and consequently

$$c_j = (4\pi)^{-1} \int_{-\pi}^{\pi} e^{ij\lambda} I_y(\lambda)\, d\lambda.$$

The procedure is equivalent therefore to locating a particular local maximiser of

$$\sum_{j=1}^{T-1} j^{-1} \cos(j\lambda) \int_{-\pi}^{\pi} e^{ij\omega} I_y(\omega)\, d\omega$$

$$= \frac{1}{2} \int_{-\pi}^{\pi} I_y(\omega) \sum_{j=1}^{T-1} j^{-1} e^{ij(\omega-\lambda)}\, d\omega + \frac{1}{2} \int_{-\pi}^{\pi} I_y(-\omega) \sum_{j=1}^{T-1} j^{-1} e^{-ij(\omega-\lambda)}\, d\omega$$

$$= \int_{-\pi}^{\pi} I_y(\omega) \sum_{j=1}^{T-1} j^{-1} \cos\{j(\lambda - \omega)\}\, d\omega,$$

since I_y is even. We are thus finding a local maximiser of the *smoothed* periodogram

$$\kappa_T\left(\lambda\right) = \int_{-\pi}^{\pi} I_y\left(\omega\right) \mu_T\left(\lambda - \omega\right) d\omega$$

where

$$\mu_T\left(\omega\right) = \sum_{k=1}^{T-1} k^{-1} \cos\left(k\omega\right).$$

In fact, since $I_y\left(\omega\right)$ does not contain the terms $\cos\left(j\omega\right)$, $j \geq T$, we also have

$$\kappa_T\left(\lambda\right) = \int_{-\pi}^{\pi} I_y\left(\omega\right) \mu\left(\lambda - \omega\right) d\omega,$$

where, for $\omega \neq 0$,

$$\mu\left(\omega\right) = \sum_{k=1}^{\infty} k^{-1} \cos\left(k\omega\right) = -\frac{1}{2} \log\left\{4\sin^2\left(\frac{\omega}{2}\right)\right\}.$$

We may thus interpret $\kappa_T\left(\omega\right)$ as a smoothed version of the periodogram, with a kernel which does not depend on T. We shall, however, use the representation in terms of μ_T, which is more convenient. Now,

$$\mu'_T(\omega) = -\sum_{k=1}^{T-1} \sin(k\omega) = -\operatorname{Im}\frac{e^{iT\omega} - 1}{e^{i\omega} - 1} = \frac{\cos\left\{(T-1/2)\omega\right\} - \cos(\omega/2)}{2\sin(\omega/2)}.$$

Thus, since

$$\mu_T(\pi) = \sum_{k=1}^{T-1} (-1)^k / k = -\log 2 + o(1),$$

it follows that

$$\mu_T(\omega) = -\log 2 + \int_{\omega}^{\pi} \frac{\cos(x/2) - \cos\left\{(T-1/2)x\right\}}{2\sin(x/2)} dx + o\left(1\right)$$

and, for all T and $\omega > 0$,

$$
\begin{aligned}
|\mu_T(\omega) + \log 2| \quad &< \quad \int_{\omega}^{\pi} \frac{dx}{\sin(x/2)} + o\left(1\right) \\
&= \quad 2\left\{\log(1 + \cos x) - \log(\sin x)\right\}_{\pi/2}^{\omega/2} + o\left(1\right) \\
&= \quad 2\log\left\{\cot(\omega/2) + \csc(\omega/2)\right\} + o\left(1\right) \\
&< \quad \infty.
\end{aligned}
$$

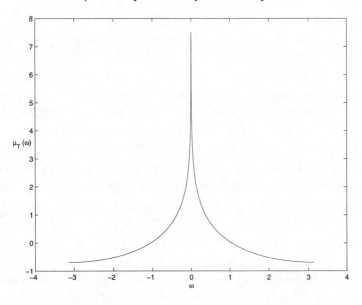

Fig. 4.1. Graph of $\mu_{1024}(\omega)$.

However,

$$
\begin{aligned}
\mu_T(0) - \mu_T(a/T) &= \sum_{k=1}^{T-1} k^{-1} - \mu_T(a/T) = \sum_{k=1}^{T-1} k^{-1}\{1 - \cos(ka/T)\} \\
&= \int_0^a T^{-1} \sum_{k=0}^{T-1} \sin(kx/T)dx \\
&= T^{-1} \int_0^a \frac{\cos\{x/(2T)\} - \cos[\{1 - 1/(2T)\}x]}{2\sin\{x/(2T)\}}dx \\
&= \int_0^a \frac{1 - \cos x}{x}dx\,\{1 + o(1)\}.
\end{aligned}
$$

Thus for fixed a,

$$
\mu_T(a/T) = \log T + \gamma - \int_0^a \frac{1 - \cos x}{x}dx + o(1)
$$

where γ (Euler's constant) ~ 0.57721. Therefore $\kappa(\lambda)$ is the convolution of I_y, which is peaky near λ_0 and of largest order T in intervals of length T^{-1} about λ_0, while μ_T, depicted in Figure 4.1 for $T = 1024$, is of order $\log T$ for arguments of order T^{-1} and of order 1 for any fixed nonzero argument.

It is thus plausible that $\kappa_T(\lambda)$ is maximised near λ_0. What is remarkable is that the maximiser of κ_T has the same central limit theorem as the periodogram maximiser, but that asymptotically $\kappa_T(\lambda)$ has no *sidelobes*. That is, $\kappa_T(\lambda)$, unlike $I_y(\lambda)$, has only one local maximiser within $O(T^{-1})$ of λ_0. The technique is thus more robust to errors in the initial estimator of λ_0. All this will be discussed in detail in Section 4.4. Figures 4.2, 4.3 and 4.4 depict the results of three simulations generated by (4.1), each with pseudo-normal independent noise and all ϕ_js zero. In Figure 4.2, where $A = 1, \lambda_0 = 1, T = 128$ and σ^2, the variance of $x(t)$, is 1, it is seen that if λ_0 were estimated by maximising κ by, say, Newton's method, a poor initial estimate of λ_0 would not prevent the correct maximiser being reached, while only slight initial inaccuracy in estimating the maximiser of the periodogram results in a sidelobe value. On the other hand, Figure 4.3, where there are two sinusoids with amplitudes 1 and 0.4, frequencies 0.3 and 0.5, $\sigma^2 = 0.25$ and $T = 128$, shows that close frequencies are not as clearly resolved by κ as by the periodogram, which is to be expected. Figure 4.4 shows the results for the same case, but with $T = 1024$, where the frequency 0.5 is clearly resolved. Finally, we show here the results of two simulations from those described at the end of Section 3.3. Figures 4.5 and 4.6, illustrate the coloured noise case with sample sizes $T = 128$ and 1024. It is easy to see why any technique will confuse the true frequency 2 with the 'pseudo-frequency' $\pi/2$. The MATLAB^{TM} code used to produce the figures is contained in the Appendix in the files `qfgen.m` and `qfpic.m`.

4.4 Asymptotic behaviour of the procedure

From what has been said in Sections 4.1 and 4.3, the analysis of the method will centre around the function $h_T(\alpha)$ defined in (4.3). Theorem 16 shows that there is, almost surely as $T \to \infty$, one and only one zero of h_T in a certain neighbourhood of $\alpha_0 = 2\cos(\lambda_0)$. It follows that there is a sequence $\{\widehat{\lambda}_T\}$, defined as the sequence of zeros of $\{h_T(\lambda)\}$ in a neighbourhood of λ_0, which converges almost surely to λ_0, and in fact it will follow that $T^\varepsilon(\widehat{\lambda}_T - \lambda_0)$ converges to 0 almost surely for any $\varepsilon < 3/2$. Theorem 17 shows that, as long as the initial estimator $\widehat{\lambda}_1$ is of a certain accuracy, only a small number of iterations is needed to guarantee that the estimator is closer in order to $\widehat{\lambda}_T$ than $\widehat{\lambda}_T$ is to λ_0. Finally, Theorem 18 describes the central limit theorem for $\widehat{\lambda}_T$, and therefore for the iterative estimator. What is shown is that $\widehat{\lambda}_T$ has the same central limit theorem as the periodogram maximiser, and thus asymptotically 'achieves the Cramér–Rao lower bound' when $\{x_t\}$ is Gaussian.

Where order notation is used in the following, the orders should, as usual, be understood as being almost surely as $T \to \infty$.

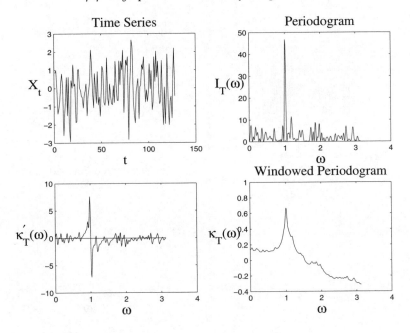

Fig. 4.2. Single sinusoid in white noise.

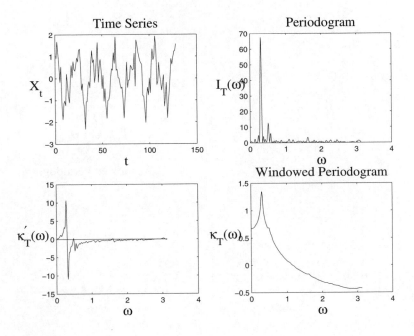

Fig. 4.3. Two sinusoids, $T = 128$.

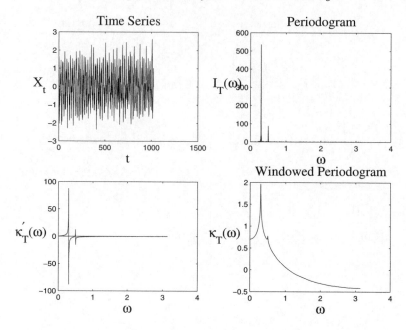

Fig. 4.4. Two sinusoids, $T = 1024$.

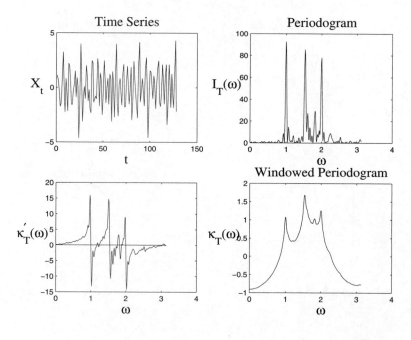

Fig. 4.5. Two sinusoids in coloured noise, $T = 128$.

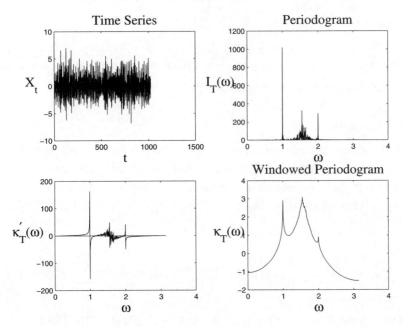

Fig. 4.6. Two sinusoids in coloured noise, $T = 1024$.

Theorem 16 *Let $A_T(\varepsilon) = \{\alpha; |\alpha - \alpha_0| < cT^{-\varepsilon}\}$, where ε and c are fixed, $1 < \varepsilon < \frac{3}{2}$, $c > 0$ and $\alpha_0 = 2\cos\lambda_0$. Then almost surely as $T \to \infty$, there exists a unique $\widehat{\alpha}_T \in A_T(\varepsilon)$ with $h_T(\widehat{\alpha}_T) = 0$. There is thus a unique solution to $h_T(\widehat{\alpha}_T) = 0$ for which $T^\varepsilon(\widehat{\lambda}_T - \lambda_0) \to 0$, almost surely, for all $\varepsilon < \frac{3}{2}$, where $\widehat{\alpha}_T = 2\cos\widehat{\lambda}_T$.*

Theorem 17 *Let $\alpha_1 \in A_T(\varepsilon)$ and put $\alpha_{j+1} = \alpha_j + 2h_T(\alpha_j)$. If $1 < \varepsilon < \frac{3}{2}$, then*

$$\alpha_{j+1} - \widehat{\alpha}_T = (\alpha_j - \widehat{\alpha}_T) \, O\left\{T^{-1/2}(\log T)^{1/2}\right\}$$

while if $\frac{1}{2} < \varepsilon \leq 1$

$$\alpha_{j+1} - \widehat{\alpha}_T = (\alpha_j - \widehat{\alpha}_T) \, O\left\{T^{1/2-\varepsilon}(\log T)^{1/2}\right\} + O\left\{T^{1/2-2\varepsilon}(\log T)^{1/2}\right\}.$$

Furthermore,

$$\widehat{\alpha}_T - \alpha_0 = O\left\{T^{-3/2}(\log T)^{1/2}\right\}$$

and

$$\alpha_k - \widehat{\alpha}_T = o\left(T^{-3/2}\right)$$

for $k \geq k\left(\varepsilon\right) = \lfloor 3 - \log\left(2\varepsilon - 1\right) / \log 2 \rfloor.$

Theorem 18 $T^{3/2}\left(\widehat{\lambda}_T - \lambda_0\right)$ *has a distribution converging to the normal with mean zero and variance* $48\pi A^{-2} f_x\left(\lambda_0\right).$

Proof (Theorem 16)

It is obvious that we need to use some specialised form of fixed point theorem, and we shall use that of Dieudonné (1969, p. 266). We must show that, almost surely as $T \to \infty$, there exists k with $0 \leq k < 1$ such that if α and $\alpha' \in A_T\left(\varepsilon\right)$, then $\left|\alpha - \alpha' + 2h_T\left(\alpha\right) - 2h_T\left(\alpha'\right)\right| \leq k\left|\alpha - \alpha'\right|$ and $\left|2h_T\left(\alpha_0\right)\right| < \left(1 - k\right)T^{-\varepsilon}$. Put

$$d_T\left(\alpha\right) = \sin\lambda \sum_{t=0}^{T-1} y\left(t\right)\xi\left(t - 1\right), \quad e_T\left(\alpha\right) = \sin^2\lambda \sum_{t=0}^{T-1} \xi^2\left(t - 1\right),$$

where $\alpha = 2\cos\lambda$ and $\xi\left(t\right) = y\left(t\right) + \alpha y\left(t - 1\right) - y\left(t - 2\right)$. Then

$$h_T\left(\alpha\right) = \sin\lambda \, d_T\left(\alpha\right) e_T^{-1}\left(\alpha\right).$$

Now

$$e_T\left(\alpha\right)$$
$$= \sum_{t=0}^{T-1}\left(\sum_{j=0}^{t} \sin\left(j\lambda\right)\left[x\left(t - j\right) + A\cos\left\{\left(t - j\right)\lambda_0 + \phi_0\right\}\right]\right)^2$$
$$= \sum_{t=0}^{T-1}\left\{\sum_{j=0}^{t} \sin\left(j\lambda\right)x\left(t - j\right)\right\}^2$$
$$+ A^2 \sum_{t=0}^{T-1}\left[\sum_{j=0}^{t} \sin\left(j\lambda\right)\cos\left\{\left(t - j\right)\lambda_0 + \phi_0\right\}\right]^2$$
$$+ 2A \sum_{t=0}^{T-1}\sum_{j=0}^{t} \sin\left(j\lambda\right)x\left(t - j\right)\sum_{k=0}^{t} \sin\left(k\lambda\right)\cos\left\{\left(t - k\right)\lambda_0 + \phi_0\right\}.$$

$$(4.4)$$

As we shall need to compare the values of $e_T\left(\alpha\right)$ for two arbitrary values in $A_T\left(\varepsilon\right)$ in order to use Dieudonné's theorem, it is natural to consider an expansion of $e_T\left(\alpha\right)$ about α_0. The first term of (4.4) is differentiable with respect to λ and may be written as

$$\sum_{t=0}^{T-1}\left\{\sum_{j=0}^{t}\sin\left(j\lambda_0\right)x\left(t-j\right)\right\}^2$$

$$+2\left(\lambda-\lambda_0\right)\sum_{t=0}^{T-1}\sum_{j=0}^{t}\sin\left(j\lambda^*\right)x\left(t-j\right)\sum_{k=0}^{t}k\cos\left(k\lambda^*\right)x\left(t-k\right),$$

where λ^* denotes, generically, a number between λ and λ_0. Now,

$$\sum_{j=0}^{t}\sin\left(j\lambda^*\right)x\left(t-j\right)=O\{(t\log t)^{1/2}\},$$

while

$$\sum_{k=0}^{t}k\cos\left(k\lambda^*\right)x\left(t-k\right)\;=\;t\sum_{k=0}^{t}\cos\left(k\lambda^*\right)x\left(t-k\right)$$

$$-\sum_{j=1}^{t}\sum_{k=0}^{t-j}\cos\left(k\lambda^*\right)x\left(t-k\right)$$

$$=\;O\left\{t^{3/2}\left(\log t\right)^{1/2}\right\},$$

uniformly in λ^*. The first term in (4.4) is thus

$$\sum_{t=0}^{T-1}\left\{\sum_{j=0}^{t}\sin\left(j\lambda_0\right)x\left(t-j\right)\right\}^2+2\left(\lambda-\lambda_0\right)O\left(T^3\log T\right),$$

which is, incidentally, $O\left(T^2\log T\right)$, uniformly in λ since $\varepsilon>1$. The third term is, using a similar argument,

$$2A\sum_{t=0}^{T-1}\sum_{j=0}^{t}\sin\left(j\lambda_0\right)x\left(t-j\right)\sum_{k=0}^{t}\sin\left(k\lambda_0\right)\cos\left\{\left(t-k\right)\lambda_0+\phi_0\right\}$$

$$+\left(\lambda-\lambda_0\right)O\left\{T^{7/2}\left(\log T\right)^{1/2}\right\},$$

which is $O\left\{T^{5/2}\left(\log T\right)^{1/2}\right\}$. The second term is the dominant one, and is

$$A^2\sum_{t=0}^{T-1}\left[\sum_{j=0}^{t}\sin\left(j\lambda_0\right)\cos\left\{\left(t-j\right)\lambda_0+\phi_0\right\}\right]^2+\left(\lambda-\lambda_0\right)O\left(T^3\right),$$

uniformly in λ, which is $\frac{A^2}{24}T^3+O\left(T^2\right)$. Thus $e_T\left(\alpha\right)$ is

$$\frac{A^2}{24}T^3+O\left\{T^{5/2}\left(\log T\right)^{1/2}\right\}$$

and also

$$e_T(\alpha) = e_T(\alpha_0) + (\lambda - \lambda_0) O\left\{T^{7/2}(\log T)^{1/2}\right\}. \tag{4.5}$$

The term $d_T(\alpha)$ may be treated similarly. Now

$$
\begin{aligned}
d_T(\alpha) \;=\; & \sum_{t=0}^{T-1} x(t) \sum_{j=0}^{t} \sin(j\lambda)\, x(t-j) \\
& + A^2 \sum_{t=0}^{T-1} \cos(t\lambda_0 + \phi_0) \sum_{j=0}^{t} \sin(j\lambda)\cos\{(t-j)\lambda_0 + \phi_0\} \\
& + A \sum_{t=0}^{T-1} x(t) \sum_{j=0}^{t} \sin(j\lambda)\cos\{(t-j)\lambda_0 + \phi_0\} \\
& + A \sum_{t=0}^{T-1} \cos(t\lambda_0 + \phi_0) \sum_{j=0}^{t} \sin(j\lambda)\, x(t-j). \tag{4.6}
\end{aligned}
$$

The first term in (4.6) may be written as

$$\sum_{t=0}^{T-1} x(t) \sum_{j=0}^{t} \sin(j\lambda_0)\, x(t-j) + (\lambda - \lambda_0) \sum_{t=0}^{T-1} x(t) \sum_{j=0}^{t} j\cos(j\lambda^*)\, x(t-j).$$

The square of the first term in this expression is less than

$$\sum_{t=0}^{T-1} x^2(t) \sum_{t=0}^{T-1}\left\{\sum_{j=0}^{t} \sin(j\lambda_0)\, x(t-j)\right\}^2$$

which is $O\left(T^3 \log T\right)$ since $T^{-1}\sum_{t=0}^{T-1} x^2(t) \to E\left\{x^2(t)\right\}$. The first expression is thus $O\left\{T^{3/2}(\log T)^{1/2}\right\}$. The second part of the expression may similarly be shown to be $(\lambda - \lambda_0) O\left\{T^{5/2}(\log T)^{1/2}\right\}$.

The second term in (4.6) is

$$A^2 \sum_{t=0}^{T-1} \cos(t\lambda_0 + \phi_0) \sum_{j=0}^{t} \sin(j\lambda_0)\cos\{(t-j)\lambda_0 + \phi_0\}$$

$$+ A^2(\lambda - \lambda_0) \sum_{t=0}^{T-1} \cos(t\lambda_0 + \phi_0) \sum_{j=0}^{t} j\cos(j\lambda^*)\cos\{(t-j)\lambda_0 + \phi_0\}$$

which is $O(T) + (\lambda - \lambda_0)\left\{A^2 T^3/24 + O(T^2)\right\}$, while the third term is

$$A \sum_{t=0}^{T-1} x(t) \sum_{j=0}^{t} \sin(j\lambda_0) \cos\{(t-j)\lambda_0 + \phi_0\}$$

$$+ A \sum_{t=0}^{T-1} x(t) \sum_{j=t}^{T-1} \sin\{(j-t)\lambda_0\} \cos(j\lambda_0 + \phi_0)$$

$$+ A(\lambda - \lambda_0) \sum_{t=0}^{T-1} x(t) \sum_{j=0}^{t} j \cos(j\lambda^*) \cos\{(t-j)\lambda_0 + \phi_0\}$$

$$+ A(\lambda - \lambda_0) \sum_{t=0}^{T-1} x(t) \sum_{j=t}^{T-1} (j-t) \cos\{(j-t)\lambda^*\} \cos(j\lambda_0 + \phi_0)$$

which can be shown to be

$$A/2 \sum_{t=0}^{T-1} x(t) \{(2t-T) \sin(t\lambda_0 + \phi_0) + O(1)\}$$

$$+ A/4(\lambda - \lambda_0) \sum_{t=0}^{T-1} x(t) \{(T^2 - 2Tt + 2t^2) \cos(t\lambda_0 + \phi_0) + O(t)\}$$

which is $O\left\{T^{3/2}(\log T)^{1/2}\right\} + (\lambda - \lambda_0) O\left\{T^{5/2}(\log T)^{1/2}\right\}$, uniformly in λ.
We thus have

$$d_T(\alpha) = d_T(\alpha_0) + (\lambda - \lambda_0)\left\{A^2 T^3/24 + O\left(T^2\right)\right\}. \tag{4.7}$$

We may now check the conditions needed for Dieudonné's theorem: let
$\alpha = 2\cos\lambda$ and $\alpha' = 2\cos\lambda'$, where $\alpha, \alpha' \in A_T(\varepsilon)$. Then

$$h_T(\alpha) - h_T(\alpha') = \sin\lambda \, d_T(\alpha)/e_T(\alpha) - \sin\lambda' \, d_T(\alpha')/e_T(\alpha')$$

and

$$\begin{aligned} e_T(\alpha) e_T(\alpha') \{h_T(\alpha) - h_T(\alpha')\} &= (\sin\lambda - \sin\lambda') d_T(\alpha) e_T(\alpha') \\ &\quad + \sin\lambda' \{d_T(\alpha) - d_T(\alpha')\} e_T(\alpha) \\ &\quad - \sin\lambda' \, d_T(\alpha) \{e_T(\alpha) - e_T(\alpha')\} \end{aligned}$$

which, because of (4.5) and (4.7), is

$$(\lambda - \lambda') O\left\{T^{9/2}(\log T)^{1/2}\right\} + (\lambda - \lambda') O\left\{T^{11/2}(\log T)^{1/2}\right\}$$

$$+ (\lambda - \lambda') \sin\lambda_0 \, (A^2/24)^2 T^6 \left[1 + O\left\{T^{-1/2}(\log T)^{1/2}\right\}\right]$$

or

$$\left(\lambda - \lambda'\right) \left[\sin \lambda_0 \ \left(A^2/24\right)^2 T^6 + O\left\{T^{11/2} \left(\log T\right)^{1/2}\right\}\right].$$

Thus

$$h_T\left(\alpha\right) - h_T\left(\alpha'\right) = \sin \lambda_0 \ \left(\lambda - \lambda'\right) \left[1 + O\left\{T^{-1/2} \left(\log T\right)^{1/2}\right\}\right]$$

and, since

$$\alpha - \alpha' = 2\cos\lambda - 2\cos\lambda' = -2\sin\lambda_0 \ \left(\lambda - \lambda'\right)\left\{1 + o\left(T^{-2}\right)\right\},$$

we have

$$\alpha - \alpha' + 2h_T\left(\alpha\right) - 2h_T\left(\alpha'\right)$$
$$= \ \alpha - \alpha' - \left(\alpha - \alpha'\right)\left[1 + O\left\{T^{-1/2}\left(\log T\right)^{1/2}\right\}\right]$$
$$= \ \left(\alpha - \alpha'\right) O\left\{T^{-1/2}\left(\log T\right)^{1/2}\right\},$$

uniformly in α and α'. Also,

$$2h_T\left(\alpha_0\right)$$
$$= \ \sin\lambda_0\, O\left\{T^{3/2}\left(\log T\right)^{1/2}\right\} \Big/ \left[A^2 T^3/24 + O\left\{T^{5/2}\left(\log T\right)^{1/2}\right\}\right]$$
$$= \ 2\sin\lambda_0\, O\left\{T^{-3/2}\left(\log T\right)^{1/2}\right\}.$$

The conditions thus hold and the first part of the theorem is proved. As the result holds for $1 < \varepsilon < 3/2$ and $A_T\left(\varepsilon_1\right) \supset A_T\left(\varepsilon_2\right)$ if $\varepsilon_1 < \varepsilon_2$, it follows that there is a unique solution $\widehat{\alpha}_T$ to $h_T\left(\alpha\right) = 0$, for which

$$T^\varepsilon\left(\widehat{\alpha}_T - \alpha_0\right) \to 0, \ \text{a.s.,} \ \text{for all } \varepsilon < \frac{3}{2}.$$

Since $\widehat{\alpha}_T = 2\cos\widehat{\lambda}_T$, it follows that $T^\varepsilon\left(\widehat{\lambda}_T - \lambda_0\right) \to 0$, almost surely. □

Proof (Theorem 17)

Let $\alpha \in A_T\left(\varepsilon\right)$, where $\varepsilon > 1$. Then the proof of the above theorem shows that

$$h_T\left(\alpha\right) \ = \ -1/2\left(\alpha - \widehat{\alpha}_T\right) + \left(\alpha - \widehat{\alpha}_T\right) O\left\{T^{-1/2}\left(\log T\right)^{1/2}\right\}$$
$$= \ -1/2\left(\alpha - \widehat{\alpha}_T\right)\left[1 + O\left\{T^{-1/2}\left(\log T\right)^{1/2}\right\}\right].$$

Thus

$$\{\alpha + 2h_T(\alpha)\} - \widehat{\alpha}_T = (\alpha - \widehat{\alpha}_T) O\left\{T^{-1/2} (\log T)^{1/2}\right\}$$

$$= O\left\{T^{-1/2-\varepsilon} (\log T)^{1/2}\right\}$$

$$= o\left(T^{-3/2}\right).$$

In particular, this holds when $\alpha = \alpha_0$, and so

$$\widehat{\alpha}_T - \alpha_0 = 2h_T(\alpha_0) + o\left(T^{-3/2}\right)$$

$$= O\left\{T^{-3/2} (\log T)^{1/2}\right\}. \tag{4.8}$$

The theorem is thus seen to be true for $\varepsilon > 1$, by noting that then

$$\lfloor 3 - \log(2\varepsilon - 1)/\log 2 \rfloor = 2.$$

If $\frac{1}{2} < \varepsilon \leq 1$, the theory developed for the proof of Theorem 16 needs to be extended. We omit the detailed proofs, but it is fairly straightforward to show that

$$e_T(\alpha) = \frac{A^2 T}{4(\lambda - \lambda_0)^2} \left[1 - \frac{\sin\{T(\lambda - \lambda_0)\}}{T(\lambda - \lambda_0)}\right] \left[1 + O\left\{T^{1/2-\varepsilon} (\log T)^{1/2}\right\}\right]$$

and

$$d_T(\alpha) = \frac{A^2 T}{4(\lambda - \lambda_0)} \left[1 - \frac{\sin\{T(\lambda - \lambda_0)\}}{T(\lambda - \lambda_0)}\right] \{1 + O(T^{-\varepsilon})\}$$

$$+ O\left\{T^{3/2} (\log T)^{1/2}\right\},$$

uniformly in λ. As $e_T^{-1}(\alpha) = O\left(T^{-1-2\varepsilon}\right)$,

$$h_T(\alpha)$$

$$= (\lambda - \lambda_0) \sin \lambda \left[1 + O\left\{T^{1/2-\varepsilon} (\log T)^{1/2}\right\}\right] + O\left\{T^{1/2-2\varepsilon} (\log T)^{1/2}\right\}$$

$$= \left(\lambda - \widehat{\lambda}_T + \widehat{\lambda}_T - \lambda_0\right) \sin \lambda \left[1 + O\left\{T^{1/2-\varepsilon} (\log T)^{1/2}\right\}\right]$$

$$+ O\left\{T^{1/2-2\varepsilon} (\log T)^{1/2}\right\}$$

$$= -\frac{1}{2}(\alpha - \widehat{\alpha}_T) \left[1 + O\left\{T^{1/2-\varepsilon} (\log T)^{1/2}\right\}\right] + O\left\{T^{1/2-2\varepsilon} (\log T)^{1/2}\right\}$$

since $\widehat{\lambda}_T - \lambda_0 = O\left\{T^{-3/2} (\log T)^{1/2}\right\}$. Thus

$$\alpha + 2h_T(\alpha) - \widehat{\alpha}_T$$

$$= (\alpha - \widehat{\alpha}_T) O\left\{T^{1/2-\varepsilon} (\log T)^{1/2}\right\} + O\left\{T^{1/2-2\varepsilon} (\log T)^{1/2}\right\},$$

and it follows that, if $\alpha_1 \in A_T(\varepsilon)$, then

$$\alpha_2 = \alpha_1 + 2h_T(\alpha_1) \in A_T(2\varepsilon - 1/2 - \delta),$$

where δ is arbitrarily small. The proof of the final part of the theorem is thus via a simple counting argument: Let $a_1 = \varepsilon$ and put $a_{j+1} = 2a_j - \frac{1}{2} - \delta$, where δ is arbitrarily small. Then

$$a_k = 2^{k-1}\left(\varepsilon - \frac{1}{2}\right) + \frac{1}{2} - \delta\left(2^{k-1} - 1\right).$$

It is easily seen that the smallest value of k for which $a_k > 1$ is

$$\lfloor 2 - \log(2\varepsilon - 1)/\log 2 \rfloor$$

and, from the earlier part of the proof, one more iteration is needed to get within $o(T^{-3/2})$ of $\widehat{\alpha}_T$. In particular, $\alpha_3 - \widehat{\alpha}_T = o\left(T^{-3/2}\right)$ if $\varepsilon = 1$, which is the case when the procedure is initialised by the maximiser of the periodogram over the Fourier frequencies $\{2\pi j/T;\ j = 1, \ldots, \lfloor (T-1)/2 \rfloor\}$.

$\square$

Proof (Theorem 18)

From (4.8), it follows that

$$\widehat{\lambda}_T - \lambda_0 = -\frac{d_T(\alpha_0)}{e_T(\alpha_0)}\{1 + o(1)\}.$$

Now, since

$$T^{-3}e_T(\omega_0) \to \frac{A^2}{24},\ \text{a.s.},$$

$T^{3/2}\frac{A^2}{24}\left(\widehat{\lambda}_T - \lambda_0\right)$ has the same asymptotic distribution as

$$-T^{-3/2}d_T(\alpha_0),$$

the dominant terms in which are, from the proof of Theorem 16,

$$T^{-3/2}\sum_{t=0}^{T-1} x(t) \sum_{j=0}^{t} \sin(j\lambda_0) x(t-j)$$

$$+\frac{1}{2}AT^{-3/2}\sum_{t=0}^{T-1} x(t)\{(2t - T)\sin(t\lambda_0 + \phi_0) + O(1)\}.$$

As it is not too hard to show that $T^{1/2}$ by the first term converges almost surely, it follows that

$$T^{-3/2}d_T\left(\lambda_0\right) \sim T^{-3/2}\frac{A}{2}\sum_{t=0}^{T-1}\left(2t-T\right)\sin\left(t\lambda_0+\phi_0\right)x\left(t\right).$$

From Theorem 4, the right side of the above is asymptotically normally distributed with mean zero and variance $2\pi f_x\left(\lambda_0\right)A^2/24$. Consequently, $T^{3/2}\left(\widehat{\lambda}_T-\lambda_0\right)$ is asymptotically distributed normally with mean zero and variance $48\pi f_x(\lambda_0)/A^2$. $\qquad\qquad\square$

We note here that the process $\{y\left(t\right)\}$ may be prefiltered by any filter whose frequency response function has no zeros at λ_0, for the net effect, if the filter is $\sum_{j=0}^{\infty}a_jz^j$, is to replace A^2 with A^2D and $f_x\left(\lambda_0\right)$ with $Df_x\left(\lambda_0\right)$ where $D=\left|\sum_{j=0}^{\infty}a_je^{-ij\lambda_0}\right|^2$. The variance in the limiting normal distribution is thus unaltered. We suppose, of course, that $D<\infty$.

4.5 More than one frequency

As in the single frequency case, the technique is motivated by the fact that a certain difference operator annihilates all of the sinusoidal components. If $\{y\left(t\right)\}$ satisfies (3.1) then, letting $\theta_j=2\cos\lambda_j$,

$$\sum_{j=0}^{2r}\alpha_jy\left(t-j\right)=\sum_{j=0}^{2r}\alpha_jx\left(t-j\right), \tag{4.9}$$

where

$$\sum_{j=0}^{2r}\alpha_jz^j=\prod_{j=1}^{r}\left(1-\theta_jz+z^2\right), \tag{4.10}$$

and so $\alpha_{2r-j}=\alpha_j;\ j=0,\ldots,r-1$, with $\alpha_0=1$. Thus $\{y\left(t\right)\}$ satisfies an ARMA(2p, 2p) equation, the zeros of the auxiliary polynomials of which are all on the unit circle. Note that there is no other polynomial of order less than or equal to $2r$ which has this property.

The algorithm for the general case is suggested by the following argument. Suppose we wish to compute the least squares estimates of the parameters $\{\alpha_j;\ j=1,\ldots,r\}$ and $\{\beta_j;\ j=1,\ldots,r\}$ in (4.9) subject to the constraints $\alpha_j=\beta_j$. We thus, for $t=0,1,\ldots,T-1$, put

$$x_{\alpha,\beta}\left(t\right)=\sum_{j=0}^{2r}\beta_jy\left(t-j\right)-\sum_{j=1}^{2r}\alpha_jx_{\alpha,\beta}\left(t-j\right),$$

where $x_{\alpha,\beta}(t)$ is zero for $t < 0$. Suppose we have estimates of the α_j. We can easily minimise $\sum_{t=0}^{T-1} x_{\alpha,\beta}^2(t)$ with respect to the β_j, since this is quadratic in the β_j. We can then replace α_j by β_j, $j = 1, \ldots, r$ and iterate until $\max_j |\alpha_j - \beta_j|$ is small enough. We thus only need to have an initial estimate of the α_j to complete the formulation. The iteration step can again be modified so as to accelerate the convergence. The complete algorithm is

Algorithm 2

(i) *Obtain* $\alpha_1, \ldots, \alpha_r$ *from* $\sum_{j=0}^{2r} \alpha_j z^j = \prod_{j=1}^{r}(1 - 2z \cos \tilde{\lambda}_j + z^2)$, *where the* $\tilde{\lambda}_j$ *are initial estimators of the* λ_j.

(ii) *Put*

$$\xi(t) = y(t) - \sum_{j=1}^{2r} \alpha_j \xi(t - j); \quad t = 0, 1, \ldots, T - 1,$$

$$\alpha = \begin{bmatrix} \alpha_1 & \cdots & \alpha_r \end{bmatrix}', \quad \beta = \begin{bmatrix} \beta_1 & \cdots & \beta_r \end{bmatrix}'$$

and

$$\eta(t - 1) = \begin{bmatrix} \tilde{\xi}(t-1) & \cdots & \tilde{\xi}(t-r+1) & \xi(t-r) \end{bmatrix}', \quad (4.11)$$

where, for $j = 1, \ldots, r - 1$,

$$\tilde{\xi}(t - j) = \xi(t - j) + \xi(t - 2r + j)$$

and $\xi(t) = 0$, $t < 0$. *Let*

$$\beta = \alpha - 2 \left\{ \sum_{t=0}^{T-1} \eta(t-1)\eta(t-1)' \right\}^{-1} \sum_{t=0}^{T-1} y(t)\eta(t-1).$$

(iii) *If* $\max_j |\beta_j - \alpha_j|$ *is suitably small, the estimators* $\widehat{\lambda}_j$ *of the* λ_j *are calculated from (4.10), with* α_j *replaced by* β_j *and* θ_j *by* $2 \cos \widehat{\lambda}_j$. *Otherwise, put* $\alpha = \beta$ *and repeat Step* (ii).

Unfortunately, the algorithm fails in practice if r and T are large, for then the matrix in step (ii) is ill-conditioned and indeed the elements of α and β may be large. An algorithm which does not exhibit this behaviour produces estimators of the θ_j directly and is obtained from the above using the following device. Putting $\theta = \begin{bmatrix} \theta_1 & \cdots & \theta_r \end{bmatrix}'$, (4.10) may be expressed symbolically as $\theta = g(\alpha)$ or $\alpha = h(\theta)$, for some one-to-one function g and its inverse h. Put $\psi = g(\beta)$. Then, if β and α are close,

$$
\begin{aligned}
\theta - \psi &= g(\alpha) - g(\beta) \\
&\sim \frac{\partial \theta}{\partial \alpha'}(\alpha - \beta) \\
&\sim 2\left\{\sum_{t=0}^{T-1} \zeta(t-1)\zeta(t-1)'\right\}^{-1} \sum_{t=0}^{T-1} y(t)\zeta(t-1)
\end{aligned}
$$

where $\zeta(t-1) = -\frac{\partial \alpha'}{\partial \theta}\eta(t-1)$. Now, differentiating (4.10) with respect to θ_j we obtain

$$
\begin{aligned}
\sum_{k=1}^{2r-1} \frac{\partial \alpha_k}{\partial \theta_j} z^k &= -z \prod_{j=1, j\neq k}^{r} (1 - \theta_j z + z^2) \\
&= -z\left(1 - \theta_j z + z^2\right)^{-1} \sum_{k=0}^{2r} \alpha_k z^k. \qquad (4.12)
\end{aligned}
$$

Thus, letting $\zeta_j(t-1)$ be the jth component of $\zeta(t-1)$, it follows that

$$
\begin{aligned}
\zeta_j(t-1) &= -\sum_{k=1}^{r-1} \frac{\partial \alpha_k}{\partial \lambda_j}\{\xi(t-k) + \xi(t-2r+k)\} - \frac{\partial \alpha_r}{\partial \lambda_j}\xi(t-r) \\
&= -\sum_{k=1}^{2r-1} \frac{\partial \alpha_k}{\partial \lambda_j}\xi(t-k),
\end{aligned}
$$

and, from (4.12), that

$$
\zeta_j(t) - \theta_j\zeta_j(t-1) + \zeta_j(t-2) = y(t), \qquad (4.13)
$$

where $\zeta_j(-1) = \zeta_j(-2) = 0$. Consequently

$$
\{\zeta_j(t); \ j = 1, \ldots, r; \ t = 0, \ldots, T-1\}
$$

may be calculated using, in parallel, r second order filters of the same type as used for the case of a single frequency. The modified algorithm is thus

Algorithm 3

(i) *Let* $\theta = \begin{bmatrix} \theta_1 & \cdots & \theta_r \end{bmatrix}'$, *where, for* $j = 1, \ldots, r$, θ_j *is an initial estimator of* $2\cos\lambda_j$.

(ii) *Calculate* $\zeta_j(t)$ *from (4.13) for* $j = 1, \ldots, r$ *and* $t = 0, \ldots, T-1$. *Let*

$$
\zeta(t) = \begin{bmatrix} \zeta_1(t) & \cdots & \zeta_r(t) \end{bmatrix}'.
$$

Put

$$\psi = \theta + 2\left\{\sum_{t=0}^{T-1}\zeta(t-1)\zeta(t-1)'\right\}^{-1}\sum_{t=0}^{T-1}y(t)\zeta(t-1).$$

(iii) *If* $\max_j |\psi_j - \theta_j|$ *is suitably small, put* $\widehat{\lambda}_j = \cos^{-1}(\psi_j/2), j = 1,\dots,r.$
Otherwise put $\theta = \psi$ *and repeat Step* (ii).

As with the case of a single frequency, a fixed number of iterations may be performed if the least order of accuracy of the initial estimators is known. For example, if all of the initial estimators used are accurate to $O(T^{-1})$ almost surely, then only two iterations are needed before the final estimator is statistically equivalent to the fixed point of the algorithm.

4.6 Asymptotic theory of the multi-frequency procedure

The first point to notice is that $\beta_j = \alpha_j$, $j = 1,\dots,r$, in Algorithm 2 if and only if

$$\sum_{t=0}^{T-1}y(t)\,\eta(t-1) = 0$$

where $\eta(t-1)$ is given by (4.11).
From the above, this is equivalent to

$$\sum_{t=0}^{T-1}y(t)\,\zeta_j(t-1) = 0$$

where $\zeta_j(t)$ satisfies (4.13). If Algorithm 2 converges, therefore, the estimators $\widehat{\lambda}_j$, $j = 1,\dots,r$, are zeros of $\sum_{t=0}^{T-1}y(t)\xi_{t-1}(2\cos\lambda)$, where $\xi_t(\alpha) = y(t) + \alpha\xi_{t-1}(\alpha) - \xi_{t-2}(\alpha)$ and $\xi_t(\alpha) = 0$, $t < 0$. The theory developed in Section 4.4 may thus be applied both to prove convergence of the algorithm and to establish the limit theory of the frequency estimators. Let λ_{0j} denote the true values of the frequencies, and define h_T as in (4.3). Then we have the following.

Theorem 19 *Let* $A_{T,j}(\varepsilon) = \{\alpha; |\alpha - 2\cos\lambda_{0j}| < cT^{-\varepsilon}\}$, *where* ε *and* c *are fixed,* $1 < \varepsilon < \frac{3}{2}$, $c > 0$. *Then almost surely as* $T \to \infty$, *and for each* $j = 1,\dots,r$, *there exists a unique* $\widehat{\lambda}_{T,j} \in \cos^{-1}\{\frac{1}{2}A_{T,j}(\varepsilon)\}$ *with the property that* $h_T(2\cos\widehat{\lambda}_{T,j}) = 0$. *There are thus unique solutions to* $h_T(2\cos\widehat{\lambda}_{T,j}) = 0$ *for which* $T^\varepsilon(\widehat{\lambda}_{T,j} - \lambda_{0j}) \to 0$, *almost surely, for all* $\varepsilon < \frac{3}{2}$.

The next theorem describes the convergence of successive iterates to the zeros of $h_T(2\cos\lambda)$.

Theorem 20 *Let* $\theta = \theta(1) = \begin{bmatrix} \theta_1 & \cdots & \theta_r \end{bmatrix}'$ *and suppose that* $\theta_j \in A_{T,j}(\varepsilon_j)$ *where* $A_{T,j}(\varepsilon)$ *is as defined in Theorem 19. Let*

$$\psi = \begin{bmatrix} \psi_1 & \cdots & \psi_r \end{bmatrix}'$$

be defined as in Algorithm 3. Then if $1 < \varepsilon_j < \frac{3}{2}$,

$$\psi_j - 2\cos\widehat{\lambda}_{T,j} = \left(\theta_j - 2\cos\widehat{\lambda}_{T,j}\right) O\left\{T^{-1/2}(\log T)^{1/2}\right\},$$

while if $\frac{1}{2} < \varepsilon_j \le 1$,

$$\begin{aligned}
\psi_j - 2\cos\widehat{\lambda}_{T,j} &= \left(\theta_j - 2\cos\widehat{\lambda}_{T,j}\right) O\left\{T^{1/2-\varepsilon_j}(\log T)^{1/2}\right\} \\
&\quad + O\left\{T^{1/2-2\varepsilon_j}(\log T)^{1/2}\right\},
\end{aligned}$$

both uniformly in θ_j. *Furthermore,* $\widehat{\lambda}_{T,j} - \lambda_{0j} = O\left\{T^{-3/2}(\log T)^{1/2}\right\}$ *and if* ε *is the minimum of* $\{\varepsilon_1, \ldots, \varepsilon_r\}$ *and* $\{\theta(k), k \ge 1\}$ *is the sequence of iterates of* θ *using Algorithm 3, then if* $\frac{1}{2} < \varepsilon < \frac{3}{2}$,

$$\cos^{-1}\{\theta(k)/2\} - \widehat{\lambda}_T = o\left(T^{-3/2}\right)$$

for $k \ge \lfloor 3 - \log(2\varepsilon - 1)/\log(2) \rfloor$, *where* $\widehat{\lambda}_T = \begin{bmatrix} \widehat{\lambda}_{T,1} & \cdots & \widehat{\lambda}_{T,r} \end{bmatrix}'$.

The final theorem provides the central limit theorem for $\widehat{\lambda}_T$.

Theorem 21 *The distribution of* $T^{3/2}\left(\widehat{\lambda}_T - \lambda_0\right)$ *converges to the normal with mean zero and diagonal covariance matrix whose* j*th diagonal element is* $48\pi f_x(\lambda_{0j})/A_j^2$.

Proof (Theorems 19–21) For any $t \ge 0$

$$\xi_t(2\cos\lambda) = (\sin\lambda)^{-1}\sum_{j=0}^{t}\sin\{(j+1)\lambda\}\, y(t-j).$$

Thus, if $\lambda \in \cos^{-1}\left\{\frac{1}{2}A_{T,k}(\varepsilon_k)\right\}$,

$$\xi_t(2\cos\lambda) = (\sin\lambda)^{-1}\sum_{j=0}^{t}\sin\{(j+1)\lambda\}\,x\,(t-j))$$

$$+A_k(\sin\lambda)^{-1}\sum_{j=0}^{t}\sin\{(j+1)\lambda\}\cos\{\lambda_{0k}(t-j)+\phi_k\}+O(1)$$

uniformly in λ. The components at different frequencies therefore do not interfere with each other and the same proofs as used in Theorems 16–18 apply with only slight modification. For example, with $\zeta(t)$ defined as in step (ii) of Algorithm 3, it is easily shown that if $\frac{1}{2} < \varepsilon_k \leq 1$ then

$$\sum_{t=0}^{T-1}\zeta_k^2(t-1) = \frac{A_k^2 T}{4\sin^2\lambda_{0k}(\psi-\lambda_{0k})^2}\left[1-\frac{\sin\{T(\psi-\lambda_{0k})\}}{T(\psi-\lambda_{0k})}\right]$$

$$\times\left[1+O\left\{T^{1/2-\varepsilon_k}(\log T)^{1/2}\right\}\right]$$

which is $O\left(T^{1+2\varepsilon_k}\right)$, while if $1 < \varepsilon_k < 3/2$,

$$\sum_{t=0}^{T-1}\zeta_k^2(t-1) = \frac{A_k^2 T^3}{24\sin^2\lambda_{0k}}\{1+o(1)\}$$

Moreover, the *cross-terms*

$$\sum_{t=0}^{T-1}\zeta_k(t-1)\,\zeta_l(t-1),\quad k\neq l,$$

are all of order $O\left(T^2\log T\right)$. Consequently, the results may be shown using essentially the same proofs as for the case of a single frequency.   □

Although the only results shown are for Algorithm 3, it is easily shown that Algorithm 2 has exactly the same properties. The results also show that there is no asymptotic benefit in estimating the frequencies jointly, as the resulting estimates are zeros of the same function of one variable $h_T(2\cos\lambda)$, a function which behaves near λ_{0j} as though $y(t)$ had only the sinusoidal component at λ_{0j}. Nevertheless, there may be some gain when the sample size is small or when the initial frequency estimates are relatively close.

5

Techniques Based on Phases and Autocovariances

5.1 Introduction

There are several types of frequency estimation techniques which we have not yet discussed. In particular, we have not paid any attention to those based on autocovariances, such as Pisarenko's technique (Pisarenko 1973), or those based on phase differences, for complex time series, such as two techniques due to Kay (1989). We have not spent much effort on these for the very reason that we have been concerned with *asymptotic theory and asymptotic optimality*. That is, for fixed system parameters, we have been interested in the behaviour of frequency estimators as the sample size T increases, with the hope that the sample size we have is large enough for the asymptotic theory to hold well enough. Moreover, we have not wished to impose conditions such as Gaussianity or whiteness on the noise process, as the latter in particular is rarely met in practice. Engineers, however, are often interested in the behaviour of estimators for *fixed* values of T, and decreasing SNR. The usual measure of this behaviour is mean square error, which may be estimated via simulations. Such properties, however, may rarely be justified theoretically, as there is no statistical limit theory which allows the mean square errors of nonlinear estimators to be calculated using what are essentially limiting distribution results. Although the methods mentioned above are *computationally* simple and *computationally* efficient, we shall see that they cannot be *statistically* asymptotically efficient and may even be inconsistent, i.e., actually converge to the wrong value. This may not be relevant if it is known *a priori* that the SNR is very large.

It is very easy to demonstrate why a technique based only on a finite number of autocovariances, as is Pisarenko's, should not be statistically asymptotically efficient: as the sample autocovariances have 'asymptotic variances' proportional to T^{-1}, any estimator which may be expressed as a nonlinear

125

function of these sample autocovariances will inherit an 'asymptotic variance' of the same order. The periodogram and likelihood approaches, on
the other hand, yield estimators with asymptotic variances of order T^{-3}.

We shall discuss in this chapter a small number of alternative frequency
estimation techniques and describe their properties and the conditions under which they hold. In Section 5.2, we demonstrate the inconsistency of
an autoregressive technique, while in Section 5.3, we show that Pisarenko's
(1973) technique, which uses the eigenvectors of an autocovariance matrix,
is consistent when the noise is white, but produces estimators which have
variances of a higher order than those of the MLE. Sections 5.4 and 5.5 discuss Kay's (1989) two estimators, which are based on relative phases from
complex time series. It is shown that the first technique has poor asymptotic properties, but that the second estimator is consistent. Finally, in
Sections 5.6 and 5.7, the MUSIC algorithm is described and analysed. Since
the real and complex cases are both of interest, we have duplicated quite a
bit of material in these sections in the interest of readability. The MUSIC
analysis is deep enough that other techniques based on the eigenvectors of
autocovariance matrices could be analysed using the same methods.

5.2 An autoregressive technique

It is well-known that the parameters in an autoregression of order 2 (AR(2))
may be chosen in such a way that a time series generated by that model
exhibits pseudo-periodic behaviour. Let $\{y_t\}$ be the real-valued stationary
process which satisfies

$$y(t) + \beta_1 y(t-1) + \beta_2 y(t-2) = x(t),$$

where $\{x(t)\}$ is a stationary process of martingale differences with

$$E\left\{x^2(t)\right\} = \sigma^2,$$

and the zeros of $1 + \beta_1 z + \beta_2 z^2$ lie outside the unit circle. Then the spectral
density of $\{y(t)\}$ is

$$f_y(\omega) = \frac{\sigma^2}{2\pi} \frac{1}{\left|1 + \beta_1 e^{-i\omega} + \beta_2 e^{-i2\omega}\right|^2}$$

and if the zeros of $1 + \beta_1 z + \beta_2 z^2$ are complex and given by $\rho^{-1} e^{\pm i\lambda}$, where
ρ and λ are real, then

$$f_y(\omega) = \frac{\sigma^2}{2\pi} \frac{1}{\left|1 - \rho^{-1} e^{-i(\lambda+\omega)}\right|^2 \left|1 - \rho^{-1} e^{i(\lambda-\omega)}\right|^2},$$

which will have very sharp peaks near $\pm\lambda$ if ρ is close to 1. Thus an obvious fast technique for estimating the frequency of a sinusoid is to fit an AR(2) model and find the arguments of the complex zeros of the auxiliary polynomial. Although the resulting estimator is not consistent, as we shall show below, the technique still appears to be in common use.

The Yule–Walker estimator of $\beta = \begin{bmatrix} \beta_1 & \beta_2 \end{bmatrix}'$ is given by

$$\widehat{\beta} = -\begin{bmatrix} C_0 & C_1 \\ C_1 & C_0 \end{bmatrix}^{-1} \begin{bmatrix} C_1 \\ C_2 \end{bmatrix},$$

where $C_j = T^{-1} \sum_{t=j}^{T-1} y(t) y(t-j)$. We may, of course, mean-correct the $y(t)$, but this does not alter the theory. Assuming that the zeros of $1 + \widehat{\beta}_1 z + \widehat{\beta}_2 z^2$ are complex, the estimator of frequency is given by

$$\widehat{\lambda} = \cos^{-1}\left(-\frac{\widehat{\beta}_1}{2\sqrt{\widehat{\beta}_2}}\right).$$

Now, if $\{y(t)\}$ satisfies (4.1), then

$$
\begin{aligned}
C_j \;=\; & A^2 \frac{1}{T} \sum_{t=j}^{T-1} \cos\left(\lambda t + \phi\right) \cos\left\{\lambda(t-j) + \phi\right\} \\
& + A \frac{1}{T} \sum_{t=j}^{T-1} \left[x(t-j)\cos\left(\lambda t + \phi\right) + x(t)\cos\left\{\lambda(t-j)+\phi\right\}\right] \\
& + \frac{1}{T} \sum_{t=j}^{T-1} x(t) x(t-j) \\
\;\rightarrow\; & \frac{A^2}{2}\cos(j\lambda) + \gamma_j,
\end{aligned}
\tag{5.1}
$$

almost surely, where $\gamma_j = E\left\{x(t)x(t-j)\right\}$. Thus $\widehat{\beta}$ converges to

$$
\begin{aligned}
& -\begin{bmatrix} \gamma_0 + \frac{A^2}{2} & \gamma_1 + \frac{A^2}{2}\cos\lambda \\ \gamma_1 + \frac{A^2}{2}\cos\lambda & \gamma_0 + \frac{A^2}{2} \end{bmatrix}^{-1} \begin{bmatrix} \gamma_1 + \frac{A^2}{2}\cos\lambda \\ \gamma_2 + \frac{A^2}{2}\cos(2\lambda) \end{bmatrix} \\
&= \begin{bmatrix} \dfrac{-\frac{A^4}{2}\sin^2\lambda\cos\lambda + \frac{A^2}{2}\left\{(\gamma_2 - \gamma_0)\cos\lambda - 2\gamma_1\sin^2\lambda\right\} + (\gamma_2 - \gamma_0)\gamma_1}{\frac{A^4}{4}\sin^2\lambda + A^2(\gamma_0 - \gamma_1\cos\lambda) + \gamma_0^2 - \gamma_1^2} \\[2ex] \dfrac{\frac{A^4}{4}\sin^2\lambda + \frac{A^2}{2}\left\{2\gamma_1\cos\lambda - \gamma_2 - \gamma_0\cos(2\lambda)\right\} + \gamma_1^2 - \gamma_0\gamma_2}{\frac{A^4}{4}\sin^2\lambda + A^2(\gamma_0 - \gamma_1\cos\lambda) + \gamma_0^2 - \gamma_1^2} \end{bmatrix}.
\end{aligned}
$$

almost surely. Now, if the frequency estimation procedure were consistent, we would have

$$\widehat{\beta} \rightarrow \left[\begin{array}{c} -2r\cos\lambda \\ r^2 \end{array} \right],$$

almost surely for some r. We would thus have both

$$r = \frac{\frac{A^4}{4}\sin^2\lambda - \frac{A^2}{4}\left\{(\gamma_2 - \gamma_0) - 2\gamma_1\sin^2\lambda\sec\lambda\right\} - \frac{1}{2}(\gamma_2 - \gamma_0)\gamma_1\sec\lambda}{\frac{A^4}{4}\sin^2\lambda + A^2(\gamma_0 - \gamma_1\cos\lambda) + \gamma_0^2 - \gamma_1^2}$$

and

$$r^2 = \frac{\frac{A^4}{4}\sin^2\lambda + \frac{A^2}{2}\left\{2\gamma_1\cos\lambda - \gamma_2 - \gamma_0\cos(2\lambda)\right\} + \gamma_1^2 - \gamma_0\gamma_2}{\frac{A^4}{4}\sin^2\lambda + A^2(\gamma_0 - \gamma_1\cos\lambda) + \gamma_0^2 - \gamma_1^2}.$$

These equations are, however, consistent only under very special circumstances. Even if $\{x(t)\}$ were white, it is easily seen that the procedure would then yield a consistent frequency estimator if and only if $\gamma_0 = 0$, that is, under noise-free conditions. So although simple to compute, the technique is useful only if there is no noise, for the 'asymptotic bias' of the frequency estimator is nearly always non-zero.

We used the program **ar2** to duplicate the simulations of Section 3.2. For SNRs other than 15 dB, there was at least one simulation out of the 100 which did not yield a valid estimate of frequency, i.e., for which the auxiliary polynomial did not have complex zeros. When the SNR was 15 dB, the statistic corresponding to the values -0.71 and -0.08 in Tables 3.3 and 3.4 was around 73, clearly indicating a severe bias problem.

5.3 Pisarenko's technique

The sample autocovariance matrix of a time series plays an important role in the understanding of ARMA processes. It is a little surprising that it may also be used to estimate frequency. Consider the matrix

$$C = \left[\begin{array}{ccc} C_0 & C_1 & C_2 \\ C_1 & C_0 & C_1 \\ C_2 & C_1 & C_0 \end{array} \right]$$

which, from the previous section, converges almost surely to

$$\frac{A^2}{2}\left[\begin{array}{ccc} 1 & \cos\lambda & \cos(2\lambda) \\ \cos\lambda & 1 & \cos\lambda \\ \cos(2\lambda) & \cos\lambda & 1 \end{array} \right] + \left[\begin{array}{ccc} \gamma_0 & \gamma_1 & \gamma_2 \\ \gamma_1 & \gamma_0 & \gamma_1 \\ \gamma_2 & \gamma_1 & \gamma_0 \end{array} \right].$$

Now, the first matrix is non-negative definite with eigenvalues 0, $2\sin^2\lambda$ and $1 + 2\cos^2\lambda$. The eigenvector corresponding to the zero eigenvalue is

$\psi = \begin{bmatrix} 1 & -2\cos\lambda & 1 \end{bmatrix}'$ and the zeros of the polynomial $1 - 2z\cos\lambda + z^2$ formed from the entries in this eigenvector are $e^{\pm i\lambda}$. This suggests the following procedure for estimating λ. First, calculate the eigenvector x of C corresponding to its smallest eigenvalue. Second, find the zeros of $x_1 + x_2 z + x_3 z^2$. Finally, assuming that these form a complex pair, estimate λ by the argument which is positive. From the above, this estimator will be consistent if and only if ψ is also an eigenvector of

$$\begin{bmatrix} \gamma_0 & \gamma_1 & \gamma_2 \\ \gamma_1 & \gamma_0 & \gamma_1 \\ \gamma_2 & \gamma_1 & \gamma_0 \end{bmatrix}.$$

This will usually only be the case when $\{x(t)\}$ is white. Consequently, Pisarenko's technique is guaranteed to be consistent only under white noise conditions. We shall assume from now on that $\{x(t)\}$ is a sequence of martingale differences, and that $\mathrm{E}\{x^2(t)|\mathcal{F}_{t-1}\} = \sigma^2 < \infty$, where $\mathcal{F}_{t-1}$ is the σ-field generated by $\{x(t-1), x(t-2), \ldots\}$. Given that the estimator is consistent, we need to determine its convergence rate by constructing a central limit theorem for $\begin{bmatrix} C_0 & C_1 & C_2 \end{bmatrix}'$. From (5.1),

$$C_j - \frac{A^2}{2}\cos(j\lambda) - \gamma_j$$

$$= A\frac{1}{T}\sum_{t=j}^{T-1}[x(t-j)\cos(\lambda t + \phi) + x(t)\cos\{\lambda(t-j) + \phi\}]$$

$$+ \frac{1}{T}\sum_{t=j}^{T-1}x(t)x(t-j) - \gamma_j$$

$$= A\frac{1}{T}\left(\sum_{t=0}^{T-1}x(t)[\cos\{\lambda(t-j) + \phi\} + \cos\{\lambda(t+j) + \phi\}]\right)$$

$$+ \frac{1}{T}\sum_{t=j}^{T-1}x(t)x(t-j) - \gamma_j + O_P(T^{-1})$$

$$= 2A\cos(j\lambda)\frac{1}{T}\sum_{t=0}^{T-1}x(t)\cos(\lambda t + \phi)$$

$$+ \frac{1}{T}\sum_{t=0}^{T-1}x(t)x(t-j) - \gamma_j + O_P(T^{-1}).$$

Thus,

$$\sqrt{T}\begin{bmatrix} C_0 - \frac{A^2}{2} - \sigma^2 \\ C_1 - \frac{A^2}{2}\cos\lambda \\ C_2 - \frac{A^2}{2}\cos(2\lambda) \end{bmatrix}$$

has the same asymptotic distribution as

$$2A\frac{1}{\sqrt{T}}\sum_{t=0}^{T-1} x(t)\cos(\lambda t + \phi)\begin{bmatrix} 1 \\ \cos\lambda \\ \cos(2\lambda) \end{bmatrix} + \begin{bmatrix} \frac{1}{\sqrt{T}}\sum_{t=0}^{T-1}\{x^2(t) - \sigma^2\} \\ \frac{1}{\sqrt{T}}\sum_{t=0}^{T-1} x(t)x(t-1) \\ \frac{1}{\sqrt{T}}\sum_{t=0}^{T-1} x(t)x(t-2) \end{bmatrix}.$$

$$(5.2)$$

But

$$T^{-1/2}\begin{bmatrix} \sum_{t=0}^{T-1} x(t)\cos(\lambda t + \phi) \\ \sum_{t=0}^{T-1}\{x^2(t) - \sigma^2\} \\ \sum_{t=0}^{T-1} x(t)x(t-1) \\ \sum_{t=0}^{T-1} x(t)x(t-2) \end{bmatrix}$$

$$(5.3)$$

is easily shown using Theorem 4 and the results of Hannan (1979) to be asymptotically normally distributed with mean zero and diagonal covariance matrix with diagonal entries $\left\{\sigma^2/2, \text{var } x^2(t), (\sigma^2)^2, (\sigma^2)^2\right\}$. Because of the structure of C, the three eigenvectors must be of one of the forms

$$\begin{bmatrix} x & 0 & x \end{bmatrix}' \quad \text{and} \quad \begin{bmatrix} 1 & x & 1 \end{bmatrix}'.$$

Solving the eigenvalue equations thus yields the eigenvalues $C_0 + \frac{C_2}{2} - \frac{1}{2}\sqrt{C_2^2 + 8C_1^2}$, $C_0 + \frac{C_2}{2} + \frac{1}{2}\sqrt{C_2^2 + 8C_1^2}$ and $C_0 - C_2$. Since the second eigenvalue is larger than the first, the contenders for the smallest eigenvalue are the first and third. But

$$(C_0 - C_2) - \left(C_0 + \frac{C_2}{2} - \frac{1}{2}\sqrt{C_2^2 + 8C_1^2}\right)$$

$$= \frac{\sqrt{C_2^2 + 8C_1^2} - 3C_2}{2}$$

$$\to \frac{\sqrt{\cos^2(2\lambda) + 8\cos^2\lambda} - 3\cos(2\lambda)}{2}$$

$$= \frac{2 - 2\cos(2\lambda)}{2} > 0$$

almost surely. Hence, with probability 1 as $T \to \infty$, the smallest eigenvalue is $C_0 + \frac{C_2}{2} - \frac{1}{2}\sqrt{C_2^2 + 8C_1^2}$ and the corresponding eigenvector is

$$\left[\begin{array}{ccc} 1 & -\frac{C_2 + \sqrt{C_2^2 + 8C_1^2}}{2C_1} & 1 \end{array}\right]'.$$

Now, since $\frac{C_2 + \sqrt{C_2^2 + 8C_1^2}}{2C_1}$ converges to $2\cos\lambda$, almost surely, the zeros of $1 - \frac{C_2 + \sqrt{C_2^2 + 8C_1^2}}{2C_1}z + z^2$, with probability 1 as $T \to \infty$, form a complex pair on the unit circle. Thus, as $T \to \infty$,

$$\cos\widehat{\lambda} = \frac{C_2 + \sqrt{C_2^2 + 8C_1^2}}{4C_1} = g(C_1, C_2),$$

say, and

$$\begin{aligned} \sqrt{T}\left(\cos\widehat{\lambda} - \cos\lambda\right) &= \sqrt{T}\left\{g(C_1, C_2) - g\left(\frac{A^2}{2}\cos\lambda, \frac{A^2}{2}\cos(2\lambda)\right)\right\} \\ &= \left[\begin{array}{cc} \widetilde{g}_1 & \widetilde{g}_2 \end{array}\right]\sqrt{T}\left[\begin{array}{c} C_1 - \frac{A^2}{2}\cos\lambda \\ C_2 - \frac{A^2}{2}\cos(2\lambda) \end{array}\right], \end{aligned}$$

where $\widetilde{g}_1$ and $\widetilde{g}_2$ are the first derivatives of g evaluated at some value of λ between λ and $\widehat{\lambda}$ and therefore converge respectively to

$$\begin{aligned} g_1 &= \frac{8\cos\lambda}{4\frac{A^2}{2}\cos\lambda\left(2\cos^2\lambda + 1\right)} - \frac{\cos(2\lambda) + 2\cos^2\lambda + 1}{4\frac{A^2}{2}\cos^2\lambda} \\ &= \frac{1}{A^2\left(1 + 2\cos^2\lambda\right)}\left\{4 - 2\left(1 + 2\cos^2\lambda\right)\right\} \\ &= -\frac{2\cos(2\lambda)}{A^2\left(1 + 2\cos^2\lambda\right)} \end{aligned}$$

and

$$g_2 = \frac{1 + \frac{\cos(2\lambda)}{2\cos^2\lambda + 1}}{4\frac{A^2}{2}\cos\lambda} = \frac{2\cos\lambda}{A^2\left(1 + 2\cos^2\lambda\right)}.$$

Now,

$$\left[\begin{array}{cc} g_1 & g_2 \end{array}\right]\left[\begin{array}{c} \cos\lambda \\ \cos(2\lambda) \end{array}\right] = 0$$

and so from (5.2) and (5.3), $\sqrt{T}\left(\cos\widehat{\lambda} - \cos\lambda\right)$ is asymptotically normally distributed with mean zero and variance

$$\left(g_1^2 + g_2^2\right)\left(\sigma^2\right)^2 = 4\left(\frac{\sigma^2}{A^2}\right)^2\frac{\cos^2\lambda + \cos^2(2\lambda)}{\left(1 + 2\cos^2\lambda\right)^2},$$

and consequently

$$\sqrt{T}\left(\widehat{\lambda} - \lambda\right) = -\frac{1}{\sin \lambda}\sqrt{T}\left(\cos \widehat{\lambda} - \cos \lambda\right) + o_P(1)$$

is asymptotically normally distributed with mean zero and variance

$$4\left(\frac{\sigma^2}{A^2}\right)^2 \frac{\cos^2 \lambda + \cos^2 (2\lambda)}{\sin^2 \lambda \ (1 + 2\cos^2 \lambda)^2}.$$

Note that this does *not* mean that the variance or mean square error of $\widehat{\lambda}$ is, to first order, $\frac{4}{T}\left(\frac{\sigma^2}{A^2}\right)^2 \frac{\cos^2 \lambda + \cos^2 (2\lambda)}{\sin^2 \lambda \ (1 + 2\cos^2 \lambda)^2}$. Although we could have used well-known formulae to bound the variances of C_1 and C_2, this would *not* have helped us bound the variance of $\widehat{\lambda}$ by using the above Taylor series arguments, for these are only valid in the sense of convergence in distribution, and *not* in mean square. Thus the above result does not allow us to compare the performance of the Pisarenko estimator with that of any other estimator, for fixed T and decreasing SNR. The above is only a 'variance' in that it is the variance of the limiting distribution.

We next show that Pisarenko's estimator is the *only* strongly consistent estimator formed only from continuous functions of C_0, C_1 and C_2. Let

$$\widehat{\lambda} = \psi(C_0, C_1, C_2),$$

where ψ is continuous in its arguments. Then

$$\widehat{\lambda} \to \psi\left(\sigma^2 + \frac{A^2}{2}, \frac{A^2}{2}\cos \lambda, \frac{A^2}{2}\cos (2\lambda)\right)$$

almost surely. Thus, for $\widehat{\lambda}$ to be consistent, we need to have

$$\lambda = \psi\left(\sigma^2 + \frac{A^2}{2}, \frac{A^2}{2}\cos \lambda, \frac{A^2}{2}\cos (2\lambda)\right),$$

or

$$\begin{aligned}
\cos \lambda &= \cos\left\{\psi\left(\sigma^2 + \frac{A^2}{2}, \frac{A^2}{2}\cos \lambda, \frac{A^2}{2}\cos (2\lambda)\right)\right\} \\
&= g\left(\sigma^2 + \frac{A^2}{2}, \frac{A^2}{2}\cos \lambda, \frac{A^2}{2}\cos (2\lambda)\right),
\end{aligned}$$

say, and this needs to hold for all σ^2, A and λ. Put

$$\begin{bmatrix} x \\ y \end{bmatrix} = \begin{bmatrix} \frac{A^2}{2}\cos \lambda \\ \frac{A^2}{2}\cos (2\lambda) \end{bmatrix}.$$

Then, it is easily seen that

$$\frac{A^2}{2} = \frac{-y + \sqrt{y^2 + 8x^2}}{2}$$

and

$$\cos \lambda = \frac{y + \sqrt{y^2 + 8x^2}}{4x}.$$

Hence

$$\frac{y + \sqrt{y^2 + 8x^2}}{4x} = g\left(\sigma^2 + \frac{-y + \sqrt{y^2 + 8x^2}}{2}, x, y\right)$$

for all $\sigma^2 > 0$ and (x, y) in a dense subset of $\mathbb{R}^2$. Consequently g does not depend on its first argument and the only strongly consistent estimator of λ is obtained from

$$\cos \lambda = g\left(C_0, C_1, C_2\right) = \frac{C_2 + \sqrt{C_2^2 + 8C_1^2}}{4C_1}.$$

We should note that Pisarenko's estimator also follows from solving the two equations

$$C_1 = \frac{\widehat{A}^2}{2} \cos \widehat{\lambda} \quad \text{and} \quad C_2 = \frac{\widehat{A}^2}{2} \cos\left(2\widehat{\lambda}\right).$$

The eigenvalue development above, however, generalises to the multi-component case. The frequency estimators when there are r components are obtained as the arguments of the complex zeros of $\sum_{j=0}^{2r} b_j z^j$, where

$$\begin{bmatrix} b_0 & b_1 & \cdots & b_{2r} \end{bmatrix}'$$

is the eigenvector corresponding to the smallest eigenvalue of

$$\begin{bmatrix} C_0 & C_1 & \cdots & C_{2r} \\ C_1 & C_0 & \ddots & \vdots \\ \vdots & \ddots & \ddots & C_1 \\ C_{2r} & \cdots & C_1 & C_0 \end{bmatrix}.$$

Although the same type of asymptotic theory may be developed, it is fairly complicated and will be omitted here. Details are given in Sakai (1984). Pisarenko's estimator is also used in the estimation of the bearing, or the direction of arrival, of a signal, given time series collected at a number of collinear receivers.

5.4 Kay's first estimator

Much of the engineering literature concerned with the estimation of frequency is associated with *complex*-valued processes. The sinusoidal process thus satisfies

$$y(t) = Ae^{i(\phi + \lambda t)} + x(t) \tag{5.4}$$

and most often the $x(t)$ are assumed to be independent and identically distributed and complex normal with means zero and variances σ^2, i.e., the real and imaginary parts are assumed to be independent and identically distributed and mutually independent and normal with means zero and variances σ^2). The frequency λ is also allowed to be in $(-\pi, \pi)$, or, equivalently, in $(0, 2\pi)$. We shall adopt the latter assumption. Such processes are used to model the combination of 'in phase and quadrature' components of a signal, and also the 'analytical signal' obtained by Hilbert-transforming a real signal and adjoining this as the imaginary part of the signal. Although there are problems associated with doing the latter, not the least of which is that this cannot possibly alter the information content of the signal, the complex model has proven popular, since problems associated with 'negative' frequencies, which arise with real signals, do not arise with complex ones. There is one important circumstance where the complex sinusoid arises naturally. If we segment a large sample from a real sinusoid, then the sequence of Fourier coefficients at a single fixed frequency near the true one, will satisfy the above equation in the asymptotic sense. In what follows, we shall not be concerned about the validity of (5.4).

If the noise term is absent in (5.4), we have, for each t,

$$\frac{y(t+1)}{y(t)} = e^{i\lambda}$$

and so, for $t = 0, \ldots, T-2$,

$$\arg \left\{ \frac{y(t+1)}{y(t)} \right\} = \lambda.$$

This suggests using the z_t in some way to estimate λ. Now, if there *is* noise,

$$
\begin{aligned}
\frac{y(t+1)}{y(t)} &= e^{i\lambda} \frac{1 + A^{-1}e^{-i\phi}e^{-i\lambda(t+1)}x(t+1)}{1 + A^{-1}e^{-i\phi}e^{-i\lambda t}x(t)} \\
&= e^{i\lambda} \frac{1 + u(t+1)}{1 + u(t)},
\end{aligned}
$$

where $u(t) = A^{-1}e^{-i\phi}e^{-i\lambda t}x(t)$ and so $u(t)$ is complex normal with mean zero and variance $\frac{\sigma^2}{A^2}$. Kay (1989) argues that if the SNR is large, then the $u(t)$ are small and so

$$\frac{y(t+1)}{y(t)} \sim e^{i\lambda}\{1 + u(t+1) - u(t)\} \tag{5.5}$$

and

$$\begin{aligned} z(t) &= \arg\frac{y(t+1)}{y(t)} \\ &\sim \lambda + \arg\{1 + u(t+1) - u(t)\} \\ &\sim \lambda + \operatorname{Im}u(t+1) - \operatorname{Im}u(t), \end{aligned} \tag{5.6}$$

so that λ may now be thought of as the mean parameter in a Gaussian MA(1) process and may therefore be estimated by minimising

$$(z - \lambda J)' \Sigma^{-1} (z - \lambda J)$$

where $z' = \begin{bmatrix} z(0) & \cdots & z(T-2) \end{bmatrix}$, J is a vector of ones and

$$\Sigma = \begin{bmatrix} 2 & -1 & 0 & 0 \\ -1 & 2 & \ddots & 0 \\ 0 & \ddots & \ddots & -1 \\ 0 & 0 & -1 & 2 \end{bmatrix}.$$

Kay's estimator of frequency is then

$$\widehat{\lambda} = \frac{J'\Sigma^{-1}z}{J'\Sigma^{-1}J} = \frac{6}{T(T^2-1)} \sum_{j=0}^{T-2} (j+1)(T-1-j)z(j), \tag{5.7}$$

since it is easily shown that the jth element of $\Sigma^{-1}J$ is $j(T-j)/2$. Kay then argues that, since the $\operatorname{Im}u(t)$ are independent and identically distributed and Gaussian with means zero and variances $\frac{\sigma^2}{A^2}$,

$$\widehat{\lambda} - \lambda = \frac{J'\Sigma^{-1}(z - \lambda J)}{J'\Sigma^{-1}J} = \frac{1}{J'\Sigma^{-1}J}J'\Sigma^{-1}\Omega e$$

where

$$\Omega = \begin{bmatrix} -1 & 1 & 0 & \cdots & 0 \\ 0 & -1 & 1 & \ddots & 0 \\ \vdots & \ddots & \ddots & \ddots & 0 \\ 0 & \cdots & 0 & -1 & 1 \end{bmatrix}$$

and

$$e = \begin{bmatrix} \operatorname{Im}u(0) & \operatorname{Im}u(1) & \cdots \operatorname{Im}u(T-1) \end{bmatrix}'.$$

Note that $\Omega\Omega' = \Sigma$. He thus claims that $\widehat{\lambda} - \lambda$ is approximately normally distributed with mean zero and variance

$$\frac{\sigma^2}{A^2} \frac{J'\Sigma^{-1}\Omega\Omega'\Sigma^{-1}J}{(J'\Sigma^{-1}J)^2} = \frac{\sigma^2}{A^2} \frac{1}{\frac{T(T^2-1)}{12}} = \frac{\sigma^2}{A^2} \frac{12}{T(T^2-1)},$$

which is precisely the CRB for the variances of unbiased estimators of frequency in the complex Gaussian white noise case. He also claims that the approximation becomes more precise as the SNR increases, for fixed T. There are, however, problems with the approximations given by (5.5) and (5.6). Firstly,

$$\frac{y(t+1)}{y(t)} - e^{i\lambda}\{1 + u(t+1) - u(t)\} = e^{i\lambda}\frac{u(t)\{u(t) - u(t+1)\}}{1 + u(t)}$$

and the range of the right side of this equation is all of the complex plane, no matter how large the SNR. Secondly, the argument of the product of $e^{i\lambda}$ and $\{1 + u(t+1) - u(t)\}$ is *not* the sum of the two arguments. Finally, if we are interested in the behaviour of the estimator as T increases, the approximations do *not* hold uniformly in t. Thus Kay's approximations cannot be used either for an assessment of the asymptotic properties of his first estimator, or to calculate an accurate expression for the bias in fixed sample sizes.

Now, from (5.7),

$$\widehat{\lambda} - \lambda = \frac{J'\Sigma^{-1}(z - \lambda J)}{J'\Sigma^{-1}J} \tag{5.8}$$

where $z' = \begin{bmatrix} z(0) & \cdots & z(T-2) \end{bmatrix}$ and $z(t) = \arg\left\{\frac{y(t+1)}{y(t)}\right\}$. Thus

$$E\left(\widehat{\lambda}\right) - \lambda = \frac{J'\Sigma^{-1}\{E(z) - \lambda J\}}{J'\Sigma^{-1}J} = E\{z(t)\} - \lambda,$$

since $\{z(t)\}$ is an independent and identically distributed sequence. But

$$z(t) - \lambda = \arg\left\{\frac{y(t+1)}{y(t)}\right\} - \lambda = \arg\left\{e^{i\lambda}\frac{1 + u(t+1)}{1 + u(t)}\right\} - \lambda$$
$$= \nu(t+1) - \nu(t) + 2\pi k(t),$$

where $\nu(t) = \arg\{1 + u(t)\}$ and

$$k(t) = \begin{cases} -1 & ; \quad \pi \le \lambda + \nu(t+1) - \nu(t) < 3\pi \\ 0 & ; \quad -\pi \le \lambda + \nu(t+1) - \nu(t) < \pi \\ 1 & ; \quad -3\pi \le \lambda + \nu(t+1) - \nu(t) < -\pi. \end{cases}$$

Hence, since $E\{\nu(t+1) - \nu(t)\} = 0$, and the distribution of $\nu(t+1) - \nu(t)$ is symmetric about zero,

$$
\begin{aligned}
E\left(\widehat{\lambda}\right) - \lambda \\
= \; & 2\pi E\{k(t)\} \\
= \; & 2\pi\left[\Pr\{k(t) = 1\} - \Pr\{k(t) = -1\}\right] \\
= \; & 2\pi\left[\Pr\{-2\pi \le \nu(t+1) - \nu(t) < -\lambda - \pi\}\right. \\
& \left. - \Pr\{\pi - \lambda \le \nu(t+1) - \nu(t) < 2\pi\}\right] \\
= \; & 2\pi\left[\Pr\{\pi + \lambda < \nu(t+1) - \nu(t) \le 2\pi\}\right. \\
& \left. - \Pr\{\pi - \lambda \le \nu(t+1) - \nu(t) < 2\pi\}\right] \\
= \; & 2\pi \left\{
\begin{array}{ll}
\Pr\{\pi + \lambda < \nu(t+1) - \nu(t) \le \pi - \lambda\} & ; \quad \lambda < 0 \\
-\Pr\{\pi - \lambda < \nu(t+1) - \nu(t) \le \pi + \lambda\} & ; \quad \lambda \ge 0
\end{array}
\right.
\end{aligned}
$$

and the bias is thus π if $\lambda = -\pi$, is 0 if and only if $\lambda = 0$ and is $-\pi$ if $\lambda = \pi$, since the range of $\nu(t+1) - \nu(t)$ is $(-2\pi, 2\pi)$. Moreover, it is easily shown, since $\{k(t)\}$ is stationary and ergodic and $\{\nu(t)\}$ is an independent and identically distributed sequence, that

$$
\widehat{\lambda} \to \lambda + 2\pi E\{k(t)\}
$$

almost surely as $T \to \infty$. That is, Kay's estimator is biased, and is therefore not strongly consistent, unless $\lambda = 0$. An explicit expression for the bias is

$$
-\frac{\sqrt{\pi}}{2} e^{-\frac{\sigma^2}{2A^2}} \sin\lambda \sum_{k=0}^{\infty} (-\cos\lambda)^k \frac{\Gamma\left(\frac{k+1}{2}\right)}{\Gamma\left(\frac{k+2}{2}\right)} \sum_{j=0}^{k} \frac{\left(\frac{\sigma^2}{2A^2}\right)^j}{j!}.
$$

The absolute value of the bias when λ is $\pm\pi/2$ is therefore $\frac{\pi}{2}\exp\left(-\frac{\sigma^2}{2A^2}\right)$ which may be acceptably small for suitably large SNR. Note that the above expression is not valid when $\lambda = \pi$, since $\sum_{k=0}^{\infty} \frac{\Gamma\left(\frac{k+1}{2}\right)}{\Gamma\left(\frac{k+2}{2}\right)} = \infty$. Nevertheless, the limit as $\lambda \to \pi$ does exist, and is $-\pi$.

The mean square error of Kay's estimator is, from (5.8),

$$
E\left\{\left(\widehat{\lambda} - \lambda\right)^2\right\} = \frac{J'\Sigma^{-1} E\left\{(z - \lambda J)(z - \lambda J)'\right\}\Sigma^{-1}J}{(J'\Sigma^{-1}J)^2}.
$$

Let

$$
\alpha = \left[\; \nu(1) - \nu(0) \quad \cdots \quad \nu(T-1) - \nu(T-2) \;\right]',
$$

$$
\beta = \left[\; 2\pi k(0) \quad \cdots \quad 2\pi k(T-2) \;\right]'
$$

and

$$\nu = \left[\begin{array}{ccc} \nu\left(0\right) & \cdots & \nu\left(T-1\right) \end{array}\right]'.$$

The mean square error of $\widehat{\lambda}$ is then

$$E\left\{\left(\widehat{\lambda}-\lambda\right)^{2}\right\}$$

$$= \frac{J'\Sigma^{-1}E\left(\alpha\alpha'\right)\Sigma^{-1}J}{\left(J'\Sigma^{-1}J\right)^{2}} + 2\frac{J'\Sigma^{-1}E\left(\alpha\beta'\right)\Sigma^{-1}J}{\left(J'\Sigma^{-1}J\right)^{2}} + \frac{J'\Sigma^{-1}E\left(\beta\beta'\right)\Sigma^{-1}J}{\left(J'\Sigma^{-1}J\right)^{2}}$$

$$= \frac{J'\Sigma^{-1}\Omega E\left(\nu\nu'\right)\Omega'\Sigma^{-1}J}{\left(J'\Sigma^{-1}J\right)^{2}} + 2\frac{J'\Sigma^{-1}\Omega E\left(\nu\beta'\right)\Sigma^{-1}J}{\left(J'\Sigma^{-1}J\right)^{2}}$$

$$+ \frac{J'\Sigma^{-1}E\left(\beta\beta'\right)\Sigma^{-1}J}{\left(J'\Sigma^{-1}J\right)^{2}}$$

$$= \frac{J'\Sigma^{-1}\Omega\Omega'\Sigma^{-1}J}{\left(J'\Sigma^{-1}J\right)^{2}}E\left\{\nu^{2}\left(t\right)\right\} - 2\frac{J'\Sigma^{-1}\Omega\Omega'\Sigma^{-1}J}{\left(J'\Sigma^{-1}J\right)^{2}}E\left\{\nu\left(t\right)\beta\left(t\right)\right\}$$

$$+ \frac{J'\Sigma^{-1}\left(\text{cov}\,\beta\right)\Sigma^{-1}J}{\left(J'\Sigma^{-1}J\right)^{2}} + \left[2\pi E\left\{k\left(t\right)\right\}\right]^{2}$$

$$= \frac{12}{T\left(T^{2}-1\right)}\left[E\left\{\nu^{2}\left(t\right)\right\} - 4\pi E\left\{\nu\left(t\right)k\left(t\right)\right\}\right] + \frac{J'\Sigma^{-1}\left(\text{cov}\,\beta\right)\Sigma^{-1}J}{\left(J'\Sigma^{-1}J\right)^{2}}$$

$$+ 4\pi^{2}\left[E\left\{k\left(t\right)\right\}\right]^{2}.$$

A little extra calculation shows that

$$\frac{J'\Sigma^{-1}\left(\text{cov}\,\beta\right)\Sigma^{-1}J}{\left(J'\Sigma^{-1}J\right)^{2}}$$

$$= \frac{24\pi^{2}}{5T\left(T^{2}-1\right)}\left[\left(T^{2}+1\right)\text{var}\,k\left(t\right) + 2\left(T^{2}-4\right)\text{cov}\left\{k\left(t\right),k\left(t+1\right)\right\}\right],$$

which is of order T^{-1}. Thus the mean square error contains an order T^{-1} term as well as the square of the bias. An expression may be found for the mean square error, which is similar to the one found for the bias, but which will not be given here.

5.5 Kay's second estimator

Kay (1989) has suggested an alternative estimator, which was shown to have a larger variance than the first estimator and therefore to be inferior. We shall now demonstrate, to the contrary, that this estimator is consistent and therefore superior, at least in the asymptotic sense. Indeed, it is possible to compute bounds on the variance of the estimator and not just the 'asymptotic variance' (see Clarkson, Kootsookos and Quinn 1994, Händel 1995 and Quinn, Clarkson and Kootsookos 1998). The alternative estimator is

$$\widehat{\lambda} = \arg\left\{\sum_{t=0}^{T-2} w_t y\,(t+1)\,\overline{y\,(t)}\right\},$$

where the w_t are real weights constructed so as to make the estimator consistent and have as small a variance as possible. Now,

$$\sum_{t=0}^{T-2} w_t y\,(t+1)\,\overline{y\,(t)}$$

$$= \sum_{t=0}^{T-2} w_t \left\{Ae^{i(\phi+\lambda t+\lambda)} + x\,(t+1)\right\}\left\{Ae^{-i(\phi+\lambda t)} + \overline{x\,(t)}\right\}$$

$$= e^{i\lambda}\left[A^2\sum_{t=0}^{T-2} w_t + A\sum_{t=0}^{T-2} w_t\left\{e^{i(\phi+\lambda t)}\overline{x\,(t)} + e^{-i(\phi+\lambda t+\lambda)}x\,(t+1)\right\}\right]$$

$$+ \sum_{t=0}^{T-2} w_t x\,(t+1)\,\overline{x\,(t)}$$

$$= A^2 e^{i\lambda}\left[\sum_{t=0}^{T-2} w_t + \sum_{t=0}^{T-2} w_t\left\{u\,(t+1) + \overline{u\,(t)} + u\,(t+1)\,\overline{u\,(t)}\right\}\right],$$

where, as before,

$$u\,(t) = A^{-1}e^{-i\phi}e^{-i\lambda t}x\,(t)\,.$$

Thus

$$\sum_{t=0}^{T-2} w_t y\,(t+1)\,\overline{y\,(t)} = C_T e^{i(\lambda+V_T)}$$

where C_T is real and positive, $V_T \in [-\pi, \pi)$ and

$$e^{iV_T} = \frac{\sum_{t=0}^{T-2} w_t + \sum_{t=0}^{T-2} w_t\left\{u\,(t+1) + \overline{u\,(t)} + u\,(t+1)\,\overline{u\,(t)}\right\}}{\left|\sum_{t=0}^{T-2} w_t + \sum_{t=0}^{T-2} w_t\left\{u\,(t+1) + \overline{u\,(t)} + u\,(t+1)\,\overline{u\,(t)}\right\}\right|}\,.$$

Therefore

$$\widehat{\lambda} = \lambda + V_T \mod(2\pi)\,.$$

Now, as the imaginary part of $\sum_{t=0}^{T-2} w_t\left\{u\,(t+1) + \overline{u\,(t)} + u\,(t+1)\,\overline{u\,(t)}\right\}$ has a symmetric probability density function, V_T also has a symmetric probability density. This does *not* ensure that $\widehat{\lambda}$ is unbiased, since $\lambda + V_T$ has range $[\lambda - \pi, \lambda + \pi)$ and is symmetrically distributed about λ. However, $\widehat{\lambda}$ will be strongly or weakly consistent if it can be shown that V_T converges almost surely or in probability to zero. Let $u(t) = \alpha(t) + i\beta(t)$. Then

$$\operatorname{Re}\left[\sum_{t=0}^{T-2} w_t + \sum_{t=0}^{T-2} w_t \left\{u(t+1) + \overline{u(t)} + u(t+1)\overline{u(t)}\right\}\right]$$

$$= \sum_{t=0}^{T-2} w_t + \sum_{t=0}^{T-2} w_t \left\{\alpha(t+1) + \alpha(t) + \alpha(t+1)\alpha(t) + \beta(t)\beta(t+1)\right\},$$

and

$$\operatorname{Im}\left[\sum_{t=0}^{T-2} w_t + \sum_{t=0}^{T-2} w_t \left\{u(t+1) + \overline{u(t)} + u(t+1)\overline{u(t)}\right\}\right]$$

$$= \sum_{t=0}^{T-2} w_t \left\{\beta(t+1) - \beta(t) + \alpha(t)\beta(t+1) - \alpha(t+1)\beta(t)\right\}.$$

The variances of each of the above are, respectively, where $\nu^2 = \sigma^2/A^2$,

$$\nu^2\left(2 + 2\nu^2\right) \sum_{t=0}^{T-2} w_t^2 + 2\nu^2 \sum_{t=1}^{T-2} w_{t-1}w_t \qquad (5.9)$$

and

$$\nu^2\left(2 + 2\nu^2\right) \sum_{t=0}^{T-2} w_t^2 - 2\nu^2 \sum_{t=1}^{T-2} w_{t-1}w_t, \qquad (5.10)$$

while the means are, respectively, $\sum_{t=0}^{T-2} w_t$ and 0. We thus have

$$e^{iV_T} = \frac{1 + A_T + iB_T}{\sqrt{(1 + A_T)^2 + B_T^2}},$$

where A_T and B_T have zero means, and variances given by (5.9) and (5.10) divided by $\left(\sum_{t=0}^{T-2} w_t\right)^2$. It follows that $\widehat{\lambda}$ converges in mean square to λ, and therefore also in probability, if the w_t are such that

$$\frac{\sum_{t=0}^{T-2} w_t^2}{\left(\sum_{t=0}^{T-2} w_t\right)^2} \quad \text{and} \quad \frac{\sum_{t=1}^{T-2} w_{t-1}w_t}{\left(\sum_{t=0}^{T-2} w_t\right)^2}$$

converge to zero with T. Since

$$\left(\sum_{t=1}^{T-2} w_{t-1}w_t\right)^2 \le \left(\sum_{t=1}^{T-2} w_t^2\right)\left(\sum_{t=0}^{T-3} w_t^2\right) \le \left(\sum_{t=0}^{T-2} w_t^2\right)^2,$$

it is only necessary that

$$\lim_{T\to\infty} \frac{\sum_{t=0}^{T-2} w_t^2}{\left(\sum_{t=0}^{T-2} w_t\right)^2} = 0. \tag{5.11}$$

Consequently, even in the trivial case where $w_t = 1$, for all t, $\widehat{\lambda}$ is weakly consistent. We shall assume henceforth that $\{w_t\}$ satisfies (5.11). Now, when $A_T > -1$, V_T is explicitly given by

$$V_T = \tan^{-1}\left(\frac{B_T}{1+A_T}\right) = B_T\left\{1 + o_p(1)\right\}.$$

Since A_T converges in probability to zero, therefore, the asymptotic variance of V_T is thus the variance of B_T, which is

$$\frac{\nu^2\left(2+2\nu^2\right)\sum_{t=0}^{T-2} w_t^2 - 2\nu^2 \sum_{t=1}^{T-2} w_{t-1}w_t}{\left(\sum_{t=0}^{T-2} w_t\right)^2}, \tag{5.12}$$

a ratio of quadratic forms in $w = \begin{bmatrix} w_0 & \cdots & w_{T-2} \end{bmatrix}'$, which is minimised when w is any multiple of

$$\widetilde{w} = \begin{bmatrix} 2+2\nu^2 & -1 & 0 & \cdots & & 0 \\ -1 & 2+2\nu^2 & \ddots & \ddots & & \vdots \\ 0 & \ddots & \ddots & \ddots & & 0 \\ \vdots & & \ddots & & 2+2\nu^2 & -1 \\ 0 & \cdots & & 0 & -1 & 2+2\nu^2 \end{bmatrix}^{-1} \begin{bmatrix} 1 \\ \vdots \\ \vdots \\ \vdots \\ 1 \end{bmatrix}.$$

Moreover, the minimum value of (5.12) is

$$\frac{\nu^2}{\sum_{t=0}^{T-2} \widetilde{w}_t}.$$

Without loss of generality, therefore, we shall choose as minimising value $w = 2\widetilde{w}$, which satisfies the difference equation

$$-w_{j-1} + \left(2+2\nu^2\right) w_j - w_{j+1} = 2$$

with boundary conditions $w_{-1} = w_{T-1} = 0$. A particular solution of this is $w_t = \frac{1}{\nu^2}$, while the general solution of the homogeneous equation is

$$w_t = c_1\rho_1^t + c_2\rho_2^t,$$

where

$$\rho_1 = 1 + \nu^2 + \sqrt{(1+\nu^2)^2 - 1}$$

and

$$\rho_2 = 1 + \nu^2 - \sqrt{(1+\nu^2)^2 - 1}\ .$$

Imposing the boundary conditions, we obtain

$$\begin{bmatrix} \rho_1^{-1} & \rho_2^{-1} \\ \rho_1^{T-1} & \rho_2^{T-1} \end{bmatrix} \begin{bmatrix} c_1 \\ c_2 \end{bmatrix} + \frac{1}{\nu^2}\begin{bmatrix} 1 \\ 1 \end{bmatrix} = 0$$

and

$$w_t = \frac{1}{\nu^2}\left(1 + \rho_1^{t+1}\frac{\rho_2^T - 1}{\rho_1^T - \rho_2^T} + \rho_2^{t+1}\frac{\rho_1^T - 1}{\rho_2^T - \rho_1^T}\right),$$

and the resulting minimum variance is therefore

$$\frac{2\nu^4}{T + \frac{2 - \rho_1^T - \rho_1^{-T}}{\rho_1^T - \rho_1^{-T}}\frac{\rho_1 + 1}{\rho_1 - 1}} = \frac{2\nu^4}{T} + O\left(T^{-2}\right),$$

which is of order $O\left(T^{-1}\right)$ as $T \to \infty$. This does not allow us to calculate an estimator, however, unless ν^2 is known. In the absence of this knowledge, therefore, we calculate the limits of the w_t as the SNR increases, that is, as $\nu^2 \to 0$. The easiest method for doing this is to solve the difference equation for the special case where $\nu^2 = 0$; i.e., to solve

$$-w_{j-1} + 2w_j - w_{j+1} = 2,$$

subject to $w_{-1} = w_{T-1} = 0$. A particular solution is $-j^2$, while the general solution of the homogeneous equation is

$$w_t = c_1 + c_2 t.$$

The solution is thus quadratic in t and must be given, because of the boundary conditions and the particular solution, by

$$w_t = (t+1)\left(T - t - 1\right).$$

Since

$$\sum_{t=0}^{T-2} w_t^2 - \sum_{t=1}^{T-2} w_{t-1}w_t = \sum_{t=0}^{T-2} w_t\left(w_t - w_{t-1}\right) = \sum_{t=1}^{T} t\left(T - t\right)\left(T - 2t + 1\right)$$

$$= \frac{T\left(T^2 - 1\right)}{6},$$

the asymptotic variance of $\widehat{\lambda}$, for this set of weights, is given by

$$\frac{\nu^2 \left(2 + 2\nu^2\right) \sum_{t=0}^{T-2} w_t^2 - 2\nu^2 \sum_{t=1}^{T-2} w_{t-1}w_t}{\left(\sum_{t=0}^{T-2} w_t\right)^2}$$

$$= \frac{\nu^2 \frac{T^3}{3} \left\{1 + o\left(1\right)\right\} + 2\nu^4 \frac{T^5}{30} \left\{1 + o\left(1\right)\right\}}{\left(\frac{T^3-T}{6}\right)^2}$$

$$= 12\frac{\nu^2}{T^3} \left\{1 + o\left(1\right)\right\} + \frac{12}{5}\frac{\nu^4}{T} \left\{1 + o\left(1\right)\right\},$$

where the $o\left(1\right)$ terms involve T and not ν^2. Thus, from an asymptotic point of view, the estimator is not very interesting – the asymptotic variance is of order $O\left(T^{-1}\right)$ rather than $O\left(T^{-3}\right)$. Engineers are, however, more often than not interested in the behaviour of the estimator for fixed T and increasing SNR, or, in this case, decreasing ν^2. The variance of B_T is dominated, as the SNR increases, by the first term of the above, namely $12\frac{\nu^2}{T^3} \left\{1 + o\left(1\right)\right\}$, which is equivalent to the CRB. In fact, the ν^2 term may be shown to be *exactly* the asymptotic CRB for the variance of unbiased estimators of λ. Thus, in the asymptotic sense and as the SNR increases, it is Kay's *second* estimator which achieves the CRB, and not the *first* estimator.

5.6 MUSIC

The MUltiple SIgnal Characterization technique of Schmidt (1981, 1986) originated in the array processing literature, and was developed to resolve and estimate the directions of arrival of a number of signals. A version of the technique uses the (complex) Fourier coefficients at the same frequency but from different sensors in a straight array and in different time blocks. The version of MUSIC for estimating frequencies is also usually presented for complex data. In this section, we shall develop the theory for the cases of real sinusoids, leaving the case of complex sinusoids until later. Since arguably the real case is of most interest to statisticians, and the complex case to engineers, we have decided to present both in full, rather than just deriving completely the results for one, and outlining the results for the other.

MUSIC makes use of the eigenvectors of the sample autocovariance matrix of order larger than twice the number of sinusoids. (Recall that in the case of Pisarenko's multiple frequency algorithm, the dimension of the autocovariance matrix used is one more than twice the number of sinusoids.) Let $\{y(t)\}$ be defined as in (2.13) with $\mu = 0$ and with $\{x(t)\}$ assumed independent and identically distributed, and Gaussian with means zero and variances σ^2. Mean-correcting the $\{y(t)\}$ will have no asymptotic effect and so we shall omit that detail here. In practice, of course, the sample will be mean-corrected before the sample autocovariances are calculated. Now

$$
\begin{aligned}
C_j &= T^{-1} \sum_{t=j}^{T-1} y(t)\, y(t-j) \\
&= T^{-1} \sum_{t=j}^{T-1} x(t)\, x(t-j) + 2 \sum_{k=1}^{r} A_k \cos(\lambda_k j)\, u_k \\
&\quad + \sum_{k=1}^{r} \frac{A_k^2}{2} \cos(\lambda_k j) + o\left(T^{-1/2}\right),
\end{aligned}
$$

almost surely as $T \to \infty$, where

$$
u_k = T^{-1} \sum_{t=0}^{T-1} x(t) \cos(\lambda_k t + \phi_k).
$$

Thus

$$
C_j \to \gamma_j = \sigma^2 \delta_{0j} + \sum_{k=1}^{r} \frac{A_k^2}{2} \cos(\lambda_k j)
$$

almost surely, and the $K \times K$ sample autocovariance matrix

$$
C = \begin{bmatrix}
C_0 & C_1 & \cdots & C_{K-1} \\
C_1 & C_0 & \cdots & C_{K-2} \\
\vdots & \ddots & \ddots & \vdots \\
C_{K-1} & \cdots & C_1 & C_0
\end{bmatrix}
$$

converges almost surely to

$$
\Gamma = \sigma^2 I_K + \sum_{k=1}^{r} \frac{A_k^2}{2} \left(c_k c_k' + s_k s_k'\right),
$$

where

$$
c_k = \begin{bmatrix} 1 & \cos(\lambda_k) & \cdots & \cos\{(K-1)\lambda_k\} \end{bmatrix}'
$$

and

$$
s_k = \begin{bmatrix} 0 & \sin(\lambda_k) & \cdots & \sin\{(K-1)\lambda_k\} \end{bmatrix}'.
$$

Now, the rank of

$$
\sum_{k=1}^{r} \frac{A_k^2}{2} \left(c_k c_k' + s_k s_k'\right)
$$

is $2r$. Thus, the eigenvectors of Γ corresponding to the largest $2r$ eigenvalues form an orthonormal basis for the space spanned by the c_j and the s_j. Also, the eigenvectors corresponding to the lowest $K - 2r$ eigenvalues of Γ (which are all equal to σ^2) form an orthonormal basis for the orthogonal complement. Thus if P is one of these eigenvectors, we have $P'c_k = P's_k = 0$ and consequently $P'e(\lambda_k) = 0$, where

$$e(\theta) = \begin{bmatrix} 1 & e^{i\theta} & \cdots & e^{i(K-1)\theta} \end{bmatrix}'.$$

The MUSIC estimators of frequency are defined to be the local minimisers of

$$\sum_{k=2r+1}^{K} \left| e^*(\theta) \widehat{P}_k \right|^2, \tag{5.13}$$

where $\widehat{P}_k$ is a normalised eigenvector of C corresponding to eigenvalue $\widehat{v}_k$, and where $\widehat{v}_1 \geq \widehat{v}_2 \geq \cdots \geq \widehat{v}_K$. The 'MUSIC spectrum' is defined to be the reciprocal of (5.13). Peaks in the function will thus correspond with estimators of frequency, but no conclusions should be drawn from the relative sizes of peaks, since essentially all amplitude information is lost. This is considered an advantage, especially when there are several separated sinusoids, with quite different amplitudes. Figure 5.1 depicts the MUSIC spectrum of a single noisy sinusoid, with amplitude and noise standard deviation both 1 and sample size 512, while Figure 5.2 shows the MUSIC spectrum of the sum of two noisy sinusoids with amplitudes both 1 and noise standard deviation 0.5. Note the two peaks, which correspond to the two frequencies. However, Figure 5.3 was constructed from another simulation for the same scenario. Note that the first peak has coalesced with the zero frequency. It is thus fairly clear, since the periodogram will have clearly pronounced peaks at both frequencies, that the behaviour of MUSIC estimators is quite different from that of the maximum likelihood estimators.

Since

$$\sum_{k=1}^{2r} \left| e^*(\theta) \widehat{P}_k \right|^2 + \sum_{k=2r+1}^{K} \left| e^*(\theta) \widehat{P}_k \right|^2 = e^*(\theta) e(\theta) = K,$$

it follows that the MUSIC estimators of frequency are also the local maximisers of

$$g_T(\theta) = \sum_{k=1}^{2r} \left| e^*(\theta) \widehat{P}_k \right|^2.$$

This representation will be preferable as the 'lower' eigenvectors of C do

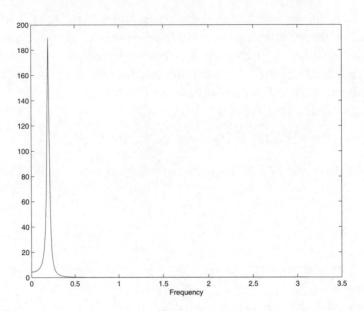

Fig. 5.1. MUSIC Spectrum

not converge. Essential to the derivation of the asymptotic properties of the MUSIC estimators of frequency, therefore, is the asymptotic behaviour of the sample eigenvectors $\widehat{P}_1, \ldots, \widehat{P}_{2r}$. Let $P_1, \ldots, P_{2r}$ be the 'ordered' eigenvectors of Γ, $P = \begin{bmatrix} P_1 & P_2 & \cdots & P_{2r} \end{bmatrix}$ and $v = \operatorname{diag}\{v_1, v_2, \ldots, v_{2r}\}$, the ordered eigenvalues. Now

$$\Gamma = \sigma^2 I_K + \sum_{k=1}^{r} \frac{A_k^2}{2} \left(c_k c_k' + s_k s_k' \right) = \sigma^2 I_K + EDE', \qquad (5.14)$$

where

$$E = \begin{bmatrix} c_1 & s_1 & c_2 & s_2 & \cdots & \cdots & c_r & s_r \end{bmatrix}$$

and

$$D = \frac{1}{2} \operatorname{diag}\left\{ A_1^2, A_1^2, A_2^2, A_2^2, \ldots, A_r^2, A_r^2 \right\}.$$

Thus

$$\Gamma = \sigma^2 I_K + P \Lambda P',$$

where $\Lambda = v - \sigma^2 I_{2r}$, and so, from (5.14), it follows that $P = EF$, for some $2r \times 2r$ matrix F. Now, E is of full column rank. To see this, suppose that

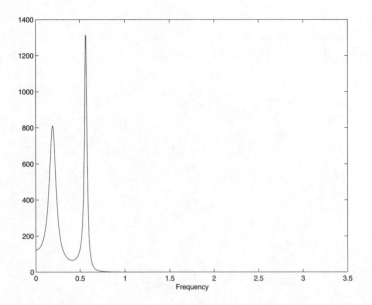

Fig. 5.2. First MUSIC spectrum for two sinusoids - 1

there exist constants $\alpha_1, \ldots, \alpha_f$ and $\beta_1, \ldots, \beta_r$, not all of which are 0, with

$$\sum_{k=1}^{r} \alpha_k c_k + \sum_{k=1}^{r} \beta_k s_k = 0.$$

We shall suppose for simplicity that $\alpha_r + i\beta_r \neq 0$. If this is not the case, the following argument is easily modified. For $j = 0, 1, \ldots, K$, we have

$$\frac{\alpha_r - i\beta_r}{2} e^{ij\lambda_r} = -\frac{\alpha_r + i\beta_r}{2} e^{-ij\lambda_r} - \sum_{k=1}^{r-1} \frac{\alpha_k - i\beta_k}{2} e^{ij\lambda_k} - \sum_{k=1}^{r-1} \frac{\alpha_k + i\beta_k}{2} e^{-ij\lambda_k}.$$

$$(5.15)$$

Now, let

$$h(z) = \left(e^{-i\lambda_r} - z\right) \prod_{k=1}^{r-1} \left(1 - 2\cos\lambda_k z + z^2\right) = \sum_{k=0}^{2r-1} \delta_k z^k,$$

say, which has zeros only from the set $\{e^{-i\lambda_r}, e^{\pm i\lambda_k}; k = 1, 2, \ldots, r-1\}$. Using (5.15) we have

$$\frac{\alpha_r - i\beta_r}{2} \sum_{j=0}^{2r-1} \delta_j e^{ij\lambda_r} = 0.$$

Thus $h\left(e^{i\lambda_r}\right) = 0$ and λ_r must be one of the elements of $\{-\lambda_r, \pm\lambda_k\}$. Since

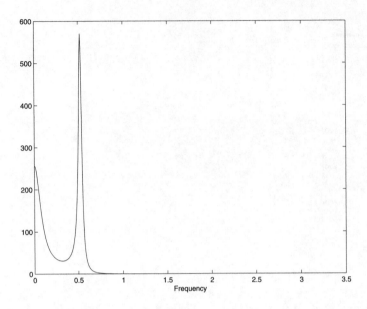

Fig. 5.3. Second MUSIC spectrum for two sinusoids - 2

the λ_k are distinct, and in $(0, \pi)$, we therefore have a contradiction. Thus E is of full column rank and F is of full rank.

Since

$$PAP' = EDE',$$

we have

$$EDE'EF = PAP'P = EF\Lambda$$

and therefore

$$D\left(E'E\right)F = F\Lambda \tag{5.16}$$

and

$$D^{1/2}\left(E'E\right)D^{1/2}\left(D^{-1/2}F\right) = \left(D^{-1/2}F\right)\Lambda.$$

Hence the columns of $D^{-1/2}F$ are eigenvectors of $D^{1/2}\left(E'E\right)D^{1/2}$ and the diagonal elements of Λ are its eigenvalues. This means, in particular, that

$$D^{-1/2}F = Rd^{1/2},$$

where $RR' = R'R = I$,

$$D^{1/2}\left(E'E\right)D^{1/2} = R\Lambda R'$$

and d is a diagonal positive definite matrix. Now, since $P'P = I$, it follows that

$$FF' = \left(E'E\right)^{-1}.$$

Thus

$$\left(E'E\right)^{-1} = D^{1/2}RdR'D^{1/2},$$

and

$$d = R'D^{-1/2}\left(E'E\right)^{-1}D^{-1/2}R = R'R\Lambda^{-1}R'R = \Lambda^{-1}$$

and so

$$P = EF$$

where

$$F = D^{1/2}R\Lambda^{-1/2}.$$

There is, of course, a lack of uniqueness, since we may multiply any normalised eigenvector by -1, and still have a normalised eigenvector. We shall need to show that this does not matter.

Since the eigenvalues are continuous functions of the elements of the matrix, it follows, since C converges almost surely to Γ, that for each k, $\widehat{v}_k$ converges almost surely to v_k. The eigenvector $\widehat{P}_k$ does not, however, converge almost surely to P_k. Since $P_1, \ldots, P_K$ form an orthogonal basis, it follows that for some $\alpha_1, \ldots, \alpha_K$,

$$\widehat{P}_k = \sum_{j=1}^{K} \alpha_j P_j.$$

In fact, since the P_j are orthonormal, it follows that $\alpha_j = P_j^* \widehat{P}_k$, whence

$$|\alpha_j|^2 \le \left(P_j'P_j\right)\left(\widehat{P}_k^*\widehat{P}_k\right) = 1.$$

Now

$$C\widehat{P}_k = \widehat{v}_k \widehat{P}_k.$$

Thus

$$(C - \Gamma + \Gamma)\sum_{j=1}^{K}\alpha_j P_j = (\widehat{v}_k - v_k + v_k)\sum_{j=1}^{K}\alpha_j P_j,$$

$$(C - \Gamma)\sum_{j=1}^{K}\alpha_j P_j + \sum_{j=1}^{K}\alpha_j v_j P_j = (\widehat{v}_k - v_k)\sum_{j=1}^{K}\alpha_j P_j + v_k\sum_{j=1}^{K}\alpha_j P_j$$

and, since $C - \Gamma$ and $\widehat{v}_k - v_k$ converge almost surely to zero, it is true that

$$\sum_{j=1}^{K} \alpha_j \left(v_j - v_k \right) P_j$$

converges almost surely to 0. Thus, for all j for which $v_j \neq v_k$, it must be the case that α_j converges almost surely to 0. Assuming for simplicity that $v_1, \ldots, v_r$ are distinct, we therefore have α_j converging to 0 for all $j \neq k$. But

$$\sum_{j=1}^{K} |\alpha_j|^2 = 1,$$

since $\widehat{P}_k^* \widehat{P}_k = 1$. Hence $|\alpha_k|$ converges almost surely to 1. Let $s_k = \mathrm{sgn}\left(\alpha_k \right)$. Then

$$
\begin{aligned}
s_k \widehat{P}_k - P_k &= s_k \sum_{j=1, j \neq k}^{K} \alpha_j P_j + |\alpha_k| P_k - P_k \\
&= s_k \sum_{j=1, j \neq k}^{K} \alpha_j P_j + \left(|\alpha_k| - 1 \right) P_k,
\end{aligned}
$$

which converges almost surely to 0. But

$$\left| e^* \left(\theta \right) s_k \widehat{P}_k \right|^2 = \left| e^* \left(\theta \right) \widehat{P}_k \right|^2.$$

We may thus act as though $\widehat{P}_k - P_k$ converges almost surely to 0. Similar results may be obtained if we drop the assumption that the $v_1, \ldots, v_r$ are distinct.

This enables us to prove the strong consistency of the MUSIC estimators. Recall that the MUSIC estimators are the maximisers of

$$g_T \left(\theta \right) = \sum_{k=1}^{2r} \left| e^* \left(\theta \right) \widehat{P}_k \right|^2 \rightarrow g \left(\theta \right) = \sum_{k=1}^{2r} \left| e^* \left(\theta \right) P_k \right|^2,$$

almost surely. Now

$$g \left(\theta \right) = e^* \left(\theta \right) P P' e \left(\theta \right)$$

and $P P'$ is of rank $2r$ and has eigenvalues 1, with multiplicity $2r$, and 0, with multiplicity $K - 2r$. Hence

$$e^* \left(\theta \right) P P' e \left(\theta \right) \leq e^* \left(\theta \right) e \left(\theta \right) = K,$$

with equality if and only if the vectors

$$c\left(\theta\right) = \left[\begin{array}{cccc} 1 & \cos\theta & \cdots & \cos\left\{\left(K-1\right)\theta\right\} \end{array}\right]'$$

and

$$s\left(\theta\right) = \left[\begin{array}{cccc} 0 & \sin\theta & \cdots & \sin\left\{\left(K-1\right)\theta\right\} \end{array}\right]'$$

are both in the space spanned by the columns of P. Thus, for some $2r \times 2$ matrix a, we have

$$\left[\begin{array}{cc} c\left(\theta\right) & s\left(\theta\right) \end{array}\right] = Pa = E\left(Fa\right) = Eb,$$

say. Now, let

$$h\left(z\right) = \prod_{k=1}^{r}\left(1 - 2\cos\lambda_{k}z + z^{2}\right) = \sum_{j=0}^{2r}\zeta_{j}z^{j},$$

the zeros of which are $e^{\pm i\lambda_{k}}$. Then

$$
\begin{aligned}
h\left(e^{i\theta}\right) &= \sum_{j=0}^{2r}\zeta_{j}\left\{\cos\left(j\theta\right) + i\sin\left(j\theta\right)\right\} \\
&= \sum_{j=0}^{2r}\zeta_{j}\left\{\sum_{k=1}^{r}\cos\left(j\lambda_{k}\right)b_{2k-1,1} + i\sum_{k=1}^{r}\sin\left(j\lambda_{k}\right)b_{2k,2}\right\} \\
&= \frac{1}{2}\sum_{k=1}^{r}\left(b_{2k-1,1} + b_{2k,2}\right)\sum_{j=0}^{2r}\zeta_{j}e^{ij\lambda_{k}} \\
&\qquad + \frac{1}{2}\sum_{k=1}^{r}\left(b_{2k-1,1} - b_{2k,2}\right)\sum_{j=0}^{2r}\zeta_{j}e^{-ij\lambda_{k}} \\
&= 0.
\end{aligned}
$$

Hence

$$g\left(\theta\right) \leq K,$$

with equality if and only if θ is one of $\lambda_{1}, \lambda_{2}, \ldots, \lambda_{k}$. Strong consistency follows using the usual Jennrich (1969) type arguments.

Now, since

$$
\begin{aligned}
T^{1/2}\left(C_{j} - \gamma_{j}\right) &= T^{-1/2}\sum_{t=j}^{T-1}\left\{x\left(t\right)x\left(t-j\right) - \sigma^{2}\delta_{0j}\right\} \\
&\quad + 2\sum_{k=1}^{r}A_{k}\cos\left(\lambda_{k}j\right)T^{1/2}u_{k} + O_{P}\left(T^{-1/2}\right), \quad (5.17)
\end{aligned}
$$

and since the terms in (5.17) for different j are asymptotically jointly normal, it follows that the elements of $\left\{ T^{1/2}\left(\widehat{P}_j - P_j\right) ; j = 1, \ldots, 2r \right\}$ are jointly asymptotically normal. It follows that the order of the central limit theorem for the estimators of frequency is also $T^{1/2}$; i.e., that the asymptotic variances of the frequency estimators are of order T^{-1}. Let

$$\widehat{P} = \left[\begin{array}{cccc} \widehat{P}_1 & \widehat{P}_2 & \cdots & \widehat{P}_{2r} \end{array}\right],$$

$Z_j = \widehat{P}_j - P_j$ and put $Z = \widehat{P} - P$. Then

$$\begin{aligned} I_{2r} &= \widehat{P}'\widehat{P} = (P+Z)'(P+Z) = P'P + P'Z + Z'P + Z'Z \\ &= I_{2r} + P'Z + Z'P + O_P\left(T^{-1}\right). \end{aligned}$$

Hence, although $Z = O_P\left(T^{-1/2}\right)$, we have

$$P'Z + Z'P = O_P\left(T^{-1}\right)$$

and in particular

$$\frac{1}{2}\left(P'_j Z_j + Z'_j P_j\right) = P'_j Z_j = O_P\left(T^{-1}\right).$$

Now

$$C\widehat{P} = \widehat{P}\widehat{v}$$

and so, putting $\xi = C - \Gamma$, we obtain

$$(\Gamma + \xi)(P + Z) = (P + Z)(v + \widehat{v} - v).$$

Thus

$$\Gamma P + \xi P + \Gamma Z + \xi Z = Pv + Zv + P\left(\widehat{v} - v\right) + Z\left(\widehat{v} - v\right).$$

Since $\Gamma P = Pv$ and ξZ and $Z\left(\widehat{v} - v\right)$ are of order $O_P\left(T^{-1}\right)$, it follows that

$$\xi P + \Gamma Z = Zv + P\left(\widehat{v} - v\right) + O_P\left(T^{-1}\right)$$

and, multiplying this on the left by P', that

$$P'\xi P + vP'Z = P'Zv + \widehat{v} - v + O_P\left(T^{-1}\right).$$

The (j, k)th component equation is, when $j \neq k$,

$$P'_j \xi P_k + v_j P'_j Z_k = P'_j Z_k v_k + O_P\left(T^{-1}\right)$$

so that

$$P'_j Z_k = -\frac{1}{v_j - v_k} P'_j \xi P_k + O_P\left(T^{-1}\right)$$

while the (j, j)th component equation is

$$P'_j \xi P_j + v_j P'_j Z_j = P'_j Z_j v_j + \widehat{v}_j - v_j + O_P\left(T^{-1}\right)$$

which shows that

$$\widehat{v}_j - v_j = P'_j \xi P_j + O_P\left(T^{-1}\right).$$

If q is any element of the orthogonal complement of the space spanned by the P_j, then, since

$$\Gamma q = \sigma^2 q,$$

it follows that

$$q' \xi P + \sigma^2 q' Z = q' Z v + O_P\left(T^{-1}\right),$$

and so

$$q' Z_k = \frac{1}{v_k - \sigma^2} q' \xi P_k + O_P\left(T^{-1}\right).$$

Now let Q be any $K \times (K - 2r)$ matrix whose columns form an orthonormal basis for the orthogonal complement of the space spanned by the P_j, $j = 1, 2, \ldots, 2r$. Then $Q'Q = I$ and

$$QQ' = I_K - PP'.$$

Hence, since $P'_k Z_k = O_P\left(T^{-1}\right)$,

$$
\begin{aligned}
Z_k &= \left(PP' + QQ'\right) Z_k = \sum_j P_j P'_j Z_k + \sum_j Q_j Q'_j Z_k \\
&= -\sum_{j \neq k} \frac{1}{v_j - v_k} P_j P'_j \xi P_k \\
&\quad + \frac{1}{v_k - \sigma^2} \sum_j Q_j Q'_j \xi P_k + O_P\left(T^{-1}\right) \\
&= -\sum_{j \neq k} \frac{1}{v_j - v_k} P_j P'_j \xi P_k \\
&\quad + \frac{1}{v_k - \sigma^2} \left(I_K - PP'\right) \xi P_k + O_P\left(T^{-1}\right).
\end{aligned}
$$
(5.18)

Using (5.17) we obtain

$$
\begin{aligned}
\xi &= C - \Gamma = R - \sigma^2 I_K + 2 \sum_{j=1}^{r} A_k u_k \left(c_k c'_k + s_k s'_k\right) \\
&= R - \sigma^2 I_K + E U E' + O_P\left(T^{-1}\right),
\end{aligned}
$$

where

$$
R = \begin{bmatrix}
R_0 & R_1 & \cdots & R_{K-1} \\
R_1 & R_0 & \cdots & R_{K-2} \\
\vdots & \ddots & \ddots & \vdots \\
R_{K-1} & \cdots & R_1 & R_0
\end{bmatrix},
$$

$$
R_j = T^{-1} \sum_{t=0}^{T-1} x(t)\, x(t-j)
$$

and $U = \operatorname{diag} \{2A_1 u_1, 2A_1 u_1, \ldots, 2A_r u_r, 2A_r u_r\}$. Thus

$$
P'\xi = P'(R - \sigma^2 I_K) + F^{-1} U E' + O_P(T^{-1})
$$

and, since $Q'E = Q'PF^{-1} = 0$, we obtain

$$
(I_K - PP')\xi = QQ'\xi = QQ'(R - \sigma^2 I_K) + O_P(T^{-1}). \qquad (5.19)
$$

The central limit theorem may now be derived using the usual Taylor series techniques. Let $\widehat{\lambda}_1, \widehat{\lambda}_2, \ldots, \widehat{\lambda}_r$ denote the arguments of θ corresponding to the r largest local maxima of $g_T(\theta)$. Then, since the $\widehat{\lambda}_j$ are consistent,

$$
0 = g_T'\left(\widehat{\lambda}_j\right) = g_T'(\lambda_j) + g_T''\left(\widetilde{\lambda}_j\right)\left(\widehat{\lambda}_j - \lambda_j\right),
$$

where $\widetilde{\lambda}_j$ is between λ_j and $\widehat{\lambda}_j$ and thus converges almost surely to λ_j. Consequently, we only need to derive a joint central limit theorem for the $g_T'(\lambda_j)$ and to show uniform convergence almost surely of $g_T''(\theta)$ in small neighbourhoods of the λ_j. Now, letting λ stand for any of the λ_j, we have

$$
g_T'(\lambda) = 2c_1'(\lambda)\, \widehat{P}\widehat{P}'c(\lambda) + 2s_1'(\lambda)\, \widehat{P}\widehat{P}'s(\lambda)
$$

where

$$
c_1(\lambda) = \frac{\partial}{\partial \lambda} c(\lambda)
$$

and

$$
s_1(\lambda) = \frac{\partial}{\partial \lambda} s(\lambda).
$$

Thus

$$
\begin{aligned}
g_T'(\lambda) =\ & 2c_1'(\lambda)\, PP'c(\lambda) + 2s_1'(\lambda)\, PP's(\lambda) \\
& + 2c_1'(\lambda)\, ZP'c(\lambda) + 2s_1'(\lambda)\, ZP's(\lambda) \\
& + 2c_1'(\lambda)\, PZ'c(\lambda) + 2s_1'(\lambda)\, PZ's(\lambda) + O_P(T^{-1}).
\end{aligned}
$$

But

$$2c_1'\left(\lambda\right)PP'c\left(\lambda\right) + 2s_1'\left(\lambda\right)PP's\left(\lambda\right)$$
$$= 2c_1'\left(\lambda\right)\left(I_K - QQ'\right)c\left(\lambda\right) + 2s_1'\left(\lambda\right)\left(I_K - QQ's\right)s\left(\lambda\right)$$
$$= 2c_1'\left(\lambda\right)c\left(\lambda\right) + 2s_1'\left(\lambda\right)s\left(\lambda\right),$$

since $Q'c\left(\lambda\right) = Q's\left(\lambda\right) = 0$. However,

$$2c_1'\left(\lambda\right)c\left(\lambda\right) + 2s_1'\left(\lambda\right)s\left(\lambda\right) = \frac{\partial}{\partial\lambda}e^*\left(\lambda\right)e\left(\lambda\right) = \frac{\partial}{\partial\lambda}K = 0.$$

Hence

$$g_T'\left(\lambda\right) = 2c_1'\left(\lambda\right)\left(ZP' + PZ'\right)c\left(\lambda\right)$$
$$+ 2s_1'\left(\lambda\right)\left(ZP' + PZ'\right)s\left(\lambda\right) + O_P\left(T^{-1}\right).$$

The above shows that we need to establish limit theorems for $ZP' + PZ'$.
Now

$$ZP' + PZ' = \sum_{k=1}^{2r}\left(Z_kP_k' + P_kZ_k'\right).$$

From (5.18),

$$Z_kP_k' + P_kZ_k'$$
$$= -\sum_{j\neq k}\frac{1}{v_j - v_k}P_jP_j'\xi P_kP_k' - \sum_{j\neq k}\frac{1}{v_j - v_k}P_kP_k'\xi P_jP_j'$$
$$+ \frac{1}{v_k - \sigma^2}\left(I_K - PP'\right)\xi P_kP_k' + \frac{1}{v_k - \sigma^2}P_kP_k'\xi\left(I_K - PP'\right) + O_P\left(T^{-1}\right).$$

Thus

$$\sum_{k=1}^{2r}\left(Z_kP_k' + P_kZ_k'\right)$$
$$= -\sum_k\sum_{j\neq k}\frac{1}{v_j - v_k}P_jP_j'\xi P_kP_k' - \sum_k\sum_{j\neq k}\frac{1}{v_j - v_k}P_kP_k'\xi P_jP_j'$$
$$+ \sum_k\frac{1}{v_k - \sigma^2}\left(I_K - PP'\right)\xi P_kP_k' + \sum_k\frac{1}{v_k - \sigma^2}P_kP_k'\xi\left(I_K - PP'\right)$$
$$+ O_P\left(T^{-1}\right).$$

But, interchanging j and k in the second double sum on the right side of the above results in a sum which exactly negates the first. Hence

$$ZP' + PZ' = \sum_k \frac{1}{v_k - \sigma^2} \left(I_K - PP'\right) \xi P_k P_k'$$
$$+ \sum_k \frac{1}{v_k - \sigma^2} P_k P_k' \xi \left(I_K - PP'\right) + O_P\left(T^{-1}\right).$$

Also

$$g_T'\left(\lambda_j\right)$$
$$= 2c_1'\left(\lambda_j\right)\left(ZP' + PZ'\right)c\left(\lambda_j\right) + 2s_1'\left(\lambda_j\right)\left(ZP' + PZ'\right)s\left(\lambda_j\right) + O_P\left(T^{-1}\right)$$
$$= 2\,\text{tr}\left\{\left(ZP' + PZ'\right)\left[\begin{array}{cc} c\left(\lambda_j\right) & s\left(\lambda_j\right) \end{array}\right]\left[\begin{array}{c} c_1'\left(\lambda_j\right) \\ s_1'\left(\lambda_j\right) \end{array}\right]\right\} + O_P\left(T^{-1}\right)$$
$$= 2\,\text{tr}\left\{\left(ZP' + PZ'\right)E\kappa_j\right\} + O_P\left(T^{-1}\right)$$
$$= 2\,\text{tr}\left\{\left(ZP' + PZ'\right)PF^{-1}\kappa_j\right\} + O_P\left(T^{-1}\right),$$

where κ_j is the $2r \times K$ matrix which has $c_1'\left(\lambda_j\right)$ and $s_1'\left(\lambda_j\right)$ as its $(2j-1)$th and $(2j)$th rows and zeros the rest. Now, since

$$\left(I_K - PP'\right)P = P - P = 0,$$

it follows that

$$\left(ZP' + PZ'\right)P = \sum_k \frac{1}{v_k - \sigma^2}\left(I_K - PP'\right)\xi P_k P_k' P$$
$$+ \sum_k \frac{1}{v_k - \sigma^2} P_k P_k' \xi \left(I_K - PP'\right)P + O_P\left(T^{-1}\right)$$
$$= \sum_k \frac{1}{v_k - \sigma^2}\left(I_K - PP'\right)\xi P_k i_k' + O_P\left(T^{-1}\right),$$

where i_k' is the kth row of I_K. Consequently,

$$\text{tr}\left\{\left(ZP' + PZ'\right)PF^{-1}\kappa_j\right\}$$
$$= \text{tr}\left\{\sum_k \frac{1}{v_k - \sigma^2}\left(I_K - PP'\right)\left(C - \Gamma\right)P_k i_k' F^{-1}\kappa_j\right\} + O_P\left(T^{-1}\right)$$
$$= \text{tr}\left\{\left(I_K - PP'\right)\xi\left(\sum_k \frac{1}{v_k - \sigma^2}P_k i_k'\right)F^{-1}\kappa_j\right\} + O_P\left(T^{-1}\right)$$
$$= \text{tr}\left\{\left(I_K - PP'\right)\xi P\Lambda^{-1}F^{-1}\kappa_j\right\} + O_P\left(T^{-1}\right).$$

Now, from (5.16)

$$P\Lambda^{-1}F^{-1} = EF\Lambda^{-1}F^{-1} = E\left(E'E\right)^{-1}D^{-1}$$

and so, using this as well as (5.19), we obtain

$$
\begin{aligned}
g_T'\left(\lambda_j\right) &= 2\operatorname{tr}\left\{\left(I_K - PP'\right)\left(R - \sigma^2 I_K\right)E\left(E'E\right)^{-1}D^{-1}\kappa_j\right\} + O_P\left(T^{-1}\right) \\
&= \frac{4}{A_j^2}\operatorname{tr}\left\{\left(I_K - PP'\right)\left(R - \sigma^2 I_K\right)E\left(E'E\right)^{-1}\kappa_j\right\} + O_P\left(T^{-1}\right) \\
&= \frac{4}{A_j^2}\operatorname{tr}\left\{H_j\left(R - \sigma^2 I_K\right)\right\} + O_P\left(T^{-1}\right),
\end{aligned}
$$

where

$$
H_j = E\left(E'E\right)^{-1}\kappa_j\left(I_K - PP'\right).
$$

Letting $H_{j,ik}$ be the (i,k)th element of H_j, we obtain

$$
\begin{aligned}
\operatorname{tr}\left\{H_j\left(R - \sigma^2 I_K\right)\right\} &= \sum_{i,k} H_{j,ik}\left(R_{k-i} - \sigma^2\delta_{ik}\right) = \sum_{l=1-K}^{K-1} R_l \sum_k H_{j,(l+k)k} \\
&= \sum_{l=1}^{K-1}\left(h_{j,l} + h_{j,-l}\right)R_l, \tag{5.20}
\end{aligned}
$$

where

$$
h_{j,l} = \begin{cases} \sum_{k=l}^{K-1} H_{j,(k-l)k} & ;\quad l > 0 \\ \sum_{k=0}^{K-1+l} H_{j,(k-l)k} & ;\quad l < 0 \end{cases}.
$$

Note that the coefficient $h_{j,0}$ of R_0 is zero, since

$$
\begin{aligned}
\sum_{k=0}^{K-1} H_{j,kk} &= \operatorname{tr}\left\{E\left(E'E\right)^{-1}\kappa_j\left(I_K - PP'\right)\right\} \\
&= \operatorname{tr}\left\{\left(E'E\right)^{-1}\kappa_j\left(I_K - PP'\right)E\right\} \\
&= \operatorname{tr}\left\{\left(E'E\right)^{-1}\kappa_j\left(I_K - PP'\right)PF^{-1}\right\} \\
&= \operatorname{tr}\left\{\left(E'E\right)^{-1}\kappa_j 0 F^{-1}\right\} \\
&= 0.
\end{aligned}
$$

The expression given in (5.20) can be simplified. We shall first show that

$$
H_{j,kl} = H_{j,K-1-k,K-1-l}.
$$

Note that, since

$$
PP' = EFF'E' = E\left(E'E\right)^{-1}E',
$$

we can write H_j as

$$H_j = A_j B_j',$$

where A_j is the matrix formed from the $(2j-1)$th and $(2j)$th columns of

$$E\left(E'E\right)^{-1},$$

and

$$B_j = \left(I - E\left(E'E\right)^{-1}E'\right)f_j,$$

where f_j is $K \times 2$, with first column $c_1\left(\lambda_j\right)$ and second column $s_1\left(\lambda_j\right)$. Denote by $\widetilde{E}$ and $\widetilde{A}_j$ the matrices formed from E and A_j by reversing their rows. Then the row vector formed from the $(2j-1)$th and $(2j)$th elements of the kth row of $\widetilde{E}$ is

$$\begin{aligned}
& \left[\ \cos\left\{(K-1-k)\lambda_j\right\}\quad \sin\left\{(K-1-k)\lambda_j\right\}\ \right] \\
=\ & \left[\ \cos\left(k\lambda_j\right)\quad \sin\left(k\lambda_j\right)\ \right]\left[\begin{array}{cc} \cos\left\{(K-1)\lambda_j\right\} & \sin\left\{(K-1)\lambda_j\right\} \\ \sin\left\{(K-1)\lambda_j\right\} & -\cos\left\{(K-1)\lambda_j\right\} \end{array}\right] \\
=\ & \left[\ \cos\left(k\lambda_j\right)\quad \sin\left(k\lambda_j\right)\ \right]C_j,
\end{aligned}$$

say. Hence

$$\widetilde{E} = EC,$$

where

$$C = \left[\begin{array}{ccc} C_1 & 0 & 0 \\ 0 & \ddots & 0 \\ 0 & 0 & C_r \end{array}\right].$$

Noting that $C^2 = I$ since $C_j^2 = I_2$, we have

$$\begin{aligned}
\widetilde{E}\left(E'E\right)^{-1} &= EC\left(E'E\right)^{-1} = E\left(E'EC\right)^{-1} = E\left(E'\widetilde{E}\right)^{-1} = E\left(\widetilde{E}'E\right)^{-1} \\
&= E\left(CE'E\right)^{-1} = E\left(E'E\right)^{-1}C
\end{aligned}$$

and

$$\left[\ A_{j,K-1-k,1}\quad A_{j,K-1-k,2}\ \right] = \left[\ A_{j,k1}\quad A_{j,k2}\ \right]C_j.$$

Thus

$$\widetilde{A}_j = A_j C_j.$$

Now,

$$B_j = f_j - E\left(E'E\right)^{-1}E'f_j.$$

Defining $\widetilde{B}_j$ and $\widetilde{f}_j$ as B_j and f_j with rows reversed, and using the result above, we obtain

$$
\begin{aligned}
\widetilde{B}_j &= \widetilde{f}_j - \widetilde{E}\left(E'E\right)^{-1}E'f_j \\
&= \widetilde{f}_j - E\left(E'E\right)^{-1}CE'f_j \\
&= \widetilde{f}_j - E\left(E'E\right)^{-1}\widetilde{E}'f_j \\
&= \widetilde{f}_j - E\left(E'E\right)^{-1}E'\widetilde{f}_j.
\end{aligned}
$$

But

$$
\begin{aligned}
\left[\begin{array}{cc} \widetilde{f}_{j,l,1} & \widetilde{f}_{j,l,2} \end{array}\right] \\
= \left[\begin{array}{cc} -(K-1-l)\sin\left\{(K-1-l)\lambda_j\right\} & (K-1-l)\cos\left\{(K-1-l)\lambda_j\right\} \end{array}\right] \\
= (K-1)\left[\begin{array}{cc} -\sin\left\{(K-1-l)\lambda_j\right\} & \cos\left\{(K-1-l)\lambda_j\right\} \end{array}\right] \\
\quad -\left[\begin{array}{cc} -l\sin\left\{(K-1-l)\lambda_j\right\} & l\cos\left\{(K-1-l)\lambda_j\right\} \end{array}\right] \\
= (K-1)\left[\begin{array}{cc} -\sin\left(l\lambda_j\right) & \cos\left(l\lambda_j\right) \end{array}\right]C_j \\
\quad +\left[\begin{array}{cc} -l\sin\left(l\lambda_j\right) & l\cos\left(l\lambda_j\right) \end{array}\right]C_j \\
= (K-1)\left[\begin{array}{cc} \cos\left(l\lambda_j\right) & \sin\left(l\lambda_j\right) \end{array}\right]S_j + \left[\begin{array}{cc} -l\sin\left(l\lambda_j\right) & l\cos\left(l\lambda_j\right) \end{array}\right]C_j,
\end{aligned}
$$

where

$$
S_j = \left[\begin{array}{cc} -\sin\left\{(K-1)\lambda_j\right\} & \cos\left\{(K-1)\lambda_j\right\} \\ \cos\left\{(K-1)\lambda_j\right\} & \sin\left\{(K-1)\lambda_j\right\} \end{array}\right].
$$

Thus, letting E_j be the $K \times 2$ matrix formed from the $(2j-1)$th and $(2j)$th columns of E, we have

$$
\widetilde{f}_j = (K-1)E_jS_j + f_jC_j.
$$

Hence, letting

$$
S = \left[\begin{array}{ccc} S_1 & 0 & 0 \\ 0 & \ddots & 0 \\ 0 & 0 & S_r \end{array}\right],
$$

it follows that

$$
\begin{aligned}
\left[\begin{array}{ccc} \widetilde{B}_1 & \cdots & \widetilde{B}_r \end{array}\right] &= (K-1)\left\{I - E\left(E'E\right)^{-1}E'\right\}ES \\
&\quad -\left\{I - E\left(E'E\right)^{-1}E'\right\}\left[\begin{array}{ccc} f_1 & \cdots & f_r \end{array}\right]C \\
&= \left\{I - E\left(E'E\right)^{-1}E'\right\}\left[\begin{array}{ccc} f_1 & \cdots & f_r \end{array}\right]C \\
&= \left[\begin{array}{ccc} B_1 & \cdots & B_r \end{array}\right]C,
\end{aligned}
$$

and so $\widetilde{B}_j = B_j C_j$. Thus

$$\widetilde{A}_j \widetilde{B}'_j = A_j C_j C'_j B'_j = A_j B_j,$$

consequently

$$H_{j,K-1-k,K-1-l} = H_{j,kl},$$

and, using (5.20), we obtain

$$\operatorname{tr} \left\{ H_j \left(R - \sigma^2 I_K \right) \right\} = 2 \sum_{l=1}^{K-1} h_{j,l} R_l.$$

Since $T^{1/2} \begin{bmatrix} R_1 & \cdots & R_{K-1} \end{bmatrix}'$ is asymptotically normally distributed with mean zero and covariance matrix

$$\left(\sigma^2 \right)^2 I_{K-1},$$

we therefore have the following result.

Lemma 1 $T^{1/2} \begin{bmatrix} g'_T (\lambda_1) & \cdots & g'_T (\lambda_r) \end{bmatrix}'$ *is asymptotically normally distributed with mean zero and covariance matrix* Ω, *where*

$$\Omega_{ij} = \frac{64 \left(\sigma^2 \right)^2}{A_i^2 A_j^2} \sum_{l=1}^{K-1} h_{i,l} h_{j,l}, \tag{5.21}$$

$$h_{j,l} = \sum_{k=0}^{K-1-l} H_{j,(l+k)k}$$

and

$$H_j = E \left(E'E \right)^{-1} \kappa_j \left\{ I_K - E \left(E'E \right)^{-1} E' \right\}.$$

Finally, we need to examine the second derivatives $g''_T (\theta)$. Now

$$\begin{aligned} g''_T (\theta) = \ & 2c'_2 (\theta) \, \widehat{P} \widehat{P}' c (\theta) + 2s'_2 (\theta) \, \widehat{P} \widehat{P}' s (\theta) \\ & + 2c'_1 (\theta) \, \widehat{P} \widehat{P}' c_1 (\theta) + 2s'_1 (\theta) \, \widehat{P} \widehat{P}' s_1 (\theta) \end{aligned}$$

which converges almost surely and uniformly in $\theta \in (0, \pi)$ to

$$\begin{aligned} g'' (\theta) = \ & 2c'_2 (\theta) \, P P' c (\theta) + 2s'_2 (\theta) \, P P' s (\theta) \\ & + 2c'_1 (\theta) \, P P' c_1 (\theta) + 2s'_1 (\theta) \, P P' s_1 (\theta). \end{aligned}$$

Hence, almost surely as $T \to \infty$, we have

$$g_T'' \left(\widetilde{\lambda}_j \right) \to g'' \left(\lambda_j \right)$$

$$= 2c_2' \left(\lambda_j \right) PP'c \left(\lambda_j \right) + 2s_2' \left(\lambda_j \right) PP's \left(\lambda_j \right)$$

$$+ 2c_1' \left(\lambda_j \right) PP'c_1 \left(\lambda_j \right) + 2s_1' \left(\lambda_j \right) PP's_1 \left(\lambda_j \right).$$

Some simplification is possible. Since c_j and s_j are columns of E, and therefore in the space spanned by the columns of P, it follows that

$$PP' \begin{bmatrix} c_j & s_j \end{bmatrix} = \begin{bmatrix} c_j & s_j \end{bmatrix}.$$

Thus

$$2c_2' \left(\lambda_j \right) PP'c \left(\lambda_j \right) + 2s_2' \left(\lambda_j \right) PP's \left(\lambda_j \right)$$

$$= 2c_2' \left(\lambda_j \right) c \left(\lambda_j \right) + 2s_2' \left(\lambda_j \right) s \left(\lambda_j \right)$$

$$= -2 \sum_{k=0}^{K-1} k^2 \cos^2 \left(k \lambda_j \right) - 2 \sum_{k=0}^{K-1} k^2 \sin^2 \left(k \lambda_j \right)$$

$$= -2c_1' \left(\lambda_j \right) c_1 \left(\lambda_j \right) - 2s_1' \left(\lambda_j \right) s_1 \left(\lambda_j \right).$$

Also

$$2c_1' \left(\lambda_j \right) PP'c_1 \left(\lambda_j \right) + 2s_1' \left(\lambda_j \right) PP's_1 \left(\lambda_j \right)$$

$$= 2c_1' \left(\lambda_j \right) E \left(E'E \right)^{-1} E'c_1 \left(\lambda_j \right) + 2s_1' \left(\lambda_j \right) E \left(E'E \right)^{-1} E's_1 \left(\lambda_j \right).$$

Hence $g_T'' \left(\widetilde{\lambda}_j \right)$ converges almost surely to

$$\chi_j = -2c_1' \left(\lambda_j \right) \left\{ I_K - E \left(E'E \right)^{-1} E' \right\} c_1 \left(\lambda_j \right)$$

$$-2s_1' \left(\lambda_j \right) \left\{ I_K - E \left(E'E \right)^{-1} E' \right\} s_1 \left(\lambda_j \right). \qquad (5.22)$$

Using this result in conjunction with the lemma, we have:

Theorem 22 *Let* $\widehat{\lambda}_1, \widehat{\lambda}_2, \ldots, \widehat{\lambda}_r$ *denote the MUSIC estimators of frequency. Then*

$$T^{1/2} \begin{bmatrix} \widehat{\lambda}_1 - \lambda_1 & \cdots & \widehat{\lambda}_r - \lambda_r \end{bmatrix}'$$

has a distribution which converges as $T \to \infty$ *to the normal with mean zero and covariance matrix* Σ, *where*

$$\Sigma_{ij} = \frac{\Omega_{ij}}{\chi_i \chi_j}$$

is given by (5.21) *and* (5.22).

It should be noted that MUSIC is just one of the so-called subspace techniques. Other techniques are obtained by weighting the terms in (5.13). A class of these, for which the asymptotic distribution is the same as that of the MUSIC estimators, is analysed for the complex case at the end of this chapter. The theory for the real case is similar and will not be reproduced here.

5.7 Complex MUSIC

We consider now the case of complex sinusoids. Suppose that

$$y(t) = \sum_{k=1}^{r} A_k e^{i\lambda_k t} + x(t),$$

where $\{x(t)\}$ is complex and Gaussian, with real and imaginary parts independent and identically distributed with means zero and variances $\sigma^2/2$, $A_1, \ldots, A_r$ are now complex constants and the λ_k are now in $(-\pi, \pi)$. Let

$$C_j = T^{-1} \sum_{t=j}^{T-1} y(t) y^*(t-j).$$

Then

$$
\begin{aligned}
C_j &= T^{-1} \sum_{t=j}^{T-1} x(t) x^*(t-j) + \sum_{k=1}^{r} A_k e^{\lambda_k j} u_k^* + \sum_{k=1}^{r} A_k^* e^{-\lambda_k j} u_k \\
&\quad + \sum_{k=1}^{r} |A_k|^2 e^{i\lambda_k j} + o\left(T^{-1/2}\right),
\end{aligned}
$$

almost surely as $T \to \infty$, where

$$u_k = T^{-1} \sum_{t=0}^{T-1} x(t) e^{-i\lambda_k t}.$$

Thus

$$C_j \to \gamma_j = \sigma^2 \delta_{0j} + \sum_{k=1}^{r} |A_k|^2 e^{i\lambda_k j}$$

almost surely and the $K \times K$ sample autocovariance matrix

$$
C = \begin{bmatrix}
C_0 & C_1^* & \cdots & C_{K-1}^* \\
C_1 & C_0 & \cdots & C_{K-2}^* \\
\vdots & \ddots & \ddots & \vdots \\
C_{K-1} & \cdots & C_1 & C_0
\end{bmatrix}
$$

converges almost surely to

$$\Gamma = \sigma^2 I_K + \sum_{k=1}^{r} |A_k|^2 \, e_k e_k^*,$$

where $e_k = e(\lambda_k)$ and

$$e(\theta) = \begin{bmatrix} 1 & e^{i\theta} & \cdots & e^{i(K-1)\theta} \end{bmatrix}'.$$

The rank of

$$\sum_{k=1}^{r} |A_k|^2 \, e_k e_k^*$$

is r. Thus, the eigenvectors of Γ corresponding to the largest r eigenvalues form an orthonormal basis for the space spanned by the e_j. Also, the eigenvectors corresponding to the lowest $K - r$ eigenvalues of Γ (which are all equal to σ^2) form an orthonormal basis for the orthogonal complement. Thus if P is one of these eigenvectors, we have $P^* e_k = 0$. The MUSIC estimators of frequency are defined to be the local minimisers of

$$\sum_{k=r+1}^{K} \left| e^*(\theta) \widehat{P}_k \right|^2,$$

where $\widehat{P}_k$ is a normalised eigenvector of C corresponding to eigenvalue $\widehat{v}_k$, and where $\widehat{v}_1 \geq \widehat{v}_2 \geq \cdots \geq \widehat{v}_K$. Since

$$\sum_{k=1}^{r} \left| e^*(\theta) \widehat{P}_k \right|^2 + \sum_{k=r+1}^{K} \left| e^*(\theta) \widehat{P}_k \right|^2 = e^*(\theta) e(\theta) = K,$$

it follows that the MUSIC estimators of frequency are also the local maximisers of

$$g_T(\theta) = \sum_{k=1}^{r} \left| e^*(\theta) \widehat{P}_k \right|^2.$$

This representation will be preferable as the 'lower' eigenvectors of C do not converge. Essential to the derivation of the asymptotic properties of the MUSIC estimators of frequency, therefore, is the asymptotic behaviour of the sample eigenvectors $\widehat{P}_1, \ldots, \widehat{P}_r$. Let $P_1, \ldots, P_r$ be any 'ordered' eigenvectors of Γ, $P = \begin{bmatrix} P_1 & P_2 & \cdots & P_r \end{bmatrix}$ and $v = \mathrm{diag}\{v_1, v_2, \ldots, v_r\}$ the ordered eigenvalues. Now

$$\Gamma = \sigma^2 I_K + \sum_{k=1}^{r} |A_k|^2 \, e_k e_k^* = \sigma^2 I_K + EDE^*,$$

where

$$E = \begin{bmatrix} e_1 & \cdots & e_r \end{bmatrix}$$

and

$$D = \text{diag}\left\{|A_1|^2, \dots, |A_r|^2\right\}.$$

Thus

$$\Gamma = \sigma^2 I_K + P\Lambda P^*,$$

where $\Lambda = v - \sigma^2 I_r$, and so $P = EF$, for some $r \times r$ matrix F. Now, E is of full column rank. To see this, suppose that there are constants $\alpha_1, \dots, \alpha_r$ not all of which are 0, with

$$\sum_{k=1}^{r} \alpha_k e_k = 0.$$

We shall suppose for simplicity that $\alpha_r \neq 0$. If this is not so, the following argument is easily modified. For $j = 0, 1, \dots, K$, we have

$$\alpha_r e^{ij\lambda_r} = -\sum_{k=1}^{r-1} \alpha_k e^{ij\lambda_k}.$$

Now, let

$$h(z) = \prod_{k=1}^{r-1}\left(1 - e^{-i\lambda_k} z\right) = \sum_{k=0}^{r-1} \delta_k z^k,$$

say, which has zeros only from the set $\{e^{i\lambda_k}; k = 1, 2, \dots, r-1\}$. Thus

$$\alpha_r \sum_{j=0}^{r-1} \delta_j e^{ij\lambda_r} = 0,$$

$h\left(e^{i\lambda_r}\right) = 0$ and λ_r must be one of the elements of $\{\lambda_k; k = 1, \dots, r-1\}$. Since the λ_k are distinct we therefore have a contradiction; thus E is of full column rank and F is of full rank.

Since

$$P\Lambda P^* = EDE^*,$$

we have

$$EDE^* EF = P\Lambda P^* P = EF\Lambda$$

and therefore

$$D\left(E^* E\right) F = F\Lambda$$

and

$$D^{1/2} \left(E^* E\right) D^{1/2} \left(D^{-1/2} F\right) = \left(D^{-1/2} F\right) \Lambda.$$

Hence the columns of $D^{-1/2} F$ are eigenvectors of $D^{1/2} \left(E^* E\right) D^{1/2}$ and the diagonal elements of Λ are its eigenvalues. This means, in particular, that

$$D^{-1/2} F = R d^{1/2},$$

where $RR^* = R^* R = I$,

$$D^{1/2} \left(E^* E\right) D^{1/2} = R \Lambda R^*$$

and d is a diagonal positive definite matrix. Now, since $P^* P = I$, it follows that

$$FF^* = \left(E^* E\right)^{-1}.$$

Thus

$$\left(E^* E\right)^{-1} = D^{1/2} R d R^* D^{1/2},$$

$$d = R^* D^{-1/2} \left(E^* E\right)^{-1} D^{-1/2} R = R^* R \Lambda^{-1} R^* R = \Lambda^{-1}$$

and so

$$P = EF,$$

where

$$F = D^{1/2} R \Lambda^{-1/2}.$$

There is, of course, a lack of uniqueness, since we may multiply any normalised eigenvector by a complex number whose modulus is 1, and still have a normalised eigenvector. We shall need to show that this does not matter.

Since the eigenvalues are continuous functions of the elements of the matrix, it follows, since C converges almost surely to Γ, that for each k, $\widehat{v}_k$ converges almost surely to v_k. The eigenvector $\widehat{P}_k$ does not, however, converge almost surely to P_k. Since $P_1, \ldots, P_K$ form an orthogonal basis, it follows that for some $\alpha_1, \ldots, \alpha_K$,

$$\widehat{P}_k = \sum_{j=1}^{K} \alpha_j P_j.$$

In fact, since the P_j are orthonormal, we have $\alpha_j = P_j^* \widehat{P}_k$, whence

$$|\alpha_j|^2 \le \left(P_j^* P_j\right) \left(\widehat{P}_k^* \widehat{P}_k\right) = 1.$$

Now

$$C\widehat{P}_k = \widehat{v}_k \widehat{P}_k.$$

Thus

$$(C - \Gamma + \Gamma) \sum_{j=1}^{K} \alpha_j P_j = (\widehat{v}_k - v_k + v_k) \sum_{j=1}^{K} \alpha_j P_j,$$

$$(C - \Gamma) \sum_{j=1}^{K} \alpha_j P_j + \sum_{j=1}^{K} \alpha_j v_j P_j = (\widehat{v}_k - v_k) \sum_{j=1}^{K} \alpha_j P_j + v_k \sum_{j=1}^{K} \alpha_j P_j$$

and, since $C - \Gamma$ and $\widehat{v}_k - v_k$ converge almost surely to zero, it follows that

$$\sum_{j=1}^{K} \alpha_j \left(v_j - v_k \right) P_j$$

converges almost surely to 0. Thus, for all j for which $v_j \neq v_k$, it must be the case that α_j converges almost surely to 0. Assuming for simplicity that $v_1, \ldots, v_r$ are distinct, we therefore have α_j converging to 0 for all $j \neq k$. But, since $\widehat{P}_k^* \widehat{P}_k = 1$, we have

$$\sum_{j=1}^{K} |\alpha_j|^2 = 1.$$

Hence $|\alpha_k|$ converges almost surely to 1. If we let $\alpha_k = r_k e^{i\phi_k}$, where r_k is real and positive, we must therefore have r_k converging almost surely to 1. Now

$$e^{-i\phi_k} \widehat{P}_k - P_k = e^{-i\phi_k} \sum_{j=1, j \neq k}^{K} \alpha_j P_j + r_k P_k - P_k$$

$$= e^{-i\phi_k} \sum_{j=1, j \neq k}^{K} \alpha_j P_j + (r_k - 1) P_k,$$

which converges almost surely to 0. But

$$\left| e^* \left(\theta \right) e^{-i\phi_k} \widehat{P}_k \right|^2 = \left| e^* \left(\theta \right) \widehat{P}_k \right|^2.$$

We may thus act as though $\widehat{P}_k - P_k$ converges almost surely to 0. Similar results may be obtained if we drop the assumption that the $v_1, \ldots, v_r$ are distinct.

This enables us to prove the strong consistency of the MUSIC estimators. Recall that the MUSIC estimators are the maximisers of

$$g_T(\theta) = \sum_{k=1}^{r} \left| e^*(\theta) \widehat{P}_k \right|^2$$

$$\rightarrow \quad g(\theta) = \sum_{k=1}^{r} |e^*(\theta) P_k|^2$$

almost surely. Now

$$g(\theta) = e^*(\theta) PP^* e(\theta)$$

and PP^* is of rank r and has eigenvalues 1, with multiplicity r and 0, with multiplicity $K - r$. Hence

$$e^*(\theta) PP^* e(\theta) \le e^*(\theta) e(\theta) = K,$$

with equality if and only if the vector

$$e(\theta) = \begin{bmatrix} 1 & e^{i\theta} & \cdots & e^{i(K-1)\theta} \end{bmatrix}'$$

is in the space spanned by the columns of P. Thus equality will hold only if for some r-dimensional vector a,

$$e(\theta) = Pa = E(Fa) = Eb,$$

say. Now, let

$$h(z) = \prod_{k=1}^{r} \left(1 - e^{-i\lambda_k} z \right) = \sum_{j=0}^{r} \zeta_j z^j,$$

the zeros of which are the $e^{i\lambda_k}$. Then

$$h\left(e^{i\theta}\right) = \sum_{j=0}^{r} \zeta_j e^{ij\theta} = \sum_{j=0}^{r} \zeta_j \left\{ \sum_{k=1}^{r} b_k e^{ij\lambda_k} \right\} = \sum_{k=1}^{r} b_k \sum_{j=0}^{r} \zeta_j e^{ij\lambda_k} = 0.$$

Hence

$$g(\theta) \le K,$$

with equality if and only if θ is one of $\lambda_1, \lambda_2, \ldots, \lambda_k$. Strong consistency follows using the usual Jennrich (1969) type arguments.

Now, since

$$T^{1/2}\left(C_j - \gamma_j\right) = T^{-1/2} \sum_{t=j}^{T-1} \left\{x\left(t\right)x\left(t-j\right) - \sigma^2 \delta_{0j}\right\}$$

$$+ \sum_{k=1}^{r} A_k e^{\lambda_k j} T^{1/2} u_k^* + \sum_{k=1}^{r} A_k^* e^{-\lambda_k j} T^{1/2} u_k + o_P\left(1\right),$$

and since the right sides of this equation for different j are asymptotically jointly normal, it follows that the elements of

$$\left\{T^{1/2}\left(\widehat{P}_j - P_j\right); j = 1, \ldots, r\right\}$$

are jointly asymptotically normal. Thus the order of the central limit theorem for the estimators of frequency will also be $T^{1/2}$, i.e., the asymptotic variances of the frequency estimators will be of order T^{-1}. Let

$$\widehat{P} = \left[\begin{array}{cccc} \widehat{P}_1 & \widehat{P}_2 & \cdots & \widehat{P}_r \end{array}\right],$$

$Z_j = \widehat{P}_j - P_j$ and put $Z = \widehat{P} - P$. Then

$$\begin{aligned} I_r &= \widehat{P}^* \widehat{P} = (P + Z)^* (P + Z) = P^* P + P^* Z + Z^* P + Z^* Z \\ &= I_r + P^* Z + Z^* P + O_P\left(T^{-1}\right). \end{aligned}$$

Hence, although $Z = O_P\left(T^{-1/2}\right)$, we have

$$P^* Z + Z^* P = O_P\left(T^{-1}\right)$$

and in particular

$$P_j^* Z_j + Z_j^* P_j = 2\operatorname{Re}\left(P_j^* Z_j\right) = O_P\left(T^{-1}\right).$$

Now

$$C\widehat{P} = \widehat{P}\widehat{v}$$

and so, putting $\xi = C - \Gamma$, we obtain

$$\left(\Gamma + \xi\right)\left(P + Z\right) = \left(P + Z\right)\left(v + \widehat{v} - v\right).$$

Thus

$$\Gamma P + \xi P + \Gamma Z + \xi Z = Pv + Zv + P\left(\widehat{v} - v\right) + Z\left(\widehat{v} - v\right).$$

Since $\Gamma P = Pv$ and ξZ and $Z\left(\widehat{v} - v\right)$ are of order $O_P\left(T^{-1}\right)$, it follows that

$$\xi P + \Gamma Z = Zv + P\left(\widehat{v} - v\right) + O_P\left(T^{-1}\right)$$

and, multiplying this on the left by P^*, we have

$$P^* \xi P + v P^* Z = P^* Z v + \widehat{v} - v + O_P \left(T^{-1} \right).$$

The (j, k)th component equation is, when $j \neq k$,

$$P_j^* \xi P_k + v_j P_j^* Z_k = P_j^* Z_k v_k + O_P \left(T^{-1} \right),$$

so that

$$P_j^* Z_k = -\frac{1}{v_j - v_k} P_j^* \xi P_k + O_P \left(T^{-1} \right),$$

while the (j, j)th component equation is

$$P_j^* \xi P_j + v_j P_j^* Z_j = P_j^* Z_j v_j + \widehat{v}_j - v_j + O_P \left(T^{-1} \right),$$

which shows that

$$\widehat{v}_j - v_j = P_j^* \xi P_j + O_P \left(T^{-1} \right).$$

If q is any element of the orthogonal complement of the space spanned by the P_j, then, since

$$\Gamma q = \sigma^2 q,$$

it follows that

$$q^* \xi P + \sigma^2 q^* Z = q^* Z v + O_P \left(T^{-1} \right),$$

and so

$$q^* Z_k = \frac{1}{v_k - \sigma^2} q^* \xi P_k + O_P \left(T^{-1} \right).$$

Now let Q be any $K \times (K - r)$ matrix whose columns form an orthonormal basis for the orthogonal complement of the space spanned by the P_j, $j = 1, 2, \ldots, r$. Then $Q^* Q = I$ and

$$QQ^* = I_K - PP^*.$$

Hence

$$
\begin{aligned}
Z_k &= (PP^* + QQ^*) Z_k = \sum_j P_j P_j^* Z_k + \sum_j Q_j Q_j^* Z_k \\
&= -\sum_{j \neq k} \frac{1}{v_j - v_k} P_j P_j^* \xi P_k + P_k P_k^* Z_k \\
&\qquad + \frac{1}{v_k - \sigma^2} \sum_j Q_j Q_j^* \xi P_k + O_P\left(T^{-1}\right) \\
&= -\sum_{j \neq k} \frac{1}{v_j - v_k} P_j P_j^* \xi P_k + P_k P_k^* Z_k \\
&\qquad + \frac{1}{v_k - \sigma^2} \left(I_K - PP^*\right) \xi P_k + O_P\left(T^{-1}\right).
\end{aligned}
$$

Now

$$
\begin{aligned}
\xi &= C - \Gamma = R - \sigma^2 I_K + \left(\sum_{k=1}^{r} A_k u_k^* e_k e_k^* + \sum_{k=1}^{r} A_k^* u_k e_k e_k^* \right) \\
&= R - \sigma^2 I_K + EUE^* + O_P\left(T^{-1}\right),
\end{aligned}
$$

where

$$
R = \begin{bmatrix}
R_0 & R_1^* & \cdots & R_{K-1}^* \\
R_1 & R_0 & \cdots & R_{K-2}^* \\
\vdots & \ddots & \ddots & \vdots \\
R_{K-1} & \cdots & R_1 & R_0
\end{bmatrix},
$$

$$
R_j = T^{-1} \sum_{t=0}^{T-1} x(t) x(t-j)
$$

and $U = 2\,\mathrm{Re}\,\mathrm{diag}\,\{A_1 u_1, A_2 u_2, \ldots, A_r u_r\}$. Thus

$$
P^* \xi = P^* \left(R - \sigma^2 I_K\right) + F^{-1} U E^* + O_P\left(T^{-1}\right)
$$

and, since $Q^* E = Q^* P F^{-1} = 0$, we obtain

$$
\left(I_K - PP^*\right) \xi = QQ^* \xi = QQ^* \left(R - \sigma^2 I_K\right) + O_P\left(T^{-1}\right).
$$

The central limit theorem may now be derived using the usual Taylor series techniques. Let $\widehat{\lambda}_1, \widehat{\lambda}_2, \ldots, \widehat{\lambda}_r$ denote the arguments of θ corresponding to the r largest local maxima of $g_T(\theta)$. Then, since the $\widehat{\lambda}_j$ are consistent,

$$
0 = g_T'\left(\widehat{\lambda}_j\right) = g_T'(\lambda_j) + g_T''\left(\widetilde{\lambda}_j\right)\left(\widehat{\lambda}_j - \lambda_j\right),
$$

where $\widetilde{\lambda}_j$ is between λ_j and $\widehat{\lambda}_j$ and thus converges almost surely to λ_j. Hence

$$\widehat{\lambda}_j - \lambda_j = -\frac{g'_T(\lambda_j)}{g''_T\left(\widetilde{\lambda}_j\right)}.$$

We thus only need a joint central limit theorem for the $g'_T(\lambda_j)$ and show the uniform convergence, almost surely, of $g''_T(\theta)$ in small neighbourhoods of the λ_j. Now, letting λ stand for any of the λ_j,

$$g'_T(\lambda) = 2\operatorname{Re}\left\{e_1^*(\lambda)\,\widehat{P}\widehat{P}^*e(\lambda)\right\}$$

where

$$e_1(\lambda) = \frac{\partial}{\partial\lambda}e(\lambda).$$

Thus

$$\begin{aligned}
g'_T(\lambda) &= 2\operatorname{Re}\{e_1^*(\lambda)\,PP^*e(\lambda)\} + 2\operatorname{Re}\{e_1^*(\lambda)\,ZP^*e(\lambda)\} \\
&\quad + 2\operatorname{Re}\{e_1^*(\lambda)\,PZ^*e(\lambda)\} + O_P\left(T^{-1}\right).
\end{aligned}$$

But

$$e_1^*(\lambda)\,PP^*e(\lambda) = e_1^*(\lambda)\,(I_K - QQ^*)\,e(\lambda) = e_1^*(\lambda)\,e(\lambda),$$

since $Q^*e(\lambda) = 0$. However,

$$2\operatorname{Re}\{e_1^*(\lambda)\,e(\lambda)\} = \frac{\partial}{\partial\lambda}e^*(\lambda)\,e(\lambda) = \frac{\partial}{\partial\lambda}K = 0.$$

Hence

$$g'_T(\lambda) = 2\operatorname{Re}\{e_1^*(\lambda)\,(ZP^* + PZ^*)\,e(\lambda)\} + O_P\left(T^{-1}\right).$$

We thus need to establish limit theorems for $ZP^* + PZ^*$. Now

$$ZP^* + PZ^* = \sum_{k=1}^{r}\left(Z_kP_k^* + P_kZ_k^*\right).$$

But

$$\begin{aligned}
&Z_kP_k^* + P_kZ_k^* \\
&= -\sum_{j\neq k}\frac{1}{v_j - v_k}P_jP_j^*\xi P_kP_k^* + P_kP_k^*Z_kP_k^* + P_kZ_k^*P_kP_k^* \\
&\quad -\sum_{j\neq k}\frac{1}{v_j - v_k}P_kP_k^*\xi P_jP_j^* + \frac{1}{v_k - \sigma^2}\left(I_K - PP^*\right)\xi P_kP_k^* \\
&\quad + \frac{1}{v_k - \sigma^2}P_kP_k^*\xi\left(I_K - PP^*\right) + O_P\left(T^{-1}\right)
\end{aligned}$$

and since

$$P_k P_k^* Z_k P_k^* + P_k Z_k^* P_k P_k^* = P_k P_k^* \left(P_k^* Z_k + Z_k^* P_k \right) = P_k P_k^* O_P \left(T^{-1} \right),$$

it follows that

$$\sum_{k=1}^{r} \left(Z_k P_k^* + P_k Z_k^* \right)$$

$$= -\sum_k \sum_{j \neq k} \frac{1}{v_j - v_k} P_j P_j^* \xi P_k P_k^* - \sum_k \sum_{j \neq k} \frac{1}{v_j - v_k} P_k P_k^* \xi P_j P_j^*$$

$$+ \sum_k \frac{1}{v_k - \sigma^2} \left(I_K - P P^* \right) \xi P_k P_k^* + \sum_k \frac{1}{v_k - \sigma^2} P_k P_k^* \xi \left(I_K - P P^* \right)$$

$$+ O_P \left(T^{-1} \right).$$

Now, interchanging j and k in the second double sum on the right side of the above results in a sum which exactly negates the first. Hence

$$ZP^* + PZ^* = \sum_k \frac{1}{v_k - \sigma^2} \left(I_K - P P^* \right) \xi P_k P_k^*$$

$$+ \sum_k \frac{1}{v_k - \sigma^2} P_k P_k^* \xi \left(I_K - P P^* \right) + O_P \left(T^{-1} \right).$$

But

$$g_T' \left(\lambda_j \right) = 2 \operatorname{Re} \left\{ e_1^* \left(\lambda_j \right) \left(ZP^* + PZ^* \right) e \left(\lambda_j \right) \right\} + O_P \left(T^{-1} \right)$$

$$= 2 \operatorname{Re} \operatorname{tr} \left\{ \left(ZP^* + PZ^* \right) e \left(\lambda_j \right) e_1^* \left(\lambda_j \right) \right\} + O_P \left(T^{-1} \right)$$

$$= 2 \operatorname{Re} \operatorname{tr} \left\{ \left(ZP^* + PZ^* \right) E \kappa_j \right\} + O_P \left(T^{-1} \right)$$

$$= 2 \operatorname{Re} \operatorname{tr} \left\{ \left(ZP^* + PZ^* \right) P F^{-1} \kappa_j \right\} + O_P \left(T^{-1} \right),$$

where κ_j is the $r \times K$ matrix which has $e_1^* \left(\lambda_j \right)$ as its jth row and zeros the rest. Now, since

$$\left(I_K - P P^* \right) P = P - P = 0,$$

we have

$$\left(ZP^* + PZ^* \right) P = \sum_k \frac{1}{v_k - \sigma^2} \left(I_K - P P^* \right) \xi P_k P_k^* P$$

$$+ \sum_k \frac{1}{v_k - \sigma^2} P_k P_k^* \xi \left(I_K - P P^* \right) P + O_P \left(T^{-1} \right)$$

$$= \sum_k \frac{1}{v_k - \sigma^2} \left(I_K - P P^* \right) \xi P_k i_k' + O_P \left(T^{-1} \right),$$

where i'_k is the kth row of I_K. Consequently,

$$
\begin{aligned}
\mathrm{tr} & \left\{ (ZP^* + PZ^*)\, PF^{-1}\kappa_j \right\} \\
&= \mathrm{tr}\Big\{ \sum_k \frac{1}{v_k - \sigma^2}\, (I_K - PP^*)\,(C - \Gamma)\, P_k i'_k F^{-1}\kappa_j \Big\} + O_P\left(T^{-1}\right) \\
&= \mathrm{tr}\Big\{ (I_K - PP^*)\,\xi \left(\sum_k \frac{1}{v_k - \sigma^2} P_k i'_k \right) F^{-1}\kappa_j \Big\} + O_P\left(T^{-1}\right) \\
&= \mathrm{tr}\left\{ (I_K - PP^*)\,\xi P\Lambda^{-1} F^{-1}\kappa_j \right\} + O_P\left(T^{-1}\right).
\end{aligned}
$$

Now

$$
P\Lambda^{-1}F^{-1} = EF\Lambda^{-1}F^{-1} = E\left(E^*E\right)^{-1}D^{-1}
$$

and so

$$
\begin{aligned}
g'_T & \left(\lambda_j\right) \\
&= 2\,\mathrm{Re}\,\mathrm{tr}\left\{ (I_K - PP^*)\left(R - \sigma^2 I_K\right) E\left(E^*E\right)^{-1}D^{-1}\kappa_j \right\} + O_P\left(T^{-1}\right) \\
&= \frac{4}{|A_j|^2}\,\mathrm{Re}\,\mathrm{tr}\left\{ (I_K - PP^*)\left(R - \sigma^2 I_K\right) E\left(E^*E\right)^{-1}\kappa_j \right\} + O_P\left(T^{-1}\right) \\
&= \frac{4}{|A_j|^2}\,\mathrm{Re}\,\mathrm{tr}\left\{ H_j \left(R - \sigma^2 I_K\right) \right\} + O_P\left(T^{-1}\right),
\end{aligned}
$$

where, since

$$
PP^* = EFF^*E^* = E\left(E^*E\right)^{-1}E^*,
$$

we have

$$
H_j = E\left(E^*E\right)^{-1}\kappa_j \left\{ I_K - E\left(E^*E\right)^{-1}E^* \right\}.
$$

Now, letting $H_{j,ik}$ be the (i,k)th element of H_j, we obtain

$$
\begin{aligned}
\mathrm{tr}\left\{ H_j \left(R - \sigma^2 I_K\right) \right\} &= \sum_{i,k} H_{j,ik}\left(R_{k-i} - \sigma^2 \delta_{ik} \right) = \sum_{l=1-K}^{K-1} R_l \sum_k H_{j,(k-l)k} \\
&= \sum_{l=1}^{K-1} \left(h_{j,l} R_l + h_{j,-l} R_l^* \right), \qquad\qquad (5.23)
\end{aligned}
$$

where

$$
h_{j,l} = \begin{cases} \sum_{k=l}^{K-1} H_{j,(k-l)k} & ;\quad l > 0 \\ \sum_{k=0}^{K-1+l} H_{j,(k-l)k} & ;\quad l < 0. \end{cases}
$$

Note that the coefficient $h_{j,0}$ of R_0 is zero, since

$$
\begin{aligned}
\sum_{k=0}^{K-1} H_{j,kk} &= \text{tr}\left\{ E\left(E^*E\right)^{-1}\kappa_j\left(I_K - PP^*\right)\right\} \\
&= \text{tr}\left\{\left(E^*E\right)^{-1}\kappa_j\left(I_K - PP^*\right)E\right\} \\
&= \text{tr}\left\{\left(E^*E\right)^{-1}\kappa_j\left(I_K - PP^*\right)PF^{-1}\right\} \\
&= \text{tr}\left\{\left(E^*E\right)^{-1}\kappa_j 0 F^{-1}\right\} \\
&= 0.
\end{aligned}
$$

Also note that the expression given in (5.23) can be simplified. We shall first show that

$$
H_{j,kl} = H^*_{j,K-1-k,K-1-l}.
$$

We can write H_j as

$$
H_j = A_j B_j^*,
$$

where

$$
A_j = E\left(E^*E\right)^{\cdot j}, \quad B_j = \left\{I - E\left(E^*E\right)^{-1}E^*\right\}e_1\left(\lambda_j\right)
$$

and $\left(E^*E\right)^{\cdot j}$ denotes the jth column of $\left(E^*E\right)^{-1}$. Now

$$
\left(E^*E\right)_{jk} = \frac{e^{iK(\lambda_k-\lambda_j)}-1}{e^{i(\lambda_k-\lambda_j)}-1} = e^{i(K-1)(\lambda_k-\lambda_j)/2}\frac{\sin\left\{K\left(\lambda_k-\lambda_j\right)/2\right\}}{\sin\left\{\left(\lambda_k-\lambda_j\right)/2\right\}}
$$

and therefore

$$
\begin{aligned}
e^{i(K-1)\lambda_j}\left(E^*E\right)_{jk} &= e^{-i(K-1)(\lambda_k-\lambda_j)/2}\frac{\sin\left\{K\left(\lambda_k-\lambda_j\right)/2\right\}}{\sin\left\{\left(\lambda_k-\lambda_j\right)/2\right\}}e^{i(K-1)\lambda_k} \\
&= \left(E'\overline{E}\right)_{jk}e^{i(K-1)\lambda_k}.
\end{aligned}
$$

If we let G be the diagonal matrix with jth diagonal element $e^{i(K-1)\lambda_j}$, we then have

$$
G\left(E^*E\right) = \left(E'\overline{E}\right)G
$$

and therefore

$$
\left(E^*E\right)^{-1}G^{-1} = G^{-1}\left(E'\overline{E}\right)^{-1}
$$

and

$$
G\left(E^*E\right)^{-1} = \left(E'\overline{E}\right)^{-1}G.
$$

Thus

$$
\begin{aligned}
A_{j,K-1-k} &= \sum_{l=1}^{r} e^{i(K-1-k)\lambda_l} \left(E^* E\right)^{lj} = \sum_{l=1}^{r} e^{-ik\lambda_l} e^{i(K-1)\lambda_l} \left(E^* E\right)^{lj} \\
&= \sum_{l=1}^{r} e^{-ik\lambda_l} \left(E' \overline{E}\right)^{lj} e^{i(K-1)\lambda_j} = A_{j,k}^* e^{i(K-1)\lambda_j},
\end{aligned}
$$

and

$$
\begin{aligned}
B_{j,K-1-k} \\
&= i\left(K-1-k\right) e^{i(K-1-k)\lambda_j} - \sum_{l,m=1}^{r} e^{i(K-1-k)\lambda_l} \left(E^* E\right)^{lm} \sum_{n=0}^{K-1} i n e^{in(\lambda_j-\lambda_m)} \\
&= -ike^{i(K-1-k)\lambda_j} + i\left(K-1\right) e^{i(K-1-k)\lambda_j} \\
&\quad - \sum_{l,m=1}^{r} e^{-ik\lambda_l} e^{i(K-1)\lambda_l} \left(E^* E\right)^{lm} \sum_{n=0}^{K-1} i\left(K-1-n\right) e^{i(K-1-n)(\lambda_j-\lambda_m)} \\
&= -ike^{i(K-1-k)\lambda_j} + i\left(K-1\right) e^{i(K-1-k)\lambda_j} \\
&\quad - \sum_{l,m=1}^{r} e^{-ik\lambda_l} \left(E' \overline{E}\right)^{lm} e^{i(K-1)\lambda_m} \sum_{n=0}^{K-1} i\left(K-1-n\right) e^{i(K-1-n)(\lambda_j-\lambda_m)} \\
&= -ike^{i(K-1-k)\lambda_j} + i\left(K-1\right) e^{i(K-1-k)\lambda_j} \\
&\quad -e^{i(K-1)\lambda_j} \sum_{l,m=1}^{r} e^{-ik\lambda_l} \left(E' \overline{E}\right)^{lm} \sum_{n=0}^{K-1} i\left(K-1-n\right) e^{-in(\lambda_j-\lambda_m)} \\
&= -ike^{i(K-1-k)\lambda_j} + i\left(K-1\right) e^{i(K-1-k)\lambda_j} \\
&\quad +e^{i(K-1)\lambda_j} \sum_{l,m=1}^{r} e^{-ik\lambda_l} \left(E' \overline{E}\right)^{lm} \sum_{n=0}^{K-1} i n e^{-in(\lambda_j-\lambda_m)} \\
&\qquad -i\left(K-1\right) e^{i(K-1)\lambda_j} \sum_{l,m=1}^{r} e^{-ik\lambda_l} \left(E' \overline{E}\right)^{lm} \sum_{n=0}^{K-1} e^{-in(\lambda_j-\lambda_m)} \\
&= e^{i(K-1)\lambda_j} B_{j,k}^* + i\left(K-1\right) e^{i(K-1-k)\lambda_j} \\
&\qquad -i\left(K-1\right) e^{i(K-1)\lambda_j} \sum_{l,m=1}^{r} e^{-ik\lambda_l} \left(E' \overline{E}\right)^{lm} \left(E' \overline{E}\right)^{mj} \\
&= e^{i(K-1)\lambda_j} B_{j,k}^* + i\left(K-1\right) e^{i(K-1-k)\lambda_j} \\
&\qquad -i\left(K-1\right) e^{i(K-1)\lambda_j} \sum_{l,m=1}^{r} e^{-ik\lambda_l} \delta_{lj} \quad = \quad e^{i(K-1)\lambda_j} B_{j,k}^*,
\end{aligned}
$$

and so

$$
\begin{aligned}
H_{j,K-1-k,K-1-l} &= A_{j,K-1-k}B_{j,K-1-l}^* \\
&= A_{j,k}^* e^{i(K-1)\lambda_j} e^{-i(K-1)\lambda_j} B_{j,l} \\
&= H_{j,kl}^*
\end{aligned}
$$

and consequently, for $l > 0$,

$$
\begin{aligned}
h_{j,-l} &= \sum_{k=0}^{K-1-l} H_{j,(k+l)k} = \sum_{n=0}^{K-1-l} H_{j,K-1-n,K-1-n+l} \\
&= \sum_{n=l}^{K-1} H_{j,K-1-n+l,K-1-n} = \sum_{n=l}^{K-1} H_{j,n-l,n}^* = h_{j,l}^*.
\end{aligned}
$$

Hence

$$
\mathrm{tr}\left\{H_j\left(R - \sigma^2 I_K\right)\right\} = \sum_{l=1}^{K-1}\left(h_{j,l}R_l + h_{j,l}^* R_l^*\right) = 2\,\mathrm{Re}\left(\sum_{l=1}^{K-1} h_{j,l}R_l\right).
$$

We thus have the following result, since the real and imaginary parts of the elements of $T^{1/2}R_j$ are asymptotically normally and independently distributed with means zero and variances $\left(\sigma^2\right)^2/2$:

Lemma 2 $T^{1/2}\left[\; g_T'\left(\lambda_1\right) \;\; \cdots \;\; g_T'\left(\lambda_r\right) \;\right]'$ *is asymptotically normally distributed with mean zero and covariance matrix* Ω, *where*

$$
\Omega_{ij} = \frac{32\left(\sigma^2\right)^2}{|A_i|^2\,|A_j|^2} \sum_{l=1}^{K-1} \mathrm{Re}\left(h_{i,l}h_{j,l}^*\right), \tag{5.24}
$$

$$
h_{j,l} = \sum_{k=0}^{K-1-l} H_{j,(l+k)k}
$$

and

$$
H_j = E\left(E^*E\right)^{-1}\kappa_j\left(I_K - E\left(E^*E\right)^{-1}E^*\right).
$$

Next, we need to examine the second derivatives $g_T''(\theta)$. Now

$$
g_T''(\theta) = 2\,\mathrm{Re}\left\{e_2^*(\theta)\,\widehat{P}\widehat{P}^*e(\theta)\right\} + 2e_1^*(\theta)\,\widehat{P}\widehat{P}^*e_1(\theta)
$$

which converges almost surely and uniformly in $\theta \in (-\pi, \pi)$ to

$$
g''(\theta) = 2\,\mathrm{Re}\left\{e_2^*(\theta)\,PP^*e(\theta)\right\} + 2e_1^*(\theta)\,PP^*e_1(\theta).
$$

Hence, almost surely as $T \to \infty$, we have

$$
\begin{aligned}
g_T'' \left(\widetilde{\lambda}_j \right) \quad &\to \quad g'' \left(\lambda_j \right) \\
&= \quad 2 \operatorname{Re} \left\{ e_2^* \left(\lambda_j \right) PP^* e \left(\lambda_j \right) \right\} + 2 e_1^* \left(\lambda_j \right) PP^* e_1 \left(\lambda_j \right).
\end{aligned}
$$

Some simplification is possible. Since e_j is a column of E, and therefore in the space spanned by the columns of P, it follows that

$$
PP^* e_j = e_j.
$$

Thus

$$
e_2^* \left(\lambda_j \right) PP^* e \left(\lambda_j \right) = e_2^* \left(\lambda_j \right) e \left(\lambda_j \right) = -2 \sum_{k=0}^{K-1} k^2 = -2 e_1^* \left(\lambda_j \right) e_1 \left(\lambda_j \right).
$$

Hence $g_T'' \left(\widetilde{\lambda}_j \right)$ converges almost surely to

$$
\chi_j = -2 e_1^* \left(\lambda_j \right) \left\{ I_K - E \left(E^* E \right)^{-1} E^* \right\} e_1 \left(\lambda_j \right). \tag{5.25}
$$

Using this result in conjunction with the lemma, we have

Theorem 23 *Let* $\widehat{\lambda}_1, \widehat{\lambda}_2, \ldots, \widehat{\lambda}_r$ *denote the complex MUSIC estimators of frequency. Then*

$$
T^{1/2} \left[\begin{array}{ccc} \widehat{\lambda}_1 - \lambda_1 & \cdots & \widehat{\lambda}_r - \lambda_r \end{array} \right]'
$$

has a distribution which converges as $T \to \infty$ *to the normal with mean zero and covariance matrix* Σ, *where*

$$
\Sigma_{ij} = \frac{\Omega_{ij}}{\chi_i \chi_j}
$$

is given by (5.24) and (5.25).

An analysis of complex MUSIC has been presented before by Stoica and Nehorai (1989). Their proofs, however, are not as rigorous as ours, and their asymptotic variances are not in the form presented here.

Finally, we look at some related subspace estimation techniques. For any $\alpha \in \mathbb{R}$, denote by $\widehat{\lambda}_1, \widehat{\lambda}_2, \ldots, \widehat{\lambda}_r$ the minimisers of

$$
\sum_{k=r+1}^{K} \widehat{v}_j^{\alpha} \left| e^* \left(\theta \right) \widehat{P}_k \right|^2.
$$

The following shows that all such techniques have the same asymptotic distribution, and so are asymptotically indistinguishable from MUSIC, which corresponds to the case $\alpha = 0$.

Theorem 24 $T^{1/2} \begin{bmatrix} \widehat{\lambda}_1 - \lambda_1 & \cdots & \widehat{\lambda}_r - \lambda_r \end{bmatrix}'$ *has the same asymptotic distribution for all* α.

Proof Let

$$
\begin{aligned}
g_{T,\alpha}(\theta) &= \sum_{k=r+1}^{K} \widehat{v}_j^{\alpha} \left| e^*(\theta) \widehat{P}_k \right|^2 \\
&= \sum_{k=r+1}^{K} \left(\widehat{v}_j^{\alpha} - \sigma^{2\alpha} \right) \left| e^*(\theta) \widehat{P}_k \right|^2 + \sigma^{2\alpha} \sum_{k=r+1}^{K} \left| e^*(\theta) \widehat{P}_k \right|^2.
\end{aligned}
$$

Then, since the $\widehat{v}_j^{\alpha} - \sigma^{2\alpha}$ converge almost surely to zero, we have almost surely as $T \to \infty$,

$$
g_{T,\alpha}(\theta) \to \sigma^{2\alpha} e^*(\theta)(I_K - PP^*) e(\theta) = \sigma^{2\alpha}(K - e^*(\theta) PP^* e(\theta)) \geq 0,
$$

with equality if and only if θ is one of the λ_j. The minimisers of $g_{T,\alpha}(\theta)$ thus converge almost surely to the true frequencies. Now, for $k > r$,

$$
0 = \widehat{P}^* \widehat{P}_k = (P + Z)^* \widehat{P}_k = P^* \widehat{P}_k + Z^* \widehat{P}_k = P^* \widehat{P}_k + O_P\left(T^{-1/2}\right).
$$

Hence

$$
P^* \widehat{P}_k = O_P\left(T^{-1/2}\right).
$$

But

$$
\begin{aligned}
g'_{T,\alpha}(\theta) &= 2 \operatorname{Re}\left\{ \sum_{k=r+1}^{K} \left(\widehat{v}_j^{\alpha} - \sigma^{2\alpha} \right) \left| e^*(\theta) \widehat{P}_k \right|^2 \right\} - \sigma^{2\alpha} g'_T(\theta) \\
&= 2 \operatorname{Re}\left\{ e_1^*(\theta) \sum_{k=r+1}^{K} \left(\widehat{v}_j^{\alpha} - \sigma^{2\alpha} \right) \widehat{P}_k \widehat{P}_k^* e(\theta) \right\} - \sigma^{2\alpha} g'_T(\theta)
\end{aligned}
$$

and so, for some b,

$$
\begin{aligned}
g'_{T,\alpha}(\lambda_j) &= 2 \operatorname{Re}\left\{ e_1^*(\lambda_j) \sum_{k=r+1}^{K} \left(\widehat{v}_j^{\alpha} - \sigma^{2\alpha} \right) \widehat{P}_k \widehat{P}_k^* P b \right\} - \sigma^{2\alpha} g'_T(\lambda_j) \\
&= -\sigma^{2\alpha} g'_T(\lambda_j) + o_P\left(T^{-1/2}\right),
\end{aligned}
$$

since the $\widehat{v}_j^\alpha - \sigma^{2\alpha}$ converge in probability to 0. Similarly,

$$
\begin{aligned}
g_{T,\alpha}''(\lambda_j) & \\
&\to\ 2\sigma^{2\alpha}\operatorname{Re}\left\{e_2^*(\lambda_j)(I_K - PP^*)e(\lambda_j) + e_1^*(\lambda_j)(I_K - PP^*)e_1(\lambda_j)\right\} \\
&=\ 2\sigma^{2\alpha}e_1^*(\lambda_j)(I_K - PP^*)e_1(\lambda_j) \\
&=\ -\sigma^{2\alpha}\chi_j,
\end{aligned}
$$

almost surely as $T \to \infty$. Hence

$$
\frac{g_{T,\alpha}'(\lambda_j)}{g_{T,\alpha}''(\lambda_j)} = \frac{g_T'(\lambda_j)}{\chi_j}\left\{1 + o_P(1)\right\}
$$

and the result follows. $\qquad\qquad\qquad\qquad\qquad\qquad\qquad\qquad\square$

Of particular interest is the case where $\alpha = -1$. The technique then is known as the eigenvector (EV) technique and was proposed by Johnson (1982).

6

Estimation using Fourier Coefficients

6.1 Introduction

In previous chapters we have seen how the Fourier transform of

$$\{y(0), \ldots, y(T-1)\}$$

and the Fourier coefficients $w_y(\omega_j)$ hold a natural place in the spectral theory of $\{y(t)\}$ and in the estimation of frequency. Often the $w_y(\omega_j)$ are available very cheaply. They may be produced, for example, by special hardware in commercial signal processing equipment. Even if implemented in software, fast Fourier transform (FFT) algorithms produce the Fourier coefficients very efficiently. As most of the statistical information concerning a frequency parameter would be expected to be concentrated in a small number of Fourier coefficients corresponding to frequencies near the true frequency, a sensible problem to pose is the estimation of frequency given only the maximiser of the periodogram over the Fourier frequencies, and a fixed number of Fourier coefficients near that frequency. In this chapter, we describe computationally efficient techniques for estimating frequency using only the Fourier coefficients from, firstly, a single time series of length T, and secondly, from consecutive series of the same length T. Although the estimators do not achieve the asymptotic CRB, their asymptotic variances may be calculated, and are not far from the asymptotic CRB.

6.2 Single series

6.2.1 Likelihood method

Let us suppose that $\{y(0), \ldots, y(T-1)\}$ is generated by (4.1). In fact, we could in what follows suppose that $\{y(0), \ldots, y(T-1)\}$ was generated by (3.1), since the sinusoidal terms do not asymptotically 'interfere' with one another. For some positive integers m, k and l, let

$$z(j) = w_y \left(2\pi \frac{j}{T} \right), \ j = m - k, \ldots, m + l$$

and suppose it is known *a priori* that for some $p \in \{1, 2, \ldots, r\}$, $\lambda_p \in \left(2\pi \frac{m-k}{T}, 2\pi \frac{m+l}{T} \right)$. Then, assuming k and l to be fixed, but allowing m to vary with T, we obtain from (2.14) and (2.15), for $j = -k, \ldots, l$,

$$z(m+j) = T^{1/2} \frac{Ae^{i\phi}}{2} \frac{e^{i2\pi\delta_T} - 1}{2\pi i (\delta_T - j)} + u(m+j) \tag{6.1}$$

where $A \equiv A_p, \phi \equiv \phi_p, \lambda \equiv \lambda_p, \delta_T = T \lambda/(2\pi) - m$, and

$$u(m+j) = w_x \left(2\pi \frac{m+j}{T} \right) + O \left(T^{-1/2} \right).$$

The dependence of the above expressions on T should be stressed here: we must have m increasing with T, and δ_T is written with a subscript to emphasise this. It will be understood, however, that the $z(m+j)$ and $u(m+j)$ depend on T, without using a subscript, to avoid clumsy notation. Note that limits are to be taken in the above when δ_T coincides with one of the integers $-k, \ldots, l$. From Theorem 4, it follows that the $u(m+j)$ are asymptotically complex Gaussian and independent, with means 0. Moreover, the real and imaginary parts are asymptotically independent with the same variances $\pi f_x(\lambda)$. If we write (6.1) as

$$z(m+j) = D_T \frac{\delta_T}{\delta_T - j} + u(m+j), \tag{6.2}$$

where

$$D_T = T^{1/2} \frac{Ae^{i\phi} \left(e^{i2\pi\delta_T} - 1 \right)}{4\pi i \delta_T},$$

then an obvious method of estimating A, $e^{i\phi}$ and δ_T is to minimise

$$\sum_{j=-k}^{l} \left| z(m+j) - D_T \frac{\delta_T}{\delta_T - j} \right|^2$$

with respect to the complex constant D_T and with respect to δ_T. Such an approach was suggested by Bartlett (1967) in his response to a discussion of his paper read to the Royal Statistical Society. His technique, however, does not appear to have been followed up at that time. Minimisation with respect to D_T is simple, as the above function is quadratic in the real and imaginary parts of D_T. The minimising value of D_T is readily computed as

$$\frac{\sum_{j=-k}^{l} z\,(m+j)\,\frac{\delta_T}{\delta_T - j}}{\sum_{j=-k}^{l}\left(\frac{\delta_T}{\delta_T - j}\right)^2}$$

and we may thus estimate δ_T by maximising

$$S_T\,(\delta_T) = \frac{\left|\sum_{j=-k}^{l} z\,(m+j)\,\frac{\delta_T}{\delta_T - j}\right|^2}{\sum_{j=-k}^{l}\left(\frac{\delta_T}{\delta_T - j}\right)^2} \tag{6.3}$$

with respect to δ_T. Although (6.3) is a highly nonlinear function of δ_T, maximisation with respect to δ_T is considerably less computationally intensive than maximisation of the periodogram over some interval, since there are only $(k+l+1)$ complex data. Care must, of course, be taken when calculating (6.3) when δ_T is near an integer. Since there is flexibility in the choice of the integers m, k and l, we may without loss of generality restrict evaluations of the above to $\delta_T \in [-1/2, 1/2]$. The 'problem value' is then $\delta_T = 0$, but $S_T\,(\delta_T)$ is seen to have the limit $|z\,(m)|^2$ as $\delta_T \to 0$. If a numerical maximisation scheme such as Newton's method is used to maximise $S_T\,(\delta_T)$, first and second derivatives are easily calculated at all values of δ_T, with limits again taken as $\delta_T \to 0$ for the case where $\delta_T = 0$.

There remains, of course, the question of the asymptotic properties of the estimators. Clearly there is insufficient information to estimate $f\,(\lambda)$. However, there *would* appear to be enough information to obtain consistent estimators of A, ϕ and λ. Proof of the consistency of $\widehat{\delta}_T$, the maximiser of $S_T\,(\delta_T)$, is straightforward: denote the 'true values' of D_T and δ_T by D_{0T} and δ_{0T}. We shall assume, without loss of generality, that $|\delta_{0T}| \leq 1/2$. Then

$$
\begin{aligned}
T^{-1}S_T\,(\delta_T) &= \frac{\left|\frac{D_{0T}}{\sqrt{T}}\sum_{j=-k}^{l}\frac{\delta_{0T}}{\delta_{0T}-j}\frac{\delta_T}{\delta_T-j} + \frac{1}{\sqrt{T}}\sum_{j=-k}^{l}\frac{\delta_T}{\delta_T-j}u\,(m+j)\right|^2}{\sum_{j=-k}^{l}\left(\frac{\delta_T}{\delta_T-j}\right)^2} \\[2ex]
&= \frac{\left|\frac{D_{0T}}{\sqrt{T}}\sum_{j=-k}^{l}\frac{\delta_{0T}}{\delta_{0T}-j}\frac{\delta_T}{\delta_T-j} + o_{\text{a.s.}}\,(1)\right|^2}{\sum_{j=-k}^{l}\left(\frac{\delta_T}{\delta_T-j}\right)^2} \\[2ex]
&= \frac{|D_{0T}|^2}{T}\frac{\left(\sum_{j=-k}^{l}\frac{\delta_{0T}}{\delta_{0T}-j}\frac{\delta_T}{\delta_T-j}\right)^2}{\sum_{j=-k}^{l}\left(\frac{\delta_T}{\delta_T-j}\right)^2} + o_{\text{a.s.}}\,(1).
\end{aligned}
$$

But

$$\frac{\left(\sum_{j=-k}^{l} \frac{\delta_{0T}}{\delta_{0T}-j} \frac{\delta_T}{\delta_T-j}\right)^2}{\sum_{j=-k}^{l} \left(\frac{\delta_T}{\delta_T-j}\right)^2} \leq \sum_{j=-k}^{l} \left(\frac{\delta_{0T}}{\delta_{0T}-j}\right)^2,$$

with equality if and only if

$$\frac{\delta_T}{\delta_T - j} = \frac{\delta_{0T}}{\delta_{0T} - j}, \quad j = -k, \ldots, l,$$

that is, when and only when $\delta_T = \delta_{0T}$. It follows using arguments originated by Jennrich (1969) that $\widehat{\delta}_T - \delta_{0T} \to 0$, almost surely, where $\widehat{\delta}_T$ is the maximiser of $T^{-1}S_T(\delta_T)$ in $\left[-\frac{1}{2}, \frac{1}{2}\right]$. It only remains to be proved that we may choose m as a function of T in such a way that $\lambda_p \in \left(2\pi\frac{m-k}{T}, 2\pi\frac{m+l}{T}\right)$. For fixed $p \in \{1, \ldots, r\}$, let m_T be the closest integer to $\frac{T\lambda_p}{2\pi}$ (if the fractional part of this is exactly $\frac{1}{2}$, take one of the integers either side). Then, for fixed j,

$$I_y\left(2\pi\frac{m_T + j}{T}\right) = \frac{TA^2}{2} \frac{\sin^2(\pi\delta_{0T})}{\pi^2(\delta_{0T} - j)^2} \{1 + o_{\text{a.s.}}(1)\}$$

where $|\delta_{0T}| \leq \frac{1}{2}$. Consequently, there is a local maximiser of $I_y\left(2\pi\frac{m_T + j}{T}\right)$ at either $j = -1, 0$ or 1, almost surely as $T \to \infty$ and it follows that almost surely as $T \to \infty$, the interval $\left(2\pi\frac{\widehat{m}_T - 1}{T}, 2\pi\frac{\widehat{m}_T + 1}{T}\right)$ contains λ, where $\widehat{m}_T$ is the local maximiser of $I_y\left(2\pi\frac{m}{T}\right)$ closest to m_T. The central limit theorem for the estimator

$$\widehat{\lambda}_T = 2\pi\frac{\widehat{m}_T + \widehat{\delta}_T}{T} \tag{6.4}$$

is nearly trivial to prove: since $\widehat{\delta}_T - \delta_{0T}$ converges almost surely to zero, it follows that

$$0 = S_T'\left(\widehat{\delta}_T\right) = S_T'(\delta_{0T}) + S_T''\left(\widetilde{\delta}_T\right)\left(\widehat{\delta}_T - \delta_{0T}\right),$$

where $\widetilde{\delta}_T$ is between δ_{0T} and $\widehat{\delta}_T$, and that

$$\frac{S_T''\left(\widetilde{\delta}_T\right)}{S_T''(\delta_{0T})} \to 1,$$

almost surely as $T \to \infty$. Thus $\widehat{\delta}_T - \delta_{0T}$ has the same asymptotic distribution as

$$-\frac{S_T'(\delta_{0T})}{S_T''(\delta_{0T})}.$$

Now

$$
S_T' \left(\delta_{0T} \right)
$$

$$
= \quad \frac{-2 \operatorname{Re} \left[\left\{ \sum_{j=-k}^{l} \overline{z} \left(m+j \right) \frac{\delta_{0T}}{\delta_{0T}-j} \right\} \left\{ \sum_{p=-k}^{l} z \left(m+p \right) \frac{p}{\left(\delta_{0T}-p \right)^2} \right\} \right]}{\sum_{j=-k}^{l} \left(\frac{\delta_{0T}}{\delta_{0T}-j} \right)^2}
$$

$$
+ 2 \frac{\left| \sum_{j=-k}^{l} z \left(m+j \right) \frac{\delta_{0T}}{\delta_{0T}-j} \right|^2 \sum_{p=-k}^{l} \frac{p \delta_{0T}}{\left(\delta_{0T}-p \right)^3}}{\left\{ \sum_{j=-k}^{l} \left(\frac{\delta_{0T}}{\delta_{0T}-j} \right)^2 \right\}^2}
$$

$$
= \quad 2 \frac{\sum_{p=-k}^{l} \frac{p \delta_{0T}}{\left(\delta_{0T}-p \right)^3}}{\left\{ \sum_{j=-k}^{l} \left(\frac{\delta_{0T}}{\delta_{0T}-j} \right)^2 \right\}^2} \operatorname{Re} \left\{ \overline{D}_{0T} \sum_{j=-k}^{l} u \left(m+j \right) \frac{\delta_{0T}}{\delta_{0T}-j} \right\}
$$

$$
- 2 \operatorname{Re} \left\{ \overline{D}_{0T} \sum_{j=-k}^{l} u \left(m+j \right) \frac{j}{\left(\delta_{0T}-j \right)^2} \right\} + o_{\text{a.s.}} \left(T^{1/2} \right).
$$

Since the $u \left(m+j \right)$ behave asymptotically as independent and identically distributed complex normal random variables with real and imaginary parts independent and with variances $\pi f_x \left(\lambda \right)$, it follows that $T^{-1/2}$ times the above is asymptotically normally distributed with mean zero and variance

$$
4 \pi f_x \left(\lambda \right) T^{-1} \left| D_{0T} \right|^2 \left[\frac{\left\{ \sum_{p=-k}^{l} \frac{p \delta_{0T}}{\left(\delta_{0T}-p \right)^3} \right\}^2}{\sum_{j=-k}^{l} \left(\frac{\delta_{0T}}{\delta_{0T}-j} \right)^2} + \sum_{j=-k}^{l} \frac{j^2}{\left(\delta_{0T}-j \right)^4} \right.
$$

$$
\left. - 2 \frac{\left\{ \sum_{p=-k}^{l} \frac{p \delta_{0T}}{\left(\delta_{0T}-p \right)^3} \right\}^2}{\sum_{j=-k}^{l} \left(\frac{\delta_{0T}}{\delta_{0T}-j} \right)^2} \right]
$$

$$
= \quad 4 \pi f_x \left(\lambda \right) T^{-1} \left| D_{0T} \right|^2 \left[\sum_{j=-k}^{l} \frac{j^2}{\left(\delta_{0T}-j \right)^4} - \frac{\left\{ \sum_{p=-k}^{l} \frac{p \delta_{0T}}{\left(\delta_{0T}-p \right)^3} \right\}^2}{\sum_{j=-k}^{l} \left(\frac{\delta_{0T}}{\delta_{0T}-j} \right)^2} \right].
$$

Also, a little messy algebra shows that

$$
S_T'' \left(\delta_{0T} \right) \sim 2 \left[\sum_{j=-k}^{l} \frac{j^2}{\left(\delta_{0T}-j \right)^4} - \frac{\left\{ \sum_{p=-k}^{l} \frac{p \delta_{0T}}{\left(\delta_{0T}-p \right)^3} \right\}^2}{\sum_{j=-k}^{l} \left(\frac{\delta_{0T}}{\delta_{0T}-j} \right)^2} \right].
$$

Consequently, we have

Theorem 25 *Let $\widehat{\lambda}_T$ be defined by (6.4). Then*

$$T^{3/2} (\log T)^{-1/2-\nu} \left(\widehat{\lambda}_T - \lambda_0\right) \to 0$$

almost surely as $T \to \infty$, for all $\nu > 0$, and

$$\Pr\left\{ T^{3/2} v_T^{-1} \left(\widehat{\lambda}_T - \lambda_0\right) \le x \right\} \to \frac{1}{\sqrt{2\pi}} \int_{-\infty}^{x} e^{-\frac{1}{2}u^2} du,$$

where

$$v_T^2 = \frac{16\pi^3 f_x(\lambda)}{A^2} \frac{\pi^2 \delta_{0T}^2}{\sin^2(\pi\delta_{0T})} \left[\sum_{j=-k}^{l} \frac{j^2}{(\delta_{0T}-j)^4} - \frac{\left\{\sum_{p=-k}^{l} \frac{p\delta_{0T}}{(\delta_{0T}-p)^3}\right\}^2}{\sum_{j=-k}^{l} \left(\frac{\delta_{0T}}{\delta_{0T}-j}\right)^2} \right]^{-1}.$$

Note from the above that the ratio of the asymptotic variance to the asymptotic variance of the periodogram maximiser is

$$\frac{\pi^2}{3} \frac{\pi^2 \delta_{0T}^2}{\sin^2(\pi\delta_{0T})} \left[\sum_{j=-k}^{l} \frac{j^2}{(\delta_{0T}-j)^4} - \frac{\left\{\sum_{p=-k}^{l} \frac{p\delta_{0T}}{(\delta_{0T}-p)^3}\right\}^2}{\sum_{j=-k}^{l} \left(\frac{\delta_{0T}}{\delta_{0T}-j}\right)^2} \right]^{-1},$$

which can be shown to converge to 1 as k and l increase.

6.2.2 Three coefficient techniques

Maximising (6.3), while easier than maximising the periodogram, must still be carried out numerically, for even in the simplest case, where $k = l = 1$, the derivative of $S_T(\delta_T)$ is the ratio of two polynomials, with numerator of sixth degree. In this subsection, we derive two techniques which provide closed form estimators of δ_T, the second of which has the same central limit theorem as the maximiser of $S_T(\delta_T)$ when $k = l = 1$. From (6.2), again assuming that m_T is known, and using the facts that $T^{-1/2}|D_{0T}|$ is bounded above and below and that the $u(m_T + j)$ are $O\left(\sqrt{\log T}\right)$, it follows that

$$\frac{z(m_T + j)}{z(m_T)} = \frac{\delta_{0T}}{\delta_{0T} - j} + D_{0T}^{-1}\left\{ u(m_T + j) - \frac{\delta_{0T}}{\delta_{0T} - j} u(m_T) \right\}$$

$$+ O_{\text{a.s.}}\left(T^{-1}\log T\right). \tag{6.5}$$

Hence

$$\text{Re}\left\{\frac{z(m_T + j)}{z(m_T)}\right\} = \frac{\delta_{0T}}{\delta_{0T} - j} + O_{\text{a.s.}}\left\{T^{-1/2}(\log T)^{1/2}\right\}$$

and two estimators of δ_{0T} are formed by solving the above equations, without the error terms. The two estimators are thus

$$\widetilde{\delta}_{jT} = \frac{j\alpha_j}{\alpha_j - 1}, \quad j = -1, 1$$

where

$$\alpha_j = \text{Re}\left\{ z\left(m_T + j\right)/z\left(m_T\right)\right\}.$$

The estimators are obviously strongly consistent. In fact, from above,

$$T^{1/2}\left(\log T\right)^{-1/2-\nu}\left(\widetilde{\delta}_{jT} - \delta_{0T}\right) \to 0,$$

almost surely, for all $\nu > 0$. That the estimators have central limit theorems follows from Theorem 4: Let

$$
\begin{aligned}
\xi_{jT} &= \text{Re}\left\{ \frac{z\left(m_T + j\right)}{z\left(m_T\right)} - \frac{\delta_{0T}}{\delta_{0T} - j}\right\} \\
&= \text{Re}\left[D_{0T}^{-1}\left\{ u\left(m_T + j\right) - \frac{\delta_{0T}}{\delta_{0T} - j}u\left(m_T\right)\right\}\right] + O_{\text{a.s.}}\left(T^{-1}\log T\right).
\end{aligned}
$$

Then

$$
\begin{aligned}
\widetilde{\delta}_{jT} - \delta_{0T} &= \frac{j\alpha_j}{\alpha_j - 1} - \delta_{0T} \\
&= \frac{j\left(\frac{\delta_{0T}}{\delta_{0T}-j} + \xi_{jT}\right)}{\frac{\delta_{0T}}{\delta_{0T}-j} + \xi_{jT} - 1} - \delta_{0T} \\
&= \frac{j\delta_{0T} + j\left(\delta_{0T} - j\right)\xi_{jT}}{\delta_{0T} + \left(\delta_{0T} - j\right)\xi_{jT} - \delta_{0T} + j} - \delta_{0T} \\
&= \frac{-\left(\delta_{0T} - j\right)^2 \xi_{jT}}{j + \left(\delta_{0T} - j\right)\xi_{jT}} \\
&= -\frac{\left(\delta_{0T} - j\right)^2}{j}\xi_{jT}\left\{ 1 + o_{\text{a.s.}}\left(1\right)\right\}. \qquad (6.6)
\end{aligned}
$$

Now, as the $u\left(m_T + j\right)$ are asymptotically complex normal, it follows that $T^{1/2}\left[\,\xi_{-1,T}\quad \xi_{1,T}\,\right]'$ is asymptotically normal with mean zero and covariance matrix

$$T\Sigma_T = \frac{\pi f_x\left(\lambda\right)}{T^{-1}\left|D_{0T}\right|^2}\begin{bmatrix} 1 + \left(\frac{\delta_{0T}}{\delta_{0T}+1}\right)^2 & \frac{\delta_{0T}^2}{\delta_{0T}^2 - 1} \\ \frac{\delta_{0T}^2}{\delta_{0T}^2 - 1} & 1 + \left(\frac{\delta_{0T}}{\delta_{0T}-1}\right)^2 \end{bmatrix}$$

or, more precisely, since the above depends on both T and δ_{0T}, that

$$\Sigma_T^{-1/2}\left[\,\xi_{-1,T}\quad \xi_{1,T}\,\right]'$$

is asymptotically normal with mean zero and covariance matrix the identity. Thus

$$\Omega_T^{-1/2} \left[\; \tilde{\delta}_{-1,T} - \delta_{0T} \quad \tilde{\delta}_{1T} - \delta_{0T} \; \right]'$$

is asymptotically normal with mean zero and covariance matrix the identity, where

$$\Omega_T$$

$$= \frac{\pi f_x(\lambda)}{|D_{0T}|^2} \left[\begin{array}{cc} (\delta_{0T} + 1)^4 \left\{ 1 + \left(\frac{\delta_{0T}}{\delta_{0T}+1} \right)^2 \right\} & -\delta_{0T}^2 \left(\delta_{0T}^2 - 1 \right) \\ -\delta_{0T}^2 \left(\delta_{0T}^2 - 1 \right) & (\delta_{0T} - 1)^4 \left\{ 1 + \left(\frac{\delta_{0T}}{\delta_{0T}-1} \right)^2 \right\} \end{array} \right].$$

$$(6.7)$$

Now, since $\delta_{0T} \in \left[-\frac{1}{2}, \frac{1}{2} \right]$, we would choose $\tilde{\delta}_{-1,T}$ to estimate δ_{0T} if it were known that $\delta_{0T} < 0$, and $\tilde{\delta}_{1,T}$ if $\delta_{0T} > 0$. The two reasons preventing us from doing this, of course, are that m_T is unknown, and that even if it *were* known, we could not determine the sign of δ_{0T}. In order to construct an estimator, therefore, we need to investigate the properties of $\hat{m}_T$ and to select between the two estimators of ω,

$$2\pi \frac{\hat{m}_T + \hat{\delta}_{jT}}{T}, \quad j = -1, 1,$$

where $\hat{\delta}_{jT} = \frac{j\hat{\alpha}_j}{\hat{\alpha}_j - 1}$ and $\hat{\alpha}_j = \mathrm{Re}\left\{ z\left(\hat{m}_T + j \right) / z\left(\hat{m}_T \right) \right\}$. First, it is fairly obvious that, along subsequences for which $\delta_{0T} \in [-a, a]$, where $0 < a < \frac{1}{2}$, we shall eventually have $\hat{m}_T = m_T$. Secondly, the worst that can occur if $|\delta_{0T}|$ is close to $\frac{1}{2}$ is that $\hat{m}_T$ differs from m_T by 1. We shall make use of these facts in what follows but first obtain the general result:

Lemma 3 $\limsup_{T \to \infty} |\hat{m}_T - m_T| \leq 1$, a.s.

Proof

$$\{ |\hat{m}_T - m_T| \leq 1 \} \supset \left\{ |z(m_T)|^2 > \max_{|j - m_T| > 1} |z(j)|^2 \right\}.$$

Now

$$|z(m_T)|^2 = |D_{0T} + u(m_T)|^2 \geq |D_{0T}|^2 - |u(m_T)|^2$$

by the triangle inequality. Also, again by the triangle inequality,

$$
\max_{|j-m_T|>1} |z(j)|^2 = \max_{|j|>1} \left| D_{0T} \frac{\delta_{0T}}{\delta_{0T}-j} + u(m_T+j) \right|^2
$$

$$
\leq \max_{|j|>1} \left| D_{0T} \frac{\delta_{0T}}{\delta_{0T}-j} \right|^2 + \max_{|j|>1} |u(m_T+j)|^2
$$

$$
\leq |D_{0T}|^2 \frac{\delta_{0T}^2}{(2-|\delta_{0T}|)^2} + \max |u(j)|^2 .
$$

Thus,

$$
\left\{ |z(m_T)|^2 > \max_{|j-m_T|>1} |z(j)|^2 \right\}
$$

$$
\supset \left\{ |D_{0T}|^2 - |u(m_T)|^2 > |D_{0T}|^2 \frac{\delta_{0T}^2}{(2-|\delta_{0T}|)^2} + \max |u(j)|^2 \right\}
$$

$$
= \left\{ \max |u(j)|^2 < |D_{0T}|^2 \left\{ 1 - \frac{\delta_{0T}^2}{(2-|\delta_{0T}|)^2} \right\} + |u(m_T)|^2 \right\}
$$

$$
\supset \left\{ \max |u(j)|^2 < |D_{0T}|^2 \left\{ 1 - \frac{\delta_{0T}^2}{(2-|\delta_{0T}|)^2} \right\} \right\} .
$$

Now,

$$
\min_{\delta_{0T}\in[-1/2,1/2]} |D_{0T}|^2 \left\{ 1 - \frac{\delta_{0T}^2}{(2-|\delta_{0T}|)^2} \right\}
$$

$$
= \frac{TA^2}{4} \min_{\delta_{0T}\in[-1/2,1/2]} \frac{\sin^2(\pi\delta_{0T})}{\pi^2\delta_{0T}^2} \left\{ 1 - \frac{\delta_{0T}^2}{(2-|\delta_{0T}|)^2} \right\}
$$

$$
= \frac{8TA^2}{9\pi^2} ,
$$

the minimum occurring when $|\delta_{0T}| = \frac{1}{2}$. Thus

$$
\{|\widehat{m}_T - m_T| \leq 1\} \supset \left\{ \max |u(j)|^2 < \frac{8TA^2}{9\pi^2} \right\} .
$$

But $\max_j |u(j)|^2 = O(\log T)$ and the result of the lemma follows.   □

We now propose an algorithm for estimating δ_{0T}. We shall refer to the frequency estimator, and the others which follow, as FTI estimators, or Fourier Transform Interpolative estimators.

Algorithm 4 *FTI 1*

(i) *Let* $\widehat{m}_T$ *be the maximiser of* $I_y\left(2\pi\frac{m}{T}\right)$, $m = 1, 2, \ldots, \left\lfloor \frac{T-1}{2} \right\rfloor$.

(ii) Let $\widehat{\delta}_{jT} = \frac{j\widehat{\alpha}_j}{\widehat{\alpha}_j - 1}$, $j = -1, 1$ where $\widehat{\alpha}_j = \mathrm{Re}\left\{ z\left(\widehat{m}_T + j\right) / z\left(\widehat{m}_T\right)\right\}$.

(iii) If $\widehat{\delta}_{-1,T} > 0$ and $\widehat{\delta}_{1,T} > 0$, put $\widehat{\delta}_T = \widehat{\delta}_{1,T}$. Otherwise, put $\widehat{\delta}_T = \widehat{\delta}_{-1,T}$.

(iv) Let $\widehat{\lambda}_T = 2\pi \left(\widehat{m}_T + \widehat{\delta}_T\right)\bigg/ T$.

At first glance, Step (iii) of the above algorithm would seem rather odd, in that we choose $\widehat{\delta}_T = \widehat{\delta}_{-1,T}$ if $\widehat{\delta}_{-1,T} > 0$ and $\widehat{\delta}_{1,T} \leq 0$, or if $\widehat{\delta}_{-1,T} \leq 0$ and $\widehat{\delta}_{1,T} > 0$, or if $\widehat{\delta}_{-1,T} \leq 0$ and $\widehat{\delta}_{1,T} \leq 0$; while we choose $\widehat{\delta}_T = \widehat{\delta}_{1,T}$ only if $\widehat{\delta}_{-1,T} > 0$ and $\widehat{\delta}_{1,T} > 0$. A careful look at the following derivation shows that only this algorithm, amongst those which choose on the basis of the signs of the $\widehat{\delta}_{jT}$, or the one which chooses $\widehat{\delta}_{-1,T}$ if and only if $\widehat{\delta}_{-1,T} \leq 0$ and $\widehat{\delta}_{1,T} \leq 0$, will yield satisfactory results.

Mindful that $\delta_{0T} \in \left[-\frac{1}{2}, \frac{1}{2}\right]$, and that in the special, although unlikely, case where $\lambda/\left(2\pi\right)$ is rational, some of the following subsets of the integers may be empty, for fixed $a > 0$ and $0 < \nu < \frac{1}{2}$, let

$$
\begin{aligned}
T_1 &= \left\{ T; \; |\delta_{0T}| \leq aT^{-\nu} \right\} \\
T_2 &= \left\{ T; \; aT^{-\nu} < |\delta_{0T}| \leq \frac{1}{2} - aT^{-\nu} \right\}, \\
T_3 &= \left\{ T; \; \frac{1}{2} - aT^{-\nu} < |\delta_{0T}| \leq \frac{1}{2} \right\}.
\end{aligned}
$$

We shall consider the behaviour of $\widehat{\lambda}_T$ along each of T_1, T_2 and T_3 separately, 'patching' the asymptotic behaviours to obtain a unified central limit theorem. Now, for $T \in T_1 \cup T_2$, and using the same ideas as used in the proof of the lemma, it follows that

$$
\begin{aligned}
\{\widehat{m}_T = m_T\} & \\
&\supset \left\{ |z\left(m_T\right)|^2 \geq \max_{j \neq m_T} |z\left(j\right)|^2 \right\} \\
&\supset \left\{ \max |u\left(j\right)|^2 < \min_{|\delta_{0T}| \leq 1/2 - aT^{-\nu}} |D_{0T}|^2 \left\{ 1 - \frac{\delta_{0T}^2}{\left(1 - |\delta_{0T}|\right)^2} \right\} \right\} \\
&= \left\{ \max |u\left(j\right)|^2 < \frac{TA^2}{4}\frac{\sin^2\left(\pi\delta\right)}{\pi^2\delta^2}\left\{ 1 - \frac{\delta^2}{\left(1 - \delta\right)^2} \right\} \right\} \\
&= \left\{ \max |u\left(j\right)|^2 < \frac{8TA^2}{\pi^2}aT^{-\nu} \right\}
\end{aligned}
$$

where $\delta = \frac{1}{2} - aT^{-\nu}$. Since $\nu < \frac{1}{2}$, we have

$$
\limsup_{T \to \infty, \, T \in T_1 \cup T_2} |\widehat{m}_T - m_T| = 0, \text{ a.s.}
$$

Let $T_2' = T_2 \cap \{T;\ \delta_{0T} > 0\}$. Then, for $T \in T_2' \cap \{T;\ \widehat{m}_T = m_T\}$,

$$\left\{\widehat{\delta}_T = \widehat{\delta}_{1,T}\right\}$$

$$= \left\{\widehat{\delta}_{-1,T} > 0, \widehat{\delta}_{1,T} > 0\right\}$$

$$= \left\{0 < \widehat{\alpha}_{-1,T} < 1\right\} \cap \left\{\{\widehat{\alpha}_{1,T} < 0\} \cup \{\widehat{\alpha}_{1,T} > 1\}\right\}$$

$$\supset \left\{0 < \widehat{\alpha}_{-1,T} < 1,\ \widehat{\alpha}_{1,T} < 0\right\}$$

$$= \left\{-\frac{\delta_{0T}}{\delta_{0T}+1} < \xi_{-1,T} < 1 - \frac{\delta_{0T}}{\delta_{0T}+1},\ \xi_{1,T} < -\frac{\delta_{0T}}{\delta_{0T}-1}\right\}$$

$$\supset \left\{-aT^{1/2-\nu}\left(\log T\right)^{-1/2}\left\{1 + o\left(1\right)\right\} < T^{1/2}\left(\log T\right)^{-1/2}\xi_{-1,T}\right.$$

$$\left. < T^{1/2}\left(\log T\right)^{-1/2} - aT^{1/2-\nu}\left(\log T\right)^{-1/2}\left\{1 + o\left(1\right)\right\}\right\}$$

$$\cap \left\{T^{1/2}\left(\log T\right)^{-1/2}\xi_{1,T} < aT^{1/2-\nu}\left(\log T\right)^{-1/2}\left\{1 + o\left(1\right)\right\}\right\}.$$

But, since $T^{1/2}\left(\log T\right)^{-1/2}\xi_{jT} = O\left(1\right)$, almost surely, and $0 < \nu < \frac{1}{2}$, the above event is true with probability 1 as T increases. Hence, since $\widehat{\delta}_{1,T} = \widetilde{\delta}_{1,T}$ almost surely as $T \to \infty$ for $T \in T_2'$, it is true that $T^{3/2}\left(\log T\right)^{-1/2-\varepsilon}\left(\widehat{\lambda}_T - \lambda\right)$ converges almost surely to zero for all $\varepsilon > 0$ as $T \to \infty$, for $T \in T_2'$. It also follows that $\frac{T}{2\pi v_{22,T}}\left(\widehat{\lambda}_T - \lambda\right)$ is asymptotically distributed normally with mean zero and variance 1, where $v_{22,T}^2$ is the $(2,2)$ entry in Ω_T, defined by (6.7). Next, if $T \in (T_2 \setminus T_2') \cap \{T;\ \widehat{m}_T = m_T\}$,

$$\left\{\widehat{\delta}_T = \widehat{\delta}_{-1,T}\right\} \supset \left\{\widehat{\delta}_{-1,T} < 0\right\} = \{\widehat{\alpha}_{-1,T} < 0\} \cup \{\widehat{\alpha}_{-1,T} > 1\} \supset \{\widehat{\alpha}_{-1,T} < 0\},$$

which is true with probability 1 as T increases for reasons similar to the above. Thus $T^{3/2}\left(\log T\right)^{-1/2-\varepsilon}\left(\widehat{\lambda}_T - \lambda\right)$ converges almost surely to zero, as $T \to \infty$ for $T \in (T_2 \setminus T_2')$, for all $\varepsilon > 0$, and $\frac{T}{2\pi v_{11,T}}\left(\widehat{\lambda}_T - \lambda\right)$ is asymptotically distributed normally with mean zero and variance 1, where $v_{11,T}^2$ is the $(1,1)$ entry in Ω_T.

Next we investigate the asymptotic behaviour along subsequences where δ_{0T} is positive and close to $\frac{1}{2}$, viz. when $T \in T_3 \cap \{T;\ \delta_{0T} > 0\}$. It follows from the above that $\widehat{m}_T$ will either be m_T or $m_T + 1$ with probability 1 as T increases along this subsequence. Now, when $\widehat{m}_T = m_T$, the above result shows that $\{\widehat{\delta}_T = \widehat{\delta}_{1,T}\}$ is true with probability 1 as T increases, while when $\widehat{m}_T = m_T + 1$, $\{\widehat{\delta}_T = \widehat{\delta}_{-1,T}\}$ is true with probability one as T increases. Thus

$$\widehat{m}_T + \widehat{\delta}_T \sim m_T + \begin{cases} \dfrac{\mathrm{Re}\{z(m_T+1)/z(m_T)\}}{\mathrm{Re}\{z(m_T+1)/z(m_T)\}-1} & ;\ \widehat{m}_T = m_T \\[2ex] 1 - \dfrac{\mathrm{Re}\{z(m_T)/z(m_T+1)\}}{\mathrm{Re}\{z(m_T)/z(m_T+1)\}-1} & ;\ \widehat{m}_T = m_T + 1. \end{cases}$$

Now,

$$\text{Re}\left\{\frac{z\left(m_T+1\right)}{z\left(m_T\right)}\right\} = \frac{\delta_{0T}}{\delta_{0T}-1} + \text{Re}\left[D_{0T}^{-1}\left\{u\left(m_T+1\right) - \frac{\delta_{0T}}{\delta_{0T}-1}u\left(m_T\right)\right\}\right]$$
$$+O_{\text{a.s.}}\left(T^{-1}\log T\right)$$

while

$$\text{Re}\left\{\frac{z\left(m_T\right)}{z\left(m_T+1\right)}\right\}$$
$$= \frac{\delta_{0T}-1}{\delta_{0T}} + \frac{\delta_{0T}-1}{\delta_{0T}}\text{Re}\left[D_{0T}^{-1}\left\{u\left(m_T\right) - \frac{\delta_{0T}-1}{\delta_{0T}}u\left(m_T+1\right)\right\}\right]$$
$$+O_{\text{a.s.}}\left(T^{-1}\log T\right).$$

Thus

$$\frac{\text{Re}\left\{z\left(m_T+1\right)/z\left(m_T\right)\right\}}{\text{Re}\left\{z\left(m_T+1\right)/z\left(m_T\right)\right\}-1}$$
$$= \delta_{0T} - \left(\delta_{0T}-1\right)^2\text{Re}\left[D_{0T}^{-1}\left\{u\left(m_T+1\right) - \frac{\delta_{0T}}{\delta_{0T}-1}u\left(m_T\right)\right\}\right]$$
$$+O_{\text{a.s.}}\left(T^{-1}\log T\right)$$

while

$$1 - \frac{\text{Re}\left\{z\left(m_T\right)/z\left(m_T+1\right)\right\}}{\text{Re}\left\{z\left(m_T\right)/z\left(m_T+1\right)\right\}-1}$$
$$= 1 - \left(1 - \delta_{0T} - \delta_{0T}^2\frac{\delta_{0T}-1}{\delta_{0T}}\text{Re}\left[D_{0T}^{-1}\left\{u\left(m_T\right) - \frac{\delta_{0T}-1}{\delta_{0T}}u\left(m_T+1\right)\right\}\right]\right)$$
$$+O_{\text{a.s.}}\left(T^{-1}\log T\right)$$
$$= \delta_{0T} + \left(\delta_{0T}-1\right)\delta_{0T}\text{Re}\left[D_{0T}^{-1}\left\{u\left(m_T\right) - \frac{\delta_{0T}-1}{\delta_{0T}}u\left(m_T+1\right)\right\}\right]$$
$$+O_{\text{a.s.}}\left(T^{-1}\log T\right)$$
$$= \delta_{0T} - \left(\delta_{0T}-1\right)^2\text{Re}\left[D_{0T}^{-1}\left\{u\left(m_T+1\right) - \frac{\delta_{0T}}{\delta_{0T}-1}u\left(m_T\right)\right\}\right]$$
$$+O_{\text{a.s.}}\left(T^{-1}\log T\right).$$

Consequently,

$$\frac{\text{Re}\left\{z\left(m_T+1\right)/z\left(m_T\right)\right\}}{\text{Re}\left\{z\left(m_T+1\right)/z\left(m_T\right)\right\}-1}$$
$$= 1 - \frac{\text{Re}\left\{z\left(m_T\right)/z\left(m_T+1\right)\right\}}{\text{Re}\left\{z\left(m_T\right)/z\left(m_T+1\right)\right\}-1} + O_{\text{a.s.}}\left(T^{-1}\log T\right) \qquad (6.8)$$

and we can see that it makes no asymptotic difference whether $\widehat{m}_T = m_T$ or $m_T + 1$. Similarly, when $T \in T_3 \cap \{T;\ \delta_{0T} < 0\}$, it makes no asymptotic difference whether $\widehat{m}_T = m_T$ or $m_T - 1$.

The final case we must consider is when $T \in T_1$, i.e. when δ_{0T} is close to 0. Although it is clear that $\widehat{m}_T = m_T$ almost surely as $T \to \infty$ for $T \in T_1$, it is equally clear that $\widehat{\delta}_T$ can be $\widetilde{\delta}_{-1,T}$ when $\delta_{0T} > 0$ and that $\widehat{\delta}_T$ can be $\widetilde{\delta}_{-1,T}$ when $\delta_{0T} > 0$. Also, along $T \in T_1$, the limit of $\frac{v_{11,T}}{v_{22,T}}$ is 1 since $\nu > 0$. Now

$$
\Pr\left\{ T^{3/2} \left(\widehat{\lambda}_T - \lambda \right) \leq 2\pi x \right\}
$$
$$
= \ \Pr\left\{ T^{1/2} \left(\widehat{\delta}_T - \delta_{0T} \right) \leq x, \widehat{m}_T = m_T \right\} + o\,(1)
$$
$$
= \ \Pr\left\{ T^{1/2} \left(\widetilde{\delta}_{1,T} - \delta_{0T} \right) \leq x, \widetilde{\delta}_{-1,T} > 0, \widetilde{\delta}_{1,T} > 0, \widehat{m}_T = m_T \right\}
$$
$$
+ \Pr\left\{ T^{1/2} \left(\widetilde{\delta}_{-1,T} - \delta_{0T} \right) \leq x, \widetilde{\delta}_{-1,T} > 0, \widetilde{\delta}_{1,T} \leq 0, \widehat{m}_T = m_T \right\}
$$
$$
+ \Pr\left\{ T^{1/2} \left(\widetilde{\delta}_{-1,T} - \delta_{0T} \right) \leq x, \widetilde{\delta}_{-1,T} \leq 0, \widehat{m}_T = m_T \right\}.
$$

Since $\Pr\{\widehat{m}_T = m_T\}$ converges to 1, the above has the same limit as

$$
\Pr\left\{ T^{1/2} \left(\widetilde{\delta}_{1,T} - \delta_{0T} \right) \leq x, \widetilde{\delta}_{-1,T} > 0, \widetilde{\delta}_{1,T} > 0 \right\}
$$
$$
+ \Pr\left\{ T^{1/2} \left(\widetilde{\delta}_{-1,T} - \delta_{0T} \right) \leq x, \widetilde{\delta}_{-1,T} > 0, \widetilde{\delta}_{1,T} \leq 0 \right\}
$$
$$
+ \Pr\left\{ T^{1/2} \left(\widetilde{\delta}_{-1,T} - \delta_{0T} \right) \leq x, \widetilde{\delta}_{-1,T} \leq 0 \right\}.
$$

Now, from (6.6),

$$
\widetilde{\delta}_{jT} = \delta_{0T} - \frac{(\delta_{0T} - j)^2}{j} \xi_{jT} \{1 + o_{\text{a.s.}}\,(1)\}
$$

where

$$
\xi_{jT} = \text{Re}\left[D_{0T}^{-1} \left\{ u\,(m_T + j) - \frac{\delta_{0T}}{\delta_{0T} - j} u\,(m_T) \right\} \right].
$$

But, since $|\delta_{0T}| \leq aT^{-\nu}$,

$$
\xi_{jT} = \frac{2T^{-1/2}}{A} \text{Re}\left\{ e^{-i\phi} u\,(m_T + j) \right\} + O_{\text{a.s.}} \left\{ T^{-1/2-\nu} (\log T)^{1/2} \right\}.
$$

Thus

$$
\widetilde{\delta}_{jT} = \delta_{0T} - j\frac{2T^{-1/2}}{A} \text{Re}\left\{ e^{-i\phi} u\,(m_T + j) \right\} + O_{\text{a.s.}} \left\{ T^{-1/2-\nu} (\log T)^{1/2} \right\}
$$

and $\widetilde{\delta}_{-1,T}$ and $\widetilde{\delta}_{1,T}$ are seen to be asymptotically independent with the same distribution. Hence

$$\Pr\left\{T^{1/2}\left(\widetilde{\delta}_{1,T} - \delta_{0T}\right) \le x, \widetilde{\delta}_{-1,T} > 0, \widetilde{\delta}_{1,T} > 0\right\}$$
$$\sim \Pr\left\{T^{1/2}\left(\widetilde{\delta}_{1,T} - \delta_{0T}\right) \le x, \widetilde{\delta}_{1,T} > 0\right\}\Pr\left\{\widetilde{\delta}_{-1,T} > 0\right\}$$

and

$$\Pr\left\{T^{1/2}\left(\widetilde{\delta}_{-1,T} - \delta_{0T}\right) \le x, \widetilde{\delta}_{-1,T} > 0, \widetilde{\delta}_{1,T} \le 0\right\}$$
$$\sim \Pr\left\{T^{1/2}\left(\widetilde{\delta}_{-1,T} - \delta_{0T}\right) \le x, \widetilde{\delta}_{-1,T} > 0\right\}\Pr\left\{\widetilde{\delta}_{1,T} \le 0\right\}.$$

Exchanging labels, which may be done since $\widetilde{\delta}_{-1,T}$ and $\widetilde{\delta}_{1,T}$ are asymptotically distributed identically, it follows that

$$\Pr\left\{T^{3/2}\left(\widehat{\lambda}_T - \lambda\right) \le 2\pi x\right\}$$
$$\sim \Pr\left\{T^{1/2}\left(\widetilde{\delta}_{-1,T} - \delta_{0T}\right) \le x, \widetilde{\delta}_{-1,T} > 0\right\}\Pr\left\{\widetilde{\delta}_{1,T} > 0\right\}$$
$$+ \Pr\left\{T^{1/2}\left(\widetilde{\delta}_{-1,T} - \delta_{0T}\right) \le x, \widetilde{\delta}_{-1,T} > 0\right\}\Pr\left\{\widetilde{\delta}_{1,T} \le 0\right\}$$
$$+ \Pr\left\{T^{1/2}\left(\widetilde{\delta}_{-1,T} - \delta_{0T}\right) \le x, \widetilde{\delta}_{-1,T} \le 0\right\}$$
$$= \Pr\left\{T^{1/2}\left(\widetilde{\delta}_{-1,T} - \delta_{0T}\right) \le x\right\}.$$

We thus have the following theorem:

Theorem 26 *Let $\widehat{\lambda}_T$ be defined as in Algorithm 4. Then*

$$T^{3/2}\left(\log T\right)^{-1/2-\nu}\left(\widehat{\lambda}_T - \lambda_0\right) \to 0$$

almost surely as $T \to \infty$, for all $\nu > 0$ and

$$\Pr\left\{T^{3/2}v_T^{-1}\left(\widehat{\lambda}_T - \lambda_0\right) \le x\right\} \to \frac{1}{\sqrt{2\pi}}\int_{-\infty}^{x} e^{-\frac{1}{2}u^2}\,du$$

where

$$v_T^2 = \frac{16\pi^3 f_x\left(\lambda\right)}{A^2}\frac{\pi^2\delta_{0T}^2}{\sin^2\left(\pi\delta_{0T}\right)}\left(1 - |\delta_{0T}|\right)^2\left\{\left(1 - |\delta_{0T}|\right)^2 + \delta_{0T}^2\right\}.$$

Note that the ratio of the asymptotic variance to the asymptotic variance of the periodogram maximiser is thus

$$\frac{\pi^2}{3}\frac{\pi^2\delta_{0T}^2}{\sin^2\left(\pi\delta_{0T}\right)}\left(1 - |\delta_{0T}|\right)^2\left(1 - 2|\delta_{0T}| + 2\delta_{0T}^2\right)$$

which has largest value $\pi^2/3 \simeq 3.2899$ when $\delta_{0T} = 0$, and smallest value $\pi^4/96 \simeq 1.0147$ when $\delta_{0T} = \pm\frac{1}{2}$.

It is obvious that the alternative algorithm which replaces Step (iii) of Algorithm 4 with

(iii)′. If $\widehat{\delta}_{-1,T} < 0$ and $\widehat{\delta}_{1,T} < 0$, put $\widehat{\delta}_T = \widehat{\delta}_{-1,T}$. Otherwise, put $\widehat{\delta}_T = \widehat{\delta}_{1,T}$;

also has the same properties. Consider, however, replacing this with

(iii)″. If $\widehat{\delta}_{1,T} > 0$, put $\widehat{\delta}_T = \widehat{\delta}_{1,T}$. Otherwise, put $\widehat{\delta}_T = \widehat{\delta}_{-1,T}$.

Clearly, there will be no problems for $T \in T_2 \cup T_3$. Suppose $T \in T_1$. Then

$$
\begin{aligned}
\Pr &\left\{ T^{3/2} \left(\widehat{\lambda}_T - \lambda \right) \le 2\pi x \right\} \\
&= \Pr \left\{ T^{1/2} \left(\widehat{\delta}_T - \delta_{0T} \right) \le x, \widehat{m}_T = m_T \right\} + o\left(1 \right) \\
&= \Pr \left\{ T^{1/2} \left(\widetilde{\delta}_{1,T} - \delta_{0T} \right) \le x, \widetilde{\delta}_{1,T} > 0 \right\} \\
&\quad + \Pr \left\{ T^{1/2} \left(\widetilde{\delta}_{-1,T} - \delta_{0T} \right) \le x, \widetilde{\delta}_{1,T} \le 0 \right\} + o\left(1 \right) \\
&\sim \Pr \left\{ T^{1/2} \left(\widetilde{\delta}_{1,T} - \delta_{0T} \right) \le x, \widetilde{\delta}_{1,T} > 0 \right\} \\
&\quad + \Pr \left\{ T^{1/2} \left(\widetilde{\delta}_{-1,T} - \delta_{0T} \right) \le x \right\} \Pr \left\{ \widetilde{\delta}_{1,T} \le 0 \right\}.
\end{aligned}
$$

Now, if $x > -T^{1/2}\delta_{0T}$,

$$
\begin{aligned}
\Pr &\left\{ T^{1/2} \left(\widetilde{\delta}_{1,T} - \delta_{0T} \right) \le x, \widetilde{\delta}_{1,T} > 0 \right\} \\
&= \Pr \left\{ T^{1/2} \left(\widetilde{\delta}_{1,T} - \delta_{0T} \right) \le x \right\} - \Pr \left\{ \widetilde{\delta}_{1,T} \le 0 \right\}
\end{aligned}
$$

and therefore

$$
\begin{aligned}
\Pr &\left\{ T^{3/2} \left(\widehat{\lambda}_T - \lambda \right) \le 2\pi x \right\} \\
&\sim \Pr \left\{ T^{1/2} \left(\widetilde{\delta}_{1,T} - \delta_{0T} \right) \le x \right\} - \Pr \left\{ \widetilde{\delta}_{1,T} \le 0 \right\} \\
&\quad + \Pr \left\{ T^{1/2} \left(\widetilde{\delta}_{-1,T} - \delta_{0T} \right) \le x \right\} \Pr \left\{ \widetilde{\delta}_{1,T} \le 0 \right\}
\end{aligned}
$$

while, if $x \le -T^{1/2}\delta_{0T}$,

$$
\begin{aligned}
\Pr &\left\{ T^{3/2} \left(\widehat{\lambda}_T - \lambda \right) \le 2\pi x \right\} \\
&\sim \Pr \left\{ T^{1/2} \left(\widetilde{\delta}_{-1,T} - \delta_{0T} \right) \le x \right\} \Pr \left\{ \widetilde{\delta}_{1,T} \le 0 \right\}.
\end{aligned}
$$

But

$$
\Pr \left\{ \widetilde{\delta}_{1,T} \le 0 \right\} = \Pr \left\{ T^{1/2} \left(\widetilde{\delta}_{1,T} - \delta_{0T} \right) \le -T^{1/2}\delta_{0T} \right\},
$$

which does not converge to zero as all that is guaranteed is that

$$T^{1/2} |\delta_{0T}| \leq aT^{1/2-\nu},$$

where $\nu < \frac{1}{2}$. In fact, if $\lambda_0/(2\pi)$ is rational, there will be a subsequence of T_1 along which $\delta_{0T} = 0$, while if $\lambda_0/(2\pi)$ is irrational, there will be a subsequence of T_1 along which $T^{1/2}\delta_{0T}$ converges to zero. Thus, along these subsequences

$$\Pr\left\{T^{3/2}\left(\widehat{\lambda}_T - \lambda\right) \leq 2\pi x\right\}$$

$$\sim \begin{cases} \frac{3}{2}\Pr\left\{T^{1/2}\left(\widetilde{\delta}_{1,T} - \delta_{0T}\right) \leq x\right\} - \frac{1}{2} & ; \quad x > 0 \\ \frac{1}{2}\Pr\left\{T^{1/2}\left(\widetilde{\delta}_{1,T} - \delta_{0T}\right) \leq x\right\} & ; \quad x \leq 0 \end{cases}$$

and the usual central limit theorem does not eventuate as the mean of a random variable with distribution given by the limit of the right side of the above is not zero. It is clearly preferable, therefore, to use Algorithm 4, or the alternative which replaces Step (iii) with Step (iii)$'$.

MacLeod (1991, 1998) has suggested yet another alternative to Step (iii).

(iii)$'''$. If $\widehat{\alpha}_{-1} > \widehat{\alpha}_1$, put $\widehat{\delta}_T = \widehat{\delta}_{-1,T}$. Otherwise, put $\widehat{\delta}_T = \widehat{\delta}_{1,T}$.

The alternative formulation is neater and can be proved to have the same asymptotic properties as Algorithm 4. Moreover, it appears to have better performance when m_T is known *a priori*.

The common asymptotic variance of the above estimators when $\delta_{0T} = 0$ is exactly twice as large as that of the best estimator using three Fourier coefficients, while it is only slightly higher when $\delta_{0T} = \pm\frac{1}{2}$. Although the average of $\widehat{\delta}_{-1,T}$ and $\widehat{\delta}_{1,T}$ produces an estimator of λ which has a reduced asymptotic variance when $\delta_{0T} = 0$, the asymptotic variance is absurdly high when $\delta_{0T} = \pm\frac{1}{2}$. Thus no linear estimator will have a uniformly lower asymptotic variance. However, the estimator described above is only one possible estimator which may be constructed from $\widehat{\delta}_{-1,T}$ and $\widehat{\delta}_{1,T}$. We shall now construct another, which has the lowest possible asymptotic variance, uniformly in δ_{0T}. Let $\widehat{\delta}_T = g\left(\widehat{\delta}_{-1,T}, \widehat{\delta}_{1,T}\right)$ where g is continuous with continuous partial derivatives. Then, assuming that $T \in T_1 \cup T_2$, above, we have

$$\begin{aligned} \widehat{\delta}_T &= g\left(\delta_{0T}, \delta_{0T}\right) + g_1\left(\delta_{0T}, \delta_{0T}\right)\left(\widehat{\delta}_{-1,T} - \delta_{0T}\right) \\ &\quad + g_2\left(\delta_{0T}, \delta_{0T}\right)\left(\widehat{\delta}_{1,T} - \delta_{0T}\right) + o\left(T^{-1/2}\right) \end{aligned}$$

almost surely, where g_1 and g_2 are the partial derivatives with respect to the first and second arguments of g. If $\widehat{\delta}_T$ is to be strongly consistent, therefore, we need to have

$$\delta_{0T} = g\left(\delta_{0T}, \delta_{0T}\right).$$

If this condition is met, the asymptotic variance of $T^{3/2}\left(\widehat{\lambda}_T - \lambda\right)$, at least along $T \in T_1 \cup T_2$, is equal to

$$\alpha\left(\delta_{0T}\right)\partial g'\begin{bmatrix} \left(\delta_{0T} + 1\right)^4\left\{1 + \frac{\delta_{0T}^2}{\left(\delta_{0T}+1\right)^2}\right\} & -\delta_{0T}^2\left(\delta_{0T}^2 - 1\right) \\ -\delta_{0T}^2\left(\delta_{0T}^2 - 1\right) & \left(\delta_{0T} - 1\right)^4\left\{1 + \frac{\delta_{0T}^2}{\left(\delta_{0T}-1\right)^2}\right\} \end{bmatrix}\partial g,$$

where

$$\partial g = \begin{bmatrix} g_1\left(\delta_{0T}, \delta_{0T}\right) & g_2\left(\delta_{0T}, \delta_{0T}\right) \end{bmatrix}'$$

and

$$\alpha\left(\delta_{0T}\right) = 16\pi^3 f_x\left(\lambda\right)\frac{\pi^2\delta_{0T}^2}{\sin^2\left(\pi^2\delta_{0T}^2\right)}.$$

We must therefore find g such that $g\left(x, x\right) = x$, for all $x \in \left(-\frac{1}{2}, \frac{1}{2}\right)$, and which minimises

$$H\left(x\right)'\begin{bmatrix} \left(2x^2 + 2x + 1\right)\left(x + 1\right)^2 & -x^2\left(x^2 - 1\right) \\ -x^2\left(x^2 - 1\right) & \left(2x^2 - 2x + 1\right)\left(x - 1\right)^2 \end{bmatrix}H\left(x\right)$$

where $H\left(x\right) = \begin{bmatrix} h_1\left(x\right) & h_2\left(x\right) \end{bmatrix}'$, $h_1\left(x\right) = g_1\left(x, x\right)$ and $h_2\left(x\right) = g_2\left(x, x\right)$. But the condition $g\left(x, x\right) = x$ implies that $g_1\left(x, x\right) + g_2\left(x, x\right) = 1$, that is, $h_2\left(x\right) = 1 - h_1\left(x\right)$. Thus, put $h = h\left(x\right) = h_1\left(x\right) = 1 - h_2\left(x\right)$. We therefore need to find, for each x, that h which minimises

$$h^2\left(2x^2 + 2x + 1\right)\left(x + 1\right)^2 + \left(1 - h\right)^2\left(2x^2 - 2x + 1\right)\left(x - 1\right)^2$$
$$- 2h\left(1 - h\right)x^2\left(x^2 - 1\right).$$

Since this is quadratic in x, we easily obtain the minimum value

$$\frac{\left(x^2 - 1\right)^2\left(3x^4 + 1\right)}{2\left(3x^4 + 6x^2 + 1\right)}$$

which occurs when $h = \frac{1}{2} - \frac{3x^3 + 2x}{3x^4 + 6x^2 + 1}$. It is easy to check that this yields an asymptotic variance which is the lowest possible. Consequently, we need only find any function g which satisfies the conditions. Because of the restriction that $h_2\left(x\right) = 1 - h_1\left(x\right)$, it is natural to consider functions g which satisfy

$$g\left(x_1, x_2\right) = \eta_1\left(x_1\right) + \eta_2\left(x_2\right).$$

With this formulation, $h_1(x) = \frac{d}{dx}\eta_1(x)$ and $h_2(x) = \frac{d}{dx}\eta_2(x)$, and consequently we may take $\eta_2(x) = x - \eta_1(x)$, without loss of generality, and solve

$$\frac{d}{dx}\eta_1(x) = \frac{1}{2} - \frac{3x^3 + 2x}{3x^4 + 6x^2 + 1}.$$

We may thus choose

$$\eta_1(x) = \frac{x}{2} - \frac{1}{4}\log\left(3x^4 + 6x^2 + 1\right) + \frac{\sqrt{6}}{24}\log\left(\frac{x^2 + 1 - \sqrt{\frac{2}{3}}}{x^2 + 1 + \sqrt{\frac{2}{3}}}\right)$$

and

$$\eta_2(x) = \frac{x}{2} + \frac{1}{4}\log\left(3x^4 + 6x^2 + 1\right) - \frac{\sqrt{6}}{24}\log\left(\frac{x^2 + 1 - \sqrt{\frac{2}{3}}}{x^2 + 1 + \sqrt{\frac{2}{3}}}\right).$$

An estimator of λ is then given by the following algorithm

Algorithm 5 *FTI 2*

(i) *Let $\widehat{m}_T$ be the maximiser of $I_y\left(2\pi\frac{m}{T}\right)$, $m = 1, 2, \ldots, \left\lfloor\frac{T-1}{2}\right\rfloor$.*

(ii) *Let $\widehat{\delta}_{jT} = \frac{j\widehat{\alpha}_j}{\widehat{\alpha}_j - 1}$, $j = -1, 1$ where $\widehat{\alpha}_j = \mathrm{Re}\left\{z\left(\widehat{m}_T + j\right)/z\left(\widehat{m}_T\right)\right\}$.*

(iii) *Let*

$$\widehat{\delta}_T = \frac{\widehat{\delta}_{-1,T} + \widehat{\delta}_{1,T}}{2} + \kappa\left(\widehat{\delta}_{1,T}^2\right) - \kappa\left(\widehat{\delta}_{-1,T}^2\right)$$

where $\kappa(x) = \frac{1}{4}\log\left(3x^2 + 6x + 1\right) - \frac{\sqrt{6}}{24}\log\left(\frac{x+1-\sqrt{\frac{2}{3}}}{x+1+\sqrt{\frac{2}{3}}}\right)$.

(iv) *Let $\widehat{\lambda}_T = 2\pi\left(\widehat{m}_T + \widehat{\delta}_T\right)\Big/T$.*

From (6.8), it follows that the result extends to $T \in T_3$, and we obtain

Theorem 27 *Let $\widehat{\lambda}_T$ be defined as in Algorithm 5. Then*

$$T^{3/2}(\log T)^{-1/2 - \nu}\left(\widehat{\lambda}_T - \lambda_0\right) \to 0$$

almost surely as $T \to \infty$, for all $\nu > 0$ and

$$\Pr\left\{T^{3/2}v_T^{-1}\left(\widehat{\lambda}_T - \lambda_0\right) \le x\right\} \to \frac{1}{\sqrt{2\pi}}\int_{-\infty}^{x} e^{-\frac{1}{2}u^2}\,du$$

where

$$v_T^2 = \frac{8\pi^3 f_x(\lambda)}{A^2}\frac{\pi^2\delta_{0T}^2}{\sin^2(\pi\delta_{0T})}\frac{\left(\delta_{0T}^2 - 1\right)^2\left(3\delta_{0T}^4 + 1\right)}{3\delta_{0T}^4 + 6\delta_{0T}^2 + 1}.$$

Note that the ratio of the asymptotic variance to the asymptotic variance of the periodogram maximiser is thus

$$\frac{\pi^2}{6} \frac{\pi^2 \delta_{0T}^2}{\sin^2 (\pi \delta_{0T})} \frac{\left(\delta_{0T}^2 - 1\right)^2 \left(3\delta_{0T}^4 + 1\right)}{3\delta_{0T}^4 + 6\delta_{0T}^2 + 1},$$

which has largest value $\pi^2/6 \simeq 1.6449$ when $\delta_{0T} = 0$, and smallest value $\frac{57}{5504}\pi^4 \simeq 1.0088$ when $\delta_{0T} = \pm\frac{1}{2}$.

Incidentally, this ratio is actually minimised *outside* the interval $\left[-\frac{1}{2}, \frac{1}{2}\right]$ and is approximately 1.0066 when δ_{0T} is ±0.5217. While this result is slightly puzzling, it cannot be known *a priori*, of course, that δ_{0T} is so close to $\frac{1}{2}$ that the slight advantage might be taken. Finally, we propose an alternative, but related, algorithm which does not involve logarithms.

Algorithm 6 *FTI 3*

   (i) *Let $\widehat{m}_T$ be the maximiser of $I_y\left(2\pi\frac{m}{T}\right)$, $m = 1, 2, \ldots, \left\lfloor\frac{T-1}{2}\right\rfloor$.*

   (ii) *Let $\widehat{\delta}_{jT} = \frac{j\widehat{\alpha}_j}{\widehat{\alpha}_j - 1}$, $j = -1, 1$ where $\widehat{\alpha}_j = \mathrm{Re}\left\{z\left(\widehat{m}_T + j\right)/z\left(\widehat{m}_T\right)\right\}$.*

   (iii) *If $\widehat{\delta}_{-1,T} > 0$ and $\widehat{\delta}_{1,T} > 0$, put $\overline{\delta}_T = \widehat{\delta}_{1,T}$. Otherwise, put $\overline{\delta}_T = \widehat{\delta}_{-1,T}$.*

   (iv) *Let $\widehat{\delta}_T = \frac{\widehat{\delta}_{-1,T} + \widehat{\delta}_{1,T}}{2} + \left(\widehat{\delta}_{1,T} - \widehat{\delta}_{-1,T}\right)\frac{3\overline{\delta}_T^3 + 2\overline{\delta}_T}{3\overline{\delta}_T^4 + 6\overline{\delta}_T^2 + 1}$.*

   (v) *Let $\widehat{\lambda}_T = 2\pi\left(\widehat{m}_T + \widehat{\delta}_T\right)\Big/ T$.*

Then we have

Corollary 1 *Let $\widehat{\lambda}_T$ be defined as in Algorithm 6. Then $\widehat{\lambda}_T$ has the same asymptotic properties as the estimator given in Algorithm 5.*

Proof Since $\widehat{\delta}_{-1,T} - \delta_{0T}$ and $\widehat{\delta}_{1,T} - \delta_{0T}$ both converge to zero, almost surely, it follows that $\widehat{\delta}_T$ from Algorithm 5 satisfies

$$\widehat{\delta}_T = \frac{\widehat{\delta}_{-1,T} + \widehat{\delta}_{1,T}}{2} + 2\left(\widehat{\delta}_{1,T} - \widehat{\delta}_{-1,T}\right)\delta_T^* \kappa'\left[(\delta_T^*)^2\right],$$

where $\delta_T^* - \delta_{0T}$ also converges almost surely to zero, since δ_T^* is between $\widehat{\delta}_{-1,T}$ and $\widehat{\delta}_{1,T}$. But, since $\widehat{\delta}_{1,T} - \widehat{\delta}_{-1,T}$ converges almost surely to zero, we may replace $\kappa'\left[(\delta_T^*)^2\right]$ with $\kappa'\left[\left(\delta_T^\#\right)^2\right]$, where $\delta_T^\# - \delta_{0T}$ converges almost surely to zero. We choose to take $\delta_T^\# = \overline{\delta}_T$, which is the estimator of δ_{0T} from Algorithm 4. We could equally well use $\delta_T^\# = \frac{\widehat{\delta}_{-1,T} + \widehat{\delta}_{1,T}}{2}$, but this is not preferable as its asymptotic variance is very large when $|\delta_{0T}|$ is near $\frac{1}{2}$.

$\square$

We note that MacLeod's (1991, 1998) ideas suggest the use of the technique which is obtained by replacing Step (iii) in Algorithm 6 with

(iii)' If $\widehat{\alpha}_{-1} > \widehat{\alpha}_1$, put $\overline{\delta}_T = \widehat{\delta}-_{1,T}$. Otherwise, put $\overline{\delta}_T = \widehat{\delta}_{1,T}$.

6.2.3 Techniques involving only the moduli of Fourier coefficients

As the storage requirement for the moduli of Fourier coefficients is only half that for the complex coefficients themselves, there will be some advantage in using a technique which needs only the moduli. Indeed, many signal processing devices store only these moduli, as their main aim is to produce displays (lofargrams, sonograms, spectrograms) which show regions of frequency where there is significant energy, and since enormous quantities of data need to be displayed frequently. The following paragraphs describe various techniques which have been used. It is fair to say that none of the techniques work well, in the sense of bias or consistency.

Quadratic Interpolation

Since the periodogram yields a good estimator of frequency, accurate to $O\left(T^{-1}\right)$, it is tempting to look for an estimation technique which uses only the $|w_y\left(\omega_j\right)|^2$. A commonly used technique fits a quadratic through the three points

$$\left(j, \left|w_y\left(2\pi\frac{\widehat{m}_T + j}{T}\right)\right|^2\right), \quad j = -1, 0, 1.$$

Letting the maximiser of this quadratic be $\widehat{\delta}_T$, the quadratic estimator of λ is defined as $\widehat{\lambda}_T = 2\pi\frac{\widehat{m}_T + \widehat{\delta}_T}{T}$. The technique is called 'quadratic interpolation' and seems reasonable at first glance. In fact, it turns out to be no better than the 'coarse periodogram maximiser' $2\pi\frac{\widehat{m}_T}{T}$: It is easily shown that

$$\widehat{\delta}_T = \frac{1}{2}\frac{\left|w_y\left(2\pi\frac{\widehat{m}_T+1}{T}\right)\right|^2 - \left|w_y\left(2\pi\frac{\widehat{m}_T-1}{T}\right)\right|^2}{2\left|w_y\left(2\pi\frac{\widehat{m}_T}{T}\right)\right|^2 - \left|w_y\left(2\pi\frac{\widehat{m}_T-1}{T}\right)\right|^2 - \left|w_y\left(2\pi\frac{\widehat{m}_T+1}{T}\right)\right|^2}.$$

But, letting $\widetilde{\delta}_T$ be the above with $\widehat{m}_T$ replaced by m_T, it follows that

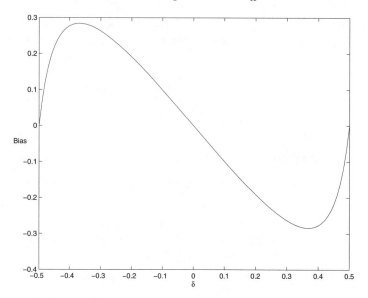

Fig. 6.1. Bias of quadratic estimator as function of δ.

$$\widetilde{\delta}_T = \frac{1}{2}\frac{\left(\frac{\delta_{0T}}{\delta_{0T}-1}\right)^2 - \left(\frac{\delta_{0T}}{\delta_{0T}+1}\right)^2}{2 - \left(\frac{\delta_{0T}}{\delta_{0T}-1}\right)^2 - \left(\frac{\delta_{0T}}{\delta_{0T}+1}\right)^2} + o\left(1\right) = \frac{\delta_{0T}^3}{1 - 3\delta_{0T}^2} + o\left(1\right)$$

$$= \delta_{0T} + \frac{4\delta_{0T}^3 - \delta_{0T}}{1 - 3\delta_{0T}^2} + o\left(1\right).$$

The bias in $\widehat{\lambda}_T$ is thus of order $O\left(T^{-1}\right)$ as $T \to \infty$, the same order as the bias in the estimator $2\pi\frac{\widehat{m}_T}{T}$, at least when $\delta_{0T} \in T_1 \cup T_2$. If $\widehat{m}_T = m_T + 1$, which is possible asymptotically when $T \in T_3$ (defined above) and $\delta_{0T} > 0$, we obtain

$$\widehat{\delta}_T = \frac{1}{2}\frac{\left(\frac{\delta_{0T}}{\delta_{0T}-2}\right)^2 - 1}{2\left(\frac{\delta_{0T}}{\delta_{0T}-1}\right)^2 - \left(\frac{\delta_{0T}}{\delta_{0T}-2}\right)^2 - 1} + o\left(1\right) = \frac{(\delta_{0T} - 1)^3}{-3\delta_{0T}^2 + 6\delta_{0T} - 2} + o\left(1\right).$$

which, since $\delta_{0T} = \frac{1}{2} + o\left(1\right)$, is $-\frac{1}{2} + o\left(1\right)$. Thus $\widehat{m}_T + \widehat{\delta}_T = m_T + \frac{1}{2} + o\left(1\right)$, which is the same as when $T \in T_3$, $\delta_{0T} > 0$ but $\widehat{m}_T = m_T$. The same can be said when $\delta_{0T} < 0$, when it is possible that $\widehat{m}_T = m_T - 1$. The bias thus disappears when $\delta_{0T} = -\frac{1}{2}$, 0 or $\frac{1}{2}$, but is of order $O\left(T^{-1}\right)$ otherwise. A plot of the bias as a function of δ_{0T} is given in Figure 6.1.

The Rife and Vincent technique

From (6.5),

$$\left|\frac{z\left(m_T + j\right)}{z\left(m_T\right)}\right|^2 = \left(\frac{\delta_{0T}}{\delta_{0T} - j}\right)^2 + o\left(1\right)$$

and therefore

$$\left|\frac{z\left(m_T + j\right)}{z\left(m_T\right)}\right| = \left|\frac{\delta_{0T}}{\delta_{0T} - j}\right| + o\left(1\right).$$

Now, if $0 \le \delta_{0T} \le \frac{1}{2}$, then $\left|\frac{\delta_{0T}}{\delta_{0T}-1}\right| \ge \left|\frac{\delta_{0T}}{\delta_{0T}+1}\right|$ and

$$\frac{\left|\frac{\delta_{0T}}{\delta_{0T}-1}\right|}{1 + \left|\frac{\delta_{0T}}{\delta_{0T}-1}\right|} = \frac{\frac{\delta_{0T}}{1-\delta_{0T}}}{1 + \frac{\delta_{0T}}{1-\delta_{0T}}} = \delta_{0T},$$

while if $-\frac{1}{2} \le \delta_{0T} \le 0$, then $\left|\frac{\delta_{0T}}{\delta_{0T}-1}\right| \le \left|\frac{\delta_{0T}}{\delta_{0T}+1}\right|$ and

$$\frac{\left|\frac{\delta_{0T}}{\delta_{0T}+1}\right|}{1 + \left|\frac{\delta_{0T}}{\delta_{0T}+1}\right|} = \frac{\frac{-\delta_{0T}}{1+\delta_{0T}}}{1 + \frac{-\delta_{0T}}{1+\delta_{0T}}} = -\delta_{0T}.$$

This suggests the following algorithm for estimating λ:

Algorithm 7 *Rife and Vincent (1970)*

   (i) *Let $\widehat{m}_T$ be the maximiser of $I_y\left(2\pi \frac{m}{T}\right)$, $m = 1, 2, \ldots, \left[\frac{T-1}{2}\right]$.*

   (ii) *Let $\widehat{\alpha}_T = 1$ if $\left|z\left(\widehat{m}_T + 1\right)\right|^2 > \left|z\left(\widehat{m}_T - 1\right)\right|^2$ and -1 otherwise.*

   (iii) *Let $\widehat{\delta}_T = \widehat{\alpha}_T \left|\frac{z(\widehat{m}_T + \widehat{\alpha}_T)}{z(\widehat{m}_T)}\right| \Big/ \left(1 + \left|\frac{z(\widehat{m}_T + \widehat{\alpha}_T)}{z(\widehat{m}_T)}\right|\right)$.*

   (iv) *Let $\widehat{\lambda}_T = 2\pi \left(\widehat{m}_T + \widehat{\delta}_T\right) \Big/ T$.*

In order to analyse this technique, we shall need a few number-theoretic results.

Lemma 4 *Let $\langle x \rangle$ denote the distance between x and its closest integer. Then, for $\theta \in (0, 1)$, and irrational, there exists an infinite sequence of integers $\{r_j\}$ for which $r_j^{1/4} \langle r_j \theta \rangle$ converges to 1.*

Proof Since θ is irrational, the jth convergent p_j/q_j in the continued fraction of θ satisfies $|\theta - p_j/q_j| < q_j^{-2}$. Let $d_j = \langle h_j q_j \theta \rangle$, for some integer $h_j > 1$, and let $\nu_j = \theta - p_j/q_j$. Then $d_j = \langle h_j q_j \nu_j \rangle$, which equals $|h_j q_j \nu_j|$ if $h_j q_j < (2|\nu_j|)^{-1}$. Let $r_j = h_j q_j$, where h_j is the integer part of $q_j^{-1} |\nu_j|^{-4/5}$. Then $\{h_j\}$ is divergent, since $q_j^{-1} |\nu_j|^{-4/5} > q_j^{-1} q_j^{8/5} = q_j^{3/5}$ and $\{q_j\}$ diverges.

Also, $h_j q_j = |\nu_j|^{-4/5} = |\nu_j|^{-1} |\nu_j|^{1/5}$ which will be less than $(2|\nu_j|)^{-1}$ for large enough j since $\{\nu_j\}$ converges to 0. Thus, for large enough j,

$$r_j^{1/4} \langle r_j \theta \rangle = (h_j q_j)^{1/4} |h_j q_j \nu_j| = \left(h_j \left| q_j \nu_j^{4/5} \right| \right)^{5/4} = \left(\frac{\left\lfloor q_j^{-1} \left| \nu_j^{-4/5} \right| \right\rfloor}{q_j^{-1} \left| \nu_j^{-4/5} \right|} \right)^{5/4},$$

which converges to 1 as $\left\{ q_j^{-1} \left| \nu_j^{-4/5} \right| \right\}$ diverges. $\square$

The next result follows similarly.

Lemma 5 *Let $\xi > 0$. Then, for $\theta \in (0,1)$, and irrational, there exists an infinite sequence of integers $\{r_j\}$ for which $r_j^{1/4} \langle r_j \theta \rangle$ converges to ξ.*

We now show that the Rife and Vincent estimator does not share the important property of the estimators in Subsection 6.2.2, namely, that

$$T^{3/2} (\log T)^{-1/2 - \nu} \left(\widehat{\lambda}_T - \lambda_0 \right) \to 0$$

almost surely as $T \to \infty$.

Theorem 28 *Let $\widehat{\lambda}_T$ be as defined in Algorithm 7, where $\lambda_0/(2\pi)$ is irrational. Then $T^{5/4} \left(\widehat{\lambda}_T - \lambda_0 \right)$ does not converge in probability to 0.*

Proof Define r_j as in Lemma 4 with $\theta = \lambda_0/(2\pi)$. Then $\lambda_0 = 2\pi \frac{m_T + \delta_{0T}}{T}$, for all T and $r_j^{1/4} \left| \delta_{0 r_j} \right|$ converges to 1, which implies that $\left| \delta_{0 r_k} \right|$ converges to 0, and so $\widehat{m}_{r_k} - m_{r_k} = 0$, almost surely as $k \to \infty$. We shall assume from now on that $T \in \{r_k\}$. Since

$$\frac{z(m_T + j)}{z(m_T)} = \frac{\delta_{0T}}{\delta_{0T} - j} + D_{0T}^{-1} \left\{ u(m_T + j) - \frac{\delta_{0T}}{\delta_{0T} - j} u(m_T) \right\}$$
$$+ O_{\text{a.s.}} \left(T^{-1} \log T \right),$$

it follows that

$$\left| \frac{z(m_T + j)}{z(m_T)} \right|^2$$
$$= \left(\frac{\delta_{0T}}{\delta_{0T} - j} \right)^2 + 2 \frac{\delta_{0T}}{\delta_{0T} - j} \operatorname{Re} \left[D_{0T}^{-1} \left\{ u(m_T + j) - \frac{\delta_{0T}}{\delta_{0T} - j} u(m_T) \right\} \right]$$
$$+ O_{\text{a.s.}} \left(T^{-1} \log T \right)$$

and, since $|\delta_{0T}| \sim T^{-1/4}$,

$$\left| \frac{z\left(m_T + j\right)}{z\left(m_T\right)} \right|$$

$$= \left| \frac{\delta_{0T}}{\delta_{0T} - j} \right| + \operatorname{sgn}\left(\frac{\delta_{0T}}{\delta_{0T} - j}\right) \operatorname{Re}\left[D_{0T}^{-1}\left\{ u\left(m_T + j\right) - \frac{\delta_{0T}}{\delta_{0T} - j} u\left(m_T\right) \right\} \right]$$

$$+ O_{\text{a.s.}}\left(T^{-3/4} \log T\right)$$

$$= |\delta_{0T}| + \operatorname{sgn}\left(\frac{\delta_{0T}}{\delta_{0T} - j}\right) \Re\left\{ D_{0T}^{-1} u\left(m_T + j\right) \right\} + O_{\text{a.s.}}\left(T^{-1/2}\right).$$

Now, if $\widehat{\alpha}_T = 1$,

$$T^{5/4} \frac{\widehat{\lambda}_{0T} - \lambda_0}{2\pi} = T^{1/4}\left(\widehat{\delta}_T - \delta_{0T}\right) = T^{1/4}\left\{ \frac{\left| \frac{z(m_T+1)}{z(m_T)} \right|}{1 + \left| \frac{z(m_T+1)}{z(m_T)} \right|} - \delta_{0T} \right\}$$

$$= T^{1/4}\left\{ |\delta_{0T}| - \delta_{0T} + O_P\left(T^{-1/2}\right) \right\},$$

while if $\widehat{\alpha}_T = -1$,

$$T^{5/4} \frac{\widehat{\lambda}_{0T} - \lambda_0}{2\pi} = T^{1/4}\left(\widehat{\delta}_T - \delta_{0T}\right) = T^{1/4}\left\{ -\frac{\left| \frac{z(m_T-1)}{z(m_T)} \right|}{1 + \left| \frac{z(m_T-1)}{z(m_T)} \right|} - \delta_{0T} \right\}$$

$$= T^{1/4}\left\{ -|\delta_{0T}| - \delta_{0T} + O_P\left(T^{-1/2}\right) \right\}.$$

Thus, if $\widehat{\alpha}_T = 1$,

$$T^{5/4}\left(\widehat{\lambda}_T - \lambda_0\right)/(2\pi) \sim \begin{cases} -2T^{1/4}\delta_{0T} & ; \quad \delta_{0T} < 0 \\ 0 & ; \quad \delta_{0T} \geq 0, \end{cases}$$

and, if $\widehat{\alpha}_T = -1$,

$$T^{5/4}\left(\widehat{\lambda}_T - \lambda_0\right)/(2\pi) \sim \begin{cases} 0 & ; \quad \delta_{0T} < 0 \\ -2T^{1/4}\delta_{0T} & ; \quad \delta_{0T} \geq 0. \end{cases}$$

But, when $\delta_{0T} < 0$,

$$\Pr\left\{\widehat{\alpha}_T = 1\right\} \sim \Pr\left\{ \left(\frac{\delta_{0T}}{\delta_{0T} - 1}\right)^2 + 2\frac{\delta_{0T}}{\delta_{0T} - 1} \operatorname{Re}\left\{ D_{0T}^{-1} u\left(m_T + 1\right) \right\} \right.$$

$$\left. > \left(\frac{\delta_{0T}}{\delta_{0T} + 1}\right)^2 + 2\frac{\delta_{0T}}{\delta_{0T} + 1} \operatorname{Re}\left\{ D_{0T}^{-1} u\left(m_T - 1\right) \right\} \right\}$$

$$\sim \ \Pr\left\{2\delta_{0T}\operatorname{Re}\left\{D_{0T}^{-1}u\left(m_{T}-1\right)-D_{0T}^{-1}u\left(m_{T}+1\right)\right\}\right.$$

$$\left. < \left(\frac{\delta_{0T}}{\delta_{0T}-1}\right)^{2}-\left(\frac{\delta_{0T}}{\delta_{0T}+1}\right)^{2}\right\}$$

$$\sim \ \Pr\left\{T^{1/2}\operatorname{Re}\left\{D_{0T}^{-1}u\left(m_{T}-1\right)-D_{0T}^{-1}u\left(m_{T}+1\right)\right\}>2\left(T^{1/4}\delta_{0T}\right)^{2}\right\}$$

$$\to \ 1-\Phi\left(\frac{|A|}{\sqrt{2\pi f\left(\lambda\right)}}\right),$$

and when $\delta_{0T}>0$,

$$\Pr\left\{\widehat{\alpha}_{T}=1\right\}$$

$$\sim \ \Pr\left\{\left(\frac{\delta_{0T}}{\delta_{0T}-1}\right)^{2}+2\frac{\delta_{0T}}{\delta_{0T}-1}\operatorname{Re}\left\{D_{0T}^{-1}u\left(m_{T}+1\right)\right\}\right.$$

$$\left. > \left(\frac{\delta_{0T}}{\delta_{0T}+1}\right)^{2}+2\frac{\delta_{0T}}{\delta_{0T}+1}\operatorname{Re}\left\{D_{0T}^{-1}u\left(m_{T}-1\right)\right\}\right\}$$

$$\sim \ \Pr\left\{2\delta_{0T}\operatorname{Re}\left\{D_{0T}^{-1}u\left(m_{T}-1\right)-D_{0T}^{-1}u\left(m_{T}+1\right)\right\}\right.$$

$$\left. < \left(\frac{\delta_{0T}}{\delta_{0T}-1}\right)^{2}-\left(\frac{\delta_{0T}}{\delta_{0T}+1}\right)^{2}\right\}$$

$$\sim \ \Pr\left\{T^{1/2}\operatorname{Re}\left\{D_{0T}^{-1}u\left(m_{T}-1\right)-D_{0T}^{-1}u\left(m_{T}+1\right)\right\}<2\left(T^{1/4}\delta_{0T}\right)^{2}\right\}$$

$$\to \ \Phi\left(\frac{|A|}{\sqrt{2\pi f\left(\lambda\right)}}\right).$$

Thus, along the subsequence of $\{r_{j}\}$ for which $\delta_{0T}<0$, $T^{5/4}\left(\widehat{\lambda}_{T}-\lambda_{0}\right)/\left(2\pi\right)$ is asymptotically 2 with probability $1-\Phi\left(\frac{|A|}{\sqrt{2\pi f(\lambda)}}\right)$, and 0 with probability $\Phi\left(\frac{|A|}{\sqrt{2\pi f(\lambda)}}\right)$, while along the subsequence of $\{r_{j}\}$ for which $\delta_{0T}>0$, $T^{5/4}\left(\widehat{\lambda}_{T}-\lambda_{0}\right)/\left(2\pi\right)$ is asymptotically -2 with probability $1-\Phi\left(\frac{|A|}{\sqrt{2\pi f(\lambda)}}\right)$, and 0 with probability $\Phi\left(\frac{|A|}{\sqrt{2\pi f(\lambda)}}\right)$. Hence $T^{5/4}\left(\widehat{\lambda}_{T}-\lambda_{0}\right)$ does not converge in probability to zero. $\qquad\square$

We include the following corollary for interest's sake. The result is of little use as we would not expect $\lambda_0/(2\pi)$ to be rational in any application where λ_0 was to be estimated.

Corollary 2 *If $\lambda_0/(2\pi)$ is rational, then $\widehat{\lambda}_T$ has 'asymptotic variance' of order T^{-3}.*

Proof Let $\lambda_0/(2\pi) = p/q$, where p and q are relatively prime. Thus δ_{0T} is of the form k/q, where k is an integer for which $|k| \le q/2$. Hence for any $\varepsilon > 0$, $T^\varepsilon |\delta_{0T}|$ is 0 along the subsequence of the integers for which δ_{0T} is 0, and diverges to ∞ off this sequence. When $\delta_{0T} = 0$,

$$T^{3/2}\left(\widehat{\lambda}_T - \lambda_0\right)/(2\pi) \sim \begin{cases} -T^{1/2}\left|D_{0T}^{-1}u\left(m_T - 1\right)\right| & ; \quad \left|\frac{u(m_T-1)}{u(m_T+1)}\right| \ge 1 \\ T^{1/2}\left|D_{0T}^{-1}u\left(m_T + 1\right)\right| & ; \quad \left|\frac{u(m_T-1)}{u(m_T+1)}\right| < 1. \end{cases} \tag{6.9}$$

When $\delta_{0T} \ne 0$, $|\delta_{0T}| > 1/q$, and so $\Pr\{\widehat{\alpha}_T = 1\}$ converges to 1 along the subsequence of the integers for which $\delta_{0T} > 0$ and to 0 along the subsequence of the integers for which $\delta_{0T} < 0$. Thus, when $\delta_{0T} > 0$,

$$T^{3/2}\left(\widehat{\lambda}_T - \lambda_0\right)/(2\pi)$$
$$\sim -(1 - \delta_{0T})^2\, T^{1/2} \operatorname{Re}\left[D_{0T}^{-1}\left\{u\left(m_T + 1\right) - \frac{\delta_{0T}}{\delta_{0T} - 1}u\left(m_T\right)\right\}\right], \tag{6.10}$$

and when $\delta_{0T} < 0$,

$$T^{3/2}\left(\widehat{\lambda}_T - \lambda_0\right)/(2\pi)$$
$$\sim (1 + \delta_{0T})^2\, T^{1/2} \operatorname{Re}\left[D_{0T}^{-1}\left\{u\left(m_T - 1\right) - \frac{\delta_{0T}}{\delta_{0T} + 1}u\left(m_T\right)\right\}\right]. \tag{6.11}$$

Since the right sides of (6.9), (6.10) and (6.11) all converge in distribution, the 'asymptotic variance' of $\widehat{\lambda}_T - \lambda_0$ is of order T^{-3}. However, the asymptotic distribution of $\widehat{\lambda}_T - \lambda_0$ is by no means standard, even though the right sides of (6.10) and (6.11) are asymptotically normal, since the right side of (6.9) is not asymptotically normal. $\qquad\square$

Finally, since the 'problem area' of Rife and Vincent's technique is the case where δ_{0T} is close to 0, where the wrong estimator is chosen with non-zero probability, a question of interest is whether the technique may be improved by using a small amount of additional information. We thus replace the Step (ii) of Algorithm 7 with one which mimics Step (ii) of Algorithm 4, obtaining

Algorithm 8 *Modified Rife and Vincent*

(i) *Let $\widehat{m}_T$ be the maximiser of $I_y\left(2\pi\frac{m}{T}\right)$, $m = 1, 2, \ldots, \left\lfloor\frac{T-1}{2}\right\rfloor$.*

(ii) *Let $s_j = \operatorname{sgn}\operatorname{Re}\left\{z\left(\widehat{m}_T + j\right)/z\left(\widehat{m}_T\right)\right\}$ and put $\widehat{\alpha}_T = 1$ if $s_{-1} = 1$ and $s_1 = -1$. Otherwise let $\widehat{\alpha}_T = -1$.*

(iii) *Let $\widehat{\delta}_T = -s_{\widehat{\alpha}_T}\widehat{\alpha}_T\left|\frac{z(\widehat{m}_T + \widehat{\alpha}_T)}{z(\widehat{m}_T)}\right|\Big/\left\{1 - s_{\widehat{\alpha}_T}\left|\frac{z(\widehat{m}_T + \widehat{\alpha}_T)}{z(\widehat{m}_T)}\right|\right\}$.*

(iv) *Let $\widehat{\lambda}_T = 2\pi\left(\widehat{m}_T + \widehat{\delta}_T\right)\Big/T$.*

We shall not prove anything about this algorithm, but note that, although $\widehat{\lambda}_T$ does not have a normal central limit theorem, it nevertheless has an asymptotic variance which is $O\left(T^{-3}\right)$.

We note here that Luginbuhl (1999) has considered the problem of estimating the frequency of a complex sinusoid using only the moduli of the Fourier coefficients, via their likelihood. In particular, he has calculated the CRB, in order to obtain lower bounds for the variance of the best estimate which uses only these moduli. The *exact* Gaussian white CRB is difficult to obtain, of course, since, although the Fourier coefficients are independent and Gaussian, the square of their moduli are non-central χ^2, and the likelihood must therefore be written in terms of the modified Bessel function I_0. However, the *asymptotic* CRB is straightforward enough, since the moduli of the Fourier coefficients, and their squares, are asymptotically normal. Using (6.2), we obtain

$$|z\left(m + j\right)|^2 = |D_T|^2\frac{\delta_T^2}{\left(\delta_T - j\right)^2} + 2\frac{\delta_T}{\delta_T - j}\operatorname{Re}\left\{\overline{D}_T u\left(m + j\right)\right\} + O_P\left(1\right),$$

and can form the *asymptotic* likelihood by using the asymptotic distribution of the $u\left(m + j\right)$. The *best* asymptotic estimator of frequency using just these periodogram values is, incidentally, found by maximising this likelihood with respect to δ_T. The asymptotic CRB is the *same* as the asymptotic CRB using the full Fourier coefficients. It is possible that this CRB also represents the asymptotic variance of the frequency estimator, but it might be difficult to prove, since the signal and error terms in the above equation will be smaller than the neglected term $|u\left(m + j\right)|^2$ when T is such that δ_T is close to zero. It may be that the technique fails, in the same way as the Rife and Vincent technique was seen to fail. The problem will be the subject of further study.

6.3 More than one series

We suppose now that $\{y\left(0\right), \ldots, y\left(RT - 1\right)\}$ is generated by (4.1), but that only a length-T Fourier transform is to be used. This often happens with real-time signal processing equipment, where the Fourier transforms are performed in hardware and are of fixed length. For $j = -k, \ldots, l$ and $r = 0, \ldots, R - 1$, put

$$z_r \left(m + j \right) = w_{y,r} \left(2\pi \frac{m + j}{T} \right)$$

where

$$w_{y,r} \left(\lambda \right) = T^{-1/2} \sum_{t=0}^{T-1} e^{-i\lambda t} y \left(rT + t \right).$$

Then

$$z_r \left(m + j \right) = T^{1/2} \frac{A e^{i\phi}}{2} e^{i2\pi r \delta_T} \frac{e^{i2\pi \delta_T} - 1}{2\pi i \left(\delta_T - j \right)} + u_r \left(m + j \right) \qquad (6.12)$$

where

$$u_r \left(m + j \right) = w_{x,r} \left(2\pi \frac{m + j}{T} \right) + O \left(T^{-1/2} \right).$$

Equation (6.12) should be compared with (6.1). Note that all that has changed is that the phase of the deterministic component is advanced by $2\pi\delta_T$ with each time-block advance. Again letting

$$D_T = T^{1/2} \frac{A e^{i\phi} \left(e^{i2\pi \delta_T} - 1 \right)}{4\pi i \delta_T},$$

and noting that the $u_r \left(m + j \right)$ are asymptotically independent and identically distributed complex Gaussian, we may thus estimate D_T and δ_T by minimising

$$E_T \left(D_T, \delta_T \right) = \sum_{r=0}^{R-1} \sum_{j=-k}^{l} \left| z_r \left(\widehat{m}_T + j \right) - D_T e^{i2\pi r \delta_T} \frac{\delta_T}{\delta_T - j} \right|^2, \qquad (6.13)$$

or, equivalently, by maximising

$$S_T \left(\delta_T \right) = \frac{\left| \sum_{r=0}^{R-1} \sum_{j=-k}^{l} z_r \left(\widehat{m}_T + j \right) e^{-i2\pi r \delta_T} \frac{\delta_T}{\delta_T - j} \right|^2}{R \sum_{j=-k}^{l} \left(\frac{\delta_T}{\delta_T - j} \right)^2} \qquad (6.14)$$

with respect to δ_T and putting

$$\widehat{D}_T = \frac{\sum_{r=0}^{R-1} \sum_{j=-k}^{l} z_r \left(\widehat{m}_T + j \right) e^{-i2\pi r \delta_T} \frac{\delta_T}{\delta_T - j}}{R \sum_{j=-k}^{l} \left(\frac{\delta_T}{\delta_T - j} \right)^2},$$

where $\widehat{\delta}_T$ is the maximiser of $S_T(\delta_T)$. Obviously $S_T(\delta_T)$ is highly nonlinear in δ_T, and must therefore be maximised using an iterative technique. Now

$$\sum_{r=0}^{R-1} \sum_{j=-k}^{l} z_r(\widehat{m}_T + j) e^{-i2\pi r \delta_T} \frac{\delta_T}{\delta_T - j}$$

$$= \sum_{j=-k}^{l} \frac{\delta_T}{\delta_T - j} \left\{ \sum_{r=0}^{R-1} z_r(\widehat{m}_T + j) e^{-i2\pi r \delta_T} \right\},$$

and $\sum_{r=0}^{R-1} z_r(\widehat{m}_T + j) e^{-i2\pi r \delta_T}$ is just the Fourier transform of the Fourier coefficients corresponding to the same frequency but from different time blocks. This may be calculated (for each fixed j) at a large number of values of $\delta_T \in \left[-\frac{1}{2}, \frac{1}{2}\right]$ by zero-padding the $z_r(\widehat{m}_T + j)$ out to a large size (of length T if the fixed-length Fourier transform is to be used) and using the FFT. Replicates of these are used when $\delta_T \notin \left[-\frac{1}{2}, \frac{1}{2}\right]$ and these may then be multiplied by the $\frac{\delta_T}{\delta_T - j}$, which should have been stored in 'lookup tables' to calculate and maximise $S_T(\delta_T)$. The estimator may then be used as an initial estimator for a Gauss–Newton technique which finds the nearest zero of the derivative of $S_T(\delta_T)$. However, as long as $R = o\left(T^{1/3}\right)$, $S_T(\delta_T)$ will have been calculated on a fine enough grid that no iteration is necessary, as the estimator of λ_0 is closer in order to the maximiser of $S_T(\delta_T)$ than the latter is to λ_0. Alternatively, estimators of δ_{0T} could be constructed from each of the time blocks using three Fourier coefficients, and averaged to produce a better estimator. Note, however, that the resulting estimator of λ_0 would have an asymptotic variance of order $O\left(R^{-1}T^{-3}\right)$, which is much larger than the CRB, which is $O\left(R^{-3}T^{-3}\right)$. Moreover, if the SNR is low, it may be that estimators from each block may be worse ('thresholding' may be evident) than the above estimator.

There remains the problem of determining the asymptotic behaviour of the estimators. The behaviour of $\widehat{\lambda}_T = 2\pi\left(\widehat{m}_T + \widehat{\delta}_T\right)\big/ T$ can be obtained either using (6.13) or (6.14). We have decided to use the former, since although we could use the latter to obtain directly the results for $\widehat{\lambda}_T$, the algebra in the former case is a little simpler. That $T\left(\widehat{\lambda}_T - \lambda_{0T}\right) \to 0$, almost surely, follows in the same way as does the result for the single series. This implies the strong consistency of the estimators of all the suitably-scaled parameters. Let $D_{Tr} = \mathrm{Re}\,(D_T)$ and $D_{Ti} = \mathrm{Im}\,(D_T)$. Put

$$\psi_{0T} = \left[\begin{array}{ccc} T^{-1/2}D_{0Tr} & T^{-1/2}D_{0Ti} & \delta_{0T} \end{array}\right]'$$

and

$$\widehat{\psi}_T = \left[\ T^{-1/2}\widehat{D}_{Tr} \quad T^{-1/2}\widehat{D}_{Ti} \quad \widehat{\delta}_T\ \right]'.$$

Then the central limit theorem follows in the usual way: letting ψ_i denote the ith element of ψ, we have

$$0 = \frac{\partial}{\partial \psi_i} E_T\left(\widehat{\psi}_T\right) = \frac{\partial}{\partial \psi_i} E_T\left(\psi_{0T}\right) + \frac{\partial^2}{\partial \psi_i \partial \psi'} E_T\left(\widetilde{\psi}_{iT}\right)\left(\widehat{\psi}_T - \psi_{0T}\right),$$

where the $\widetilde{\psi}_{iT}$ lie on the line segment joining ψ_{0T} and $\widehat{\psi}_T$. Since $\left|\widehat{\psi}_T - \psi_{0T}\right|$ converges almost surely to zero, it follows that

$$\widehat{\psi}_T - \psi_{0T} \sim -\left\{\frac{\partial^2}{\partial \psi \partial \psi'} E_T\left(\psi_{0T}\right)\right\}^{-1} \frac{\partial}{\partial \psi} E_T\left(\psi_{0T}\right).$$

Also, it clearly does not matter if we assume that $\widehat{m}_T = m_T$. Now,

$$\frac{\partial}{\partial D_{Tr}} E_T\left(\psi\right)$$
$$= -2\sum_{r=0}^{R-1}\sum_{j=-k}^{l} \text{Re}\left[\left\{z_r\left(m_T+j\right) - D_T e^{i2\pi r\delta_T}\frac{\delta_T}{\delta_T - j}\right\} e^{-i2\pi r\delta_T}\frac{\delta_T}{\delta_T - j}\right],$$

$$\frac{\partial}{\partial D_{Ti}} E_T\left(\psi\right)$$
$$= 2\sum_{r=0}^{R-1}\sum_{j=-k}^{l} \text{Re}\left[\left\{z_r\left(m_T+j\right) - D_T e^{i2\pi r\delta_T}\frac{\delta_T}{\delta_T - j}\right\} i e^{-i2\pi r\delta_T}\frac{\delta_T}{\delta_T - j}\right]$$

and

$$\frac{\partial}{\partial \delta_T} E_T\left(\psi\right) = -2\sum_{r=0}^{R-1}\sum_{j=-k}^{l} \text{Re}\left[\left\{z_r\left(m_T+j\right) - D_T e^{i2\pi r\delta_T}\frac{\delta_T}{\delta_T - j}\right\}\overline{D}_T \right.$$
$$\left. \times\left\{-i2\pi r e^{-i2\pi r\delta_T}\frac{\delta_T}{\delta_T - j} - e^{-i2\pi r\delta_T}\frac{j}{(\delta_T - j)^2}\right\}\right].$$

Thus

$$\frac{\partial}{\partial D_{Tr}} E_T\left(\psi_{0T}\right) = -2\sum_{r=0}^{R-1}\sum_{j=-k}^{l} \text{Re}\left\{u_r\left(m_T+j\right) e^{-i2\pi r\delta_{0T}}\frac{\delta_{0T}}{\delta_{0T} - j}\right\},$$

$$\frac{\partial}{\partial D_{Ti}} E_T\left(\psi_{0T}\right) = 2\sum_{r=0}^{R-1}\sum_{j=-k}^{l} \text{Re}\left\{u_r\left(m_T+j\right) i e^{-i2\pi r\delta_{0T}}\frac{\delta_{0T}}{\delta_{0T} - j}\right\}$$

and

$$\frac{\partial}{\partial \delta_T} E_T \left(\psi_{0T} \right) = -2 \sum_{r=0}^{R-1} \sum_{j=-k}^{l} \mathrm{Re} \Bigg[u_r \left(m_T + j \right) \overline{D}_{0T}$$

$$\times \left\{ -i2\pi r e^{-i2\pi r \delta_{0T}} \frac{\delta_{0T}}{\delta_{0T} - j} - e^{-i2\pi r \delta_{0T}} \frac{j}{(\delta_{0T} - j)^2} \right\} \Bigg].$$

Let $v_r \left(m_T + j \right) = u_r \left(m_T + j \right) e^{-i2\pi r \delta_{0T}}$, $v_{rr} \left(m_T + j \right) = \mathrm{Re} \left\{ v_r \left(m_T + j \right) \right\}$ and $v_{ri} \left(m_T + j \right) = \mathrm{Im} \left\{ v_r \left(m_T + j \right) \right\}$. Then, for $j = -k, \ldots, l$ and $r = 0, \ldots, R-1$, the $v_{rr} \left(m_T + j \right)$ and $v_{ri} \left(m_T + j \right)$ are asymptotically independent and Gaussian with means zero and variances $\pi f_x \left(\lambda_0 \right)$,

$$\frac{\partial}{\partial D_{Tr}} E_T \left(\psi_{0T} \right) = -2 \sum_{r=0}^{R-1} \sum_{j=-k}^{l} v_{rr} \left(m_T + j \right) \frac{\delta_{0T}}{\delta_{0T} - j},$$

$$\frac{\partial}{\partial D_{Ti}} E_T \left(\psi_{0T} \right) = -2 \sum_{r=0}^{R-1} \sum_{j=-k}^{l} v_{ri} \left(m_T + j \right) \frac{\delta_{0T}}{\delta_{0T} - j}$$

and

$$\frac{\partial}{\partial \delta_T} E_T \left(\psi_{0T} \right)$$

$$= -2 \sum_{r=0}^{R-1} \sum_{j=-k}^{l} \mathrm{Re} \left[v_r \left(m_T + j \right) \overline{D}_{0T} \left\{ -i2\pi r \frac{\delta_{0T}}{\delta_{0T} - j} - \frac{j}{(\delta_{0T} - j)^2} \right\} \right]$$

$$= -2 \sum_{r=0}^{R-1} \sum_{j=-k}^{l} \Bigg[2\pi r \left\{ D_{0Tr} v_{ri} \left(m_T + j \right) - D_{0Ti} v_{rr} \left(m_T + j \right) \right\} \frac{\delta_{0T}}{\delta_{0T} - j}$$

$$- \left\{ D_{0Tr} v_{rr} \left(m_T + j \right) + D_{0Ti} v_{ri} \left(m_T + j \right) \right\} \frac{j}{(\delta_{0T} - j)^2} \Bigg].$$

Hence, $T^{1/2} \Omega_T^{-1/2} \frac{1}{2} \frac{\partial}{\partial \psi} E_T \left(\psi_{0T} \right)$ is asymptotically normally distributed with mean zero and covariance matrix $\pi f_x \left(\lambda \right)$ times the identity, where

$$\Omega_{T,11} = \Omega_{T,22} = R \sum_{j=-k}^{l} \left(\frac{\delta_{0T}}{\delta_{0T} - j} \right)^2 = RU_1, \quad \Omega_{T,12} = 0,$$

$$\Omega_{T,13} = T^{-1/2} \sum_{r=0}^{R-1} \sum_{j=-k}^{l} \left\{ -2\pi r D_{0Ti} \left(\frac{\delta_{0T}}{\delta_{0T} - j} \right)^2 - D_{0Tr} \frac{j \delta_{0T}}{(\delta_{0T} - j)^3} \right\}$$

$$= -\pi R \left(R - 1 \right) T^{-1/2} D_{0Ti} U_1 - R T^{-1/2} D_{0Tr} U_2,$$

$$\Omega_{T,23} = T^{-1/2} \sum_{r=0}^{R-1} \sum_{j=-k}^{l} \left\{ 2\pi r D_{0Tr} \left(\frac{\delta_{0T}}{\delta_{0T} - j} \right)^2 - D_{0Ti} \frac{j\delta_{0T}}{(\delta_{0T} - j)^3} \right\}$$

$$= \pi R(R-1) T^{-1/2} D_{0Tr} U_1 - R T^{-1/2} D_{0Ti} U_2,$$

$$\Omega_{T,33} = T^{-1} |D_{0T}|^2 \sum_{r=0}^{R-1} \sum_{j=-k}^{l} \left\{ 4\pi^2 r^2 \left(\frac{\delta_{0T}}{\delta_{0T} - j} \right)^2 + \frac{j^2}{(\delta_{0T} - j)^4} \right\}$$

$$= \frac{2}{3} \pi^2 R(R-1)(2R-1) T^{-1} |D_{0T}|^2 U_1 + R T^{-1} |D_{0T}|^2 U_3,$$

and

$$U_1 = \sum_{j=-k}^{l} \left(\frac{\delta_{0T}}{\delta_{0T} - j} \right)^2, \quad U_2 = \sum_{j=-k}^{l} \frac{j\delta_{0T}}{(\delta_{0T} - j)^3}, \quad U_3 = \sum_{j=-k}^{l} \frac{j^2}{(\delta_{0T} - j)^4}.$$

Also,

$$\frac{\partial^2}{(\partial D_{Tr})^2} E_T(\psi_{0T}) = -2 \sum_{r=0}^{R-1} \sum_{j=-k}^{l} \mathrm{Re} \left(-e^{i2\pi r \delta_{0T}} \frac{\delta_{0T}}{\delta_{0T} - j} e^{-i2\pi r \delta_{0T}} \frac{\delta_{0T}}{\delta_{0T} - j} \right)$$

$$= 2R U_1,$$

$$\frac{\partial^2}{(\partial D_{Ti})^2} E_T(\psi_{0T}) = 2 \sum_{r=0}^{R-1} \sum_{j=-k}^{l} \mathrm{Re} \left(-i e^{i2\pi r \delta_{0T}} \frac{\delta_{0T}}{\delta_{0T} - j} i e^{-i2\pi r \delta_{0T}} \frac{\delta_{0T}}{\delta_{0T} - j} \right)$$

$$= 2R U_1,$$

$$\frac{\partial^2}{\partial D_{Tr} \partial D_{Ti}} E_T(\psi_{0T})$$

$$= -2 \sum_{r=0}^{R-1} \sum_{j=-k}^{l} \mathrm{Re} \left(-i e^{i2\pi r \delta_{0T}} \frac{\delta_{0T}}{\delta_{0T} - j} e^{-i2\pi r \delta_{0T}} \frac{\delta_{0T}}{\delta_{0T} - j} \right)$$

$$= 0,$$

$$\frac{\partial^2}{\partial D_{Tr} \partial \delta_T} E_T(\psi_{0T})$$

$$= -2 \sum_{r=0}^{R-1} \sum_{j=-k}^{l} \mathrm{Re} \left[\left\{ u_r(m_T + j) - \overline{D}_{0T} e^{i2\pi r \delta_{0T}} \frac{\delta_{0T}}{\delta_{0T} - j} \right\} \right.$$

$$\left. \times \left\{ -i2\pi r e^{-i2\pi r \delta_{0T}} \frac{\delta_{0T}}{\delta_{0T} - j} - e^{-i2\pi r \delta_{0T}} \frac{j}{(\delta_{0T} - j)^2} \right\} \right]$$

$$= 2 \sum_{r=0}^{R-1} \sum_{j=-k}^{l} \left\{ -2\pi r D_{0Ti} \left(\frac{\delta_{0T}}{\delta_{0T} - j} \right)^2 - D_{0Tr} \frac{j\delta_{0T}}{(\delta_{0T} - j)^3} \right\} \{1 + o_P(1)\}$$

$$= \left\{ -2\pi R(R-1) T^{-1/2} D_{0Ti} U_1 - 2R T^{-1/2} D_{0Tr} U_2 \right\} \{1 + o_P(1)\},$$

$$\frac{\partial^2}{\partial D_{Ti}\partial\delta_T}E_T\left(\psi_{0T}\right)$$

$$= -2\sum_{r=0}^{R-1}\sum_{j=-k}^{l}\text{Re}\left[-i\left\{u_r\left(m_T+j\right)+\overline{D}_{0T}e^{i2\pi r\delta_{0T}}\frac{\delta_{0T}}{\delta_{0T}-j}\right\}\right.$$

$$\left.\times\left\{-i2\pi r e^{-i2\pi r\delta_{0T}}\frac{\delta_{0T}}{\delta_{0T}-j}-e^{-i2\pi r\delta_{0T}}\frac{j}{(\delta_{0T}-j)^2}\right\}\right]$$

$$= 2\sum_{r=0}^{R-1}\sum_{j=-k}^{l}\left\{2\pi r D_{0Tr}\left(\frac{\delta_{0T}}{\delta_{0T}-j}\right)^2-D_{0Ti}\frac{j\delta_{0T}}{(\delta_{0T}-j)^3}\right\}\left\{1+o_P\left(1\right)\right\}$$

$$= \left\{2\pi R\left(R-1\right)T^{-1/2}D_{0Tr}U_1-2RT^{-1/2}D_{0Ti}U_2\right\}\left\{1+o_P\left(1\right)\right\}$$

and

$$\frac{\partial^2}{(\partial\delta_T)^2}E_T\left(\psi_{0T}\right)$$

$$= -2\sum_{r=0}^{R-1}\sum_{j=-k}^{l}\text{Re}\left[\left\{z_r\left(m_T+j\right)-D_{0T}e^{i2\pi r\delta_{0T}}\frac{\delta_{0T}}{\delta_{0T}-j}\right\}\overline{D}_{0T}\right.$$

$$\times\frac{\partial}{\partial\delta_{0T}}\left\{-i2\pi r e^{-i2\pi r\delta_T}\frac{\delta_{0T}}{\delta_{0T}-j}-e^{-i2\pi r\delta_T}\frac{j}{(\delta_{0T}-j)^2}\right\}\right]$$

$$\times\left[-D_{0T}e^{i2\pi r\delta_{0T}}\left\{i2\pi r\frac{\delta_{0T}}{\delta_{0T}-j}-\frac{j}{(\delta_{0T}-j)^2}\right\}\overline{D}_{0T}\right.$$

$$\left.\times\left\{-i2\pi r e^{-i2\pi r\delta_T}\frac{\delta_{0T}}{\delta_{0T}-j}-e^{-i2\pi r\delta_T}\frac{j}{(\delta_{0T}-j)^2}\right\}\right]$$

$$= 2\left|D_{0T}\right|^2\sum_{r=0}^{R-1}\sum_{j=-k}^{l}\left\{4\pi^2r^2\left(\frac{\delta_{0T}}{\delta_{0T}-j}\right)^2+\frac{j^2}{(\delta_{0T}-j)^4}\right\}\left\{1+o_P\left(1\right)\right\}$$

$$= \left\{\frac{4}{3}\pi^2R\left(R-1\right)\left(2R-1\right)T^{-1}\left|D_{0T}\right|^2U_1+2RT^{-1}\left|D_{0T}\right|^2U_3\right\}\left\{1+o_P\left(1\right)\right\}.$$

Thus

$$T^{-1}\frac{\partial^2}{\partial\psi\partial\psi'}E_T\left(\psi_{0T}\right)=2\Omega_T\left\{1+o_P\left(1\right)\right\}.$$

Hence, $T^{1/2}\Omega_T^{1/2}\left(\widehat{\psi}_T-\psi_{0T}\right)$ is asymptotically normally distributed with mean zero and covariance matrix $\pi f_x\left(\lambda_0\right)$ times the identity. The asymptotic covariance matrix of

$$T^{1/2}\left(\widehat{\theta}_T-\theta_{0T}\right)=T^{1/2}\left[\begin{array}{ccc}\widehat{A}_T-A_0 & \widehat{\phi}_T-\phi_0 & T\left(\widehat{\lambda}_T-\lambda_0\right)\end{array}\right]'$$

is thus $\pi f_x (\lambda_0) \Sigma$, where

$$\Sigma = \left(\frac{\partial \psi'}{\partial \theta} \Omega_T \frac{\partial \psi}{\partial \theta'} \right)^{-1},$$

and we have

Theorem 29 $\widehat{\theta}_T - \theta_{0T}$ *converges almost surely to 0, and the distribution of* $T^{1/2} \Sigma^{-1/2} \left(\widehat{\theta}_T - \theta_{0T} \right)$ *converges to the normal with mean zero and covariance matrix* $\pi f_x (\lambda_0)$ *times the identity.*

A little algebra shows that Σ^{-1} is given by

$$v_T \begin{bmatrix} RU_1/A_0^2 & 0 & 2\pi R (\tau U_1 - U_2)/A_0^2 \\ 0 & RU_1 & 2\pi^2 R^2 U_1 \\ 2\pi R (\tau U_1 - U_2)/A_0^2 & 2\pi^2 R^2 U_1 & 4\pi^2 RV \end{bmatrix}$$

where

$$v_T = T^{-1} |D_{0T}|^2, \; \tau = \frac{\dfrac{d}{d\delta_{0T}} \dfrac{\sin(\pi \delta_{0T})}{\pi \delta_{0T}}}{\dfrac{\sin(\pi \delta_{0T})}{\pi \delta_{0T}}} = \pi \cot(\pi \delta_{0T}) - \frac{1}{\delta_{0T}}$$

and

$$V = \frac{\pi^2}{3} \left(4R^2 - 1 \right) U_1 + \tau^2 U_1 - 2\tau U_2 + U_3.$$

Thus $\Sigma = (\Sigma_{ij})$, where

$$\begin{aligned} \Sigma_{33} &= 4\pi^2 R^{-1} v_T^{-1} \left\{ \frac{\pi^2}{3} \left(4R^2 - 1 \right) U_1 + \tau^2 U_1 - 2\tau U_2 + U_3 \right. \\ & \left. - \begin{bmatrix} \frac{\tau U_1 - U_2}{A_0} & \pi R U_1 \end{bmatrix} \begin{bmatrix} \frac{U_1}{A_0^2} & 0 \\ 0 & U_1 \end{bmatrix}^{-1} \begin{bmatrix} \frac{\tau U_1 - U_2}{A_0} \\ \pi R U_1 \end{bmatrix} \right\}^{-1} \\ &= \frac{4\pi^2 R^{-1} v_T^{-1}}{\frac{\pi^2}{3} \left(R^2 - 1 \right) U_1 + U_3 - \frac{U_2^2}{U_1}}, \end{aligned}$$

$$\begin{bmatrix} \Sigma_{13} \\ \Sigma_{23} \end{bmatrix} = -\frac{2\pi R^{-1} v_T^{-1}}{\frac{\pi^2}{3} \left(R^2 - 1 \right) U_1 + U_3 - \frac{U_2^2}{U_1}} \begin{bmatrix} A_0 \frac{\tau U_1 - U_2}{U_1} \\ \pi R \end{bmatrix},$$

and

$$\begin{bmatrix} \Sigma_{11} & \Sigma_{12} \\ \Sigma_{21} & \Sigma_{22} \end{bmatrix} =$$

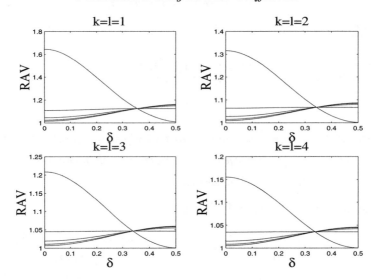

Fig. 6.2. Comparison of relative asymptotic variances for $R = 1, \ldots, 5$

$$R^{-1}v_T^{-1}\left\{\begin{bmatrix} \frac{A_0^2}{U_1} & 0 \\ 0 & \frac{1}{U_1} \end{bmatrix}\right.$$

$$+ \frac{1}{\frac{\pi^2}{3}(R^2-1)U_1 + U_3 - \frac{U_2^2}{U_1}}\left.\begin{bmatrix} \frac{A_0^2(\tau U_1 - U_2)^2}{U_1^2} & \pi R\frac{A_0(\tau U_1 - U_2)}{U_1} \\ \pi R\frac{A_0(\tau U_1 - U_2)}{U_1} & \pi^2 R^2 \end{bmatrix}\right\}.$$

The asymptotic variance of the frequency estimator, relative to the maximiser of the periodogram of $\{y(0), \ldots, y(TR-1)\}$ is thus

$$\frac{4\pi^2 R^{-1} v_T^{-1} A_0^2 R^3}{48\left\{\frac{\pi^2}{3}(R^2-1)U_1 + U_3 - \frac{U_2^2}{U_1}\right\}}$$

$$= \frac{\pi^2 R^2}{\pi^2(R^2-1)U_1 + 3\left(U_3 - \frac{U_2^2}{U_1}\right)} \frac{\pi^2 \delta_{0T}^2}{\sin^2(\pi\delta_{0T})}.$$

Figure 6.2 depicts this ratio for the cases where $k = l = K$, $K = 1, \ldots, 4$. Each subfigure contains the plots for $R = 1, \ldots, 5$. For each subfigure, the relative asymptotic variance (RAV) decreases with R at $\delta = \delta_{0T} = 0$ but increases with R at $\delta = \delta_{0T} = 0.5$. Of special interest is the fact that, while the asymptotic variance decreases with δ for $R = 1$, for all K, it increases with δ for $R \geq 3$, for all K, and for the case $R = 2$, for which it is flattest, it increases and then decreases. The code which generated the figure is contained in the Appendix in the files **asvsc.m** and **asvar.m**,

7

___

Tracking Frequency in Low SNR Conditions

7.1 Introduction

Chapter 6 was concerned with the estimation of frequency using only the Fourier coefficients of a single time series or the Fourier coefficients of consecutive segments of time series. Implicit was the assumption that the frequency was fixed. Of course, the frequency may wander in real applications. If the instantaneous frequency changes substantially under conditions of low SNR, there is not much that can be done to track the frequency as it changes. However, if the frequency is changing so slowly that it may be considered fixed within time blocks that are large enough to estimate it, the instantaneous frequency could be tracked simply by estimating it independently over time blocks, or groups of time blocks, using the methods of Chapter 6. We could thus form, for example, for

$$j = -k, \ldots, l; \ r = 0, \ldots, R-1; \ q = 0, \ldots, Q-1,$$

the quantities

$$z_{r,q}\left(\widehat{m}_{T,q} + j\right) = w_{y,r,q}\left(2\pi \frac{\widehat{m}_{T,q} + j}{T}\right)$$

where

$$w_{y,r,q}\left(\lambda\right) = T^{-1/2} \sum_{t=0}^{T-1} e^{-i\lambda t} y\left(qRT + rT + t\right)$$

and

$$\widehat{m}_{T,q} = \arg\max_m \sum_{r=0}^{R-1} \left|w_{y,r,q}\left(2\pi \frac{m}{T}\right)\right|^2$$

and track λ through time by estimating $\{\lambda_q, \ q = 0, \ldots, Q-1\}$ separately from the sets $\{z_{r,q}\left(\widehat{m}_{T,q} + j\right); \ j = -k, \ldots, l; \ r = 0, \ldots, R-1\}$. Here we im-

215

plicitly assume that the frequency is fixed within each of Q groups of R time blocks, each of length T. Such methods, however, would not take into account the transfer of phase between time blocks or the constancy of the amplitude A. It is to be expected, therefore, that taking these factors into account will improve the accuracy of the track and decrease the chance of 'thresholding'. We shall discuss such a technique in this chapter. But even this may not be sufficient for very low SNR. What is not built into the likelihood function is a device which prefers small changes in frequency to large ones. If it were known *a priori* that any change in frequency was likely to be small, we could maximise the likelihood under constraints, or penalise the likelihood for large relative excursions. We might accomplish the latter by constructing the likelihood under the assumption that the frequencies were dependent random variables. Such a likelihood would be extremely difficult to maximise unless these assumptions were simple. If the frequencies are assumed *Markovian* and discrete-valued, then our model can be expressed as a Hidden Markov Model (HMM), on which there is a wealth of literature. Moreover, there are quite general computationally efficient techniques for the estimation of the parameters in HMMs, and for the tracking of the hidden Markov process. After discussing the simple maximum likelihood tracker in Section 7.2, we introduce the general theory of hidden Markov model estimation and tracking in Section 7.3, and apply this to the tracking of frequency in Section 7.4. In Section 7.5, we apply the technique to real data where there is an evident Doppler effect, while in Section 7.6, we examine a number of simulations where the frequency tracks have been simulated using a Markov model.

7.2 Maximum likelihood tracking

Let $\lambda_q = 2\pi \frac{m_T + \delta_{T,q}}{T}$, where the $|\delta_{T,q}|$ are small, but not necessarily $\leq 1/2$. Then

$$
\begin{aligned}
z_{r,q}&(m_T + j) \\
&= T^{1/2} \frac{Ae^{i\phi}}{2} e^{2\pi i \{R \sum_{s=0}^{q} \delta_{T,s} + (r-R)\delta_{T,q}\}} \frac{e^{i2\pi\delta_{T,q}} - 1}{2\pi i (\delta_{T,q} - j)} + u_{r,q}(m_T + j) \\
&= D_T e^{2\pi i \{R \sum_{s=0}^{q-1} \delta_{T,s} + (r+\frac{1}{2})\delta_{T,q}\}} \frac{\sin(\pi\delta_{T,q})}{\pi(\delta_{T,q} - j)} + u_{r,q}(m_T + j), \qquad (7.1)
\end{aligned}
$$

where for any sequence $\{x_j\}$, $\sum_{j=0}^{-1} x_j$ is zero by definition,

$$
u_{r,q}(m_T + j) = w_{x,r,q}\left(2\pi \frac{m_T + j}{T}\right) + O\left(T^{-1/2}\right)
$$

and

$$D_T = T^{1/2}\frac{Ae^{i\phi}}{2}.$$

The (asymptotic) maximum likelihood estimators of the λ_q may thus be found by minimising

$$E_T\left(D_T, \delta_T\right) = \sum_{q=0}^{Q-1}\sum_{r=0}^{R-1}\sum_{j=-k}^{l}\left|z_{r,q}\left(\widehat{m}_T + j\right) - D_T e^{2\pi i\phi(\delta_T, q, r)}\frac{\sin\left(\pi\delta_{T,q}\right)}{\pi\left(\delta_{T,q} - j\right)}\right|^2,$$

where

$$\phi\left(\delta_T, q, r\right) = R\sum_{s=0}^{q-1}\delta_{T,s} + \left(r + \frac{1}{2}\right)\delta_{T,q},$$

$\delta_T = \begin{bmatrix} \delta_{T,1} & \cdots & \delta_{T,q} \end{bmatrix}'$ and $\widehat{m}_T$ is an estimator of m_T. This may be done, first by fixing δ_T and minimising $E_T\left(D_T, \delta_T\right)$ with respect to D_T, and then maximising

$$S_T\left(\delta_T\right) = \frac{\left|\sum_{q=0}^{Q-1}\sum_{r=0}^{R-1}\sum_{j=-k}^{l} z_{r,q}\left(\widehat{m}_T + j\right) e^{-2\pi i\phi(\delta_T, q, r)}\frac{\sin\left(\pi\delta_{T,q}\right)}{\pi\left(\delta_{T,q} - j\right)}\right|^2}{R\sum_{q=0}^{Q-1}\sum_{j=-k}^{l}\left\{\frac{\sin\left(\pi\delta_{T,q}\right)}{\pi\left(\delta_{T,q} - j\right)}\right\}^2}.$$

(7.2)

This appears to be a fairly complicated function of the $\delta_{T,q}$, but could be maximised using standard numerical techniques. For example, with initial estimates given by the values of δ_T obtained by maximising (6.14), separately for each q, we could numerically solve

$$\frac{\partial}{\partial\delta_T}S_T\left(\delta_T\right) = 0$$

using a Gauss–Newton type of method or some search technique. We have been unable, however, to achieve the (numerical) convergence of such techniques consistently, and especially with low SNR and high QR. We may think of this problem as a special case of the HMM problem, with uniform initial probabilities and transition probability matrix, the details of which will be discussed later. Note, again, that the replacement of m_T by $\widehat{m}_T$ does not matter asymptotically as long as the variation in instantaneous frequency is fairly small, i.e. stays within an interval $\left(2\pi\frac{m_T-k}{T}, 2\pi\frac{m_T+l}{T}\right)$.

7.3 Hidden Markov models

We digress here to discuss the general theory of HMMs. These have been used successfully in speech processing, and extended over the past decade to problems in tracking and communications. They have also appeared relatively recently in the statistical literature. The articles of Ferguson (1980) and Rabiner and Juang (1986) are recommended, although the following discussion is fairly self-contained. It is supposed that the system we wish to model is generated by a discrete time, discrete state space homogeneous Markov chain $\{X_t\}$, which is not observable. What *is* observed is a sequence of random vectors $\{Y_t\}$ which has the property that, conditional on $\{X_0, \ldots, X_{T-1}\}$, $\{Y_0, \ldots, Y_{T-1}\}$ is an independent sequence with the distribution of Y_j depending only on X_j and a finite-dimensional parameter θ_1. Let θ_2 be the vector of unknown parameters describing the distribution of $\{X_t\}$. This could be, for example, the initial probability distribution π for $\{X_t\}$ and its matrix of transition probabilities A. The conditional probability density function or probability function of Y given X is thus given by

$$f_{Y|X}\left(y \mid x; \theta_1\right) = \prod_{j=0}^{T-1} f_{Y_j|X_j}\left(y_j \mid x_j; \theta_1\right),$$

where $Y = \left[\begin{array}{ccc} Y_0' & \cdots & Y_{T-1}' \end{array}\right]'$ and $X = \left[\begin{array}{ccc} X_0 & \cdots & X_{T-1} \end{array}\right]'$.

We note here that there has been recent work on the development of continuous state hidden Markov models for Gaussian and Gaussian-mixture distributions, by Ainsleigh, Kehtarnavaz and Streit (2000). The continuous analogues of the discrete-state algorithms are presented, and their relationship with existing Kalman smoothing techniques and EM algorithms described. The techniques should be readily applicable to the low SNR frequency tracking scenario we describe in Section 7.4, since the Fourier coefficients are asymptotically Gaussian, and since a Gaussian Markov chain or random walk is suitable as a model for the frequency states. In fact, one of the models we describe there, and the one we have used in practice, assumes that the discrete frequencies form a random walk with innovations which are discretised normal random variables.

7.3.1 The Viterbi track and the Viterbi algorithm

The Viterbi, or HMM, track is defined as that value of

$$x = \left[\begin{array}{ccc} x_0 & \cdots & x_{T-1} \end{array}\right]'$$

which maximises

$$f_{X|Y}(x \mid y; \theta),$$

where $\theta = \begin{bmatrix} \theta_1' & \theta_2' \end{bmatrix}'$, or, equivalently, which maximises

$$f_{X,Y}(x, y; \theta). \tag{7.3}$$

For θ considered known, and supposing that the dimension of the state space is d, the domain over which (7.3) is maximised has d^T points, which soon becomes extremely large for even moderately large values of d and T. Advantage may be taken of the peculiar structure, however, and the Viterbi algorithm, which originated as the solution of a dynamic programming problem in communications theory, may be used to compute the Viterbi track. The details follow for the sake of completeness. Denote by $\{s_k\}$ the states of $\{X_t\}$, let $\pi_i = f_{X_0}(s_i)$, $A_{ij} = f_{X_{t+1}|X_t}(s_j \mid s_i)$ and $b_{it} = f_{Y_t|X_t}(y_t \mid s_i; \theta_1)$, with $\{y_t\}$ the observed data.

Algorithm 9 *The Viterbi Algorithm*

(i) *For $i = 1, \ldots, d$, let $\delta_0(i) = \log \pi_i + \log b_{i0}$.*

(ii) *For $t = 1, \ldots, T-1$ calculate, for $i = 1, \ldots, d$,*

$$z_t(i) = \arg\max_j \{\delta_{t-1}(j) + \log A_{ji}\}$$

$$\delta_t(i) = \delta_{t-1}\{z_t(i)\} + \log A_{z_t(i),i} + \log b_{it}.$$

(iii) *Put*

$$q_{T-1} = \arg\max_j \delta_{T-1}(j)$$

and, for $t = T-2, \ldots, 0$, let

$$q_t = z_{t+1}(q_{t+1}).$$

The Viterbi track is then given by $\begin{bmatrix} s_{q_0} & \cdots & s_{q_{T-1}} \end{bmatrix}'$.

Thus, instead of the anticipated d^T comparisons, the number of operations is seen to be of order at most Td^2 operations. If d is large, this can be reduced to order $Td\log_2 d$ by binary search for each maximisation. It may be seen from the nature of the operation that a crucial property for the model to satisfy is that, conditional on $\{X_0, \ldots, X_{T-1}\}$, $\{Y_0, \ldots, Y_{T-1}\}$ is an independent sequence with the distribution of Y_t depending only on X_t, and not on any other X_s. This will be important when we formulate the HMM model for frequency tracking.

7.3.2 The EM algorithm

We can use the Viterbi algorithm once we know θ. However, θ is not known and must therefore be estimated. This estimation problem fits into the framework in which the rather more general Expectation-Maximisation (EM) algorithm may be used, but the twist here is that the hidden states are from a Markov chain and the problem is therefore more complicated than usual. The *forward–backward* equations have been developed to reduce the computational load. The algorithm is often referred to in the signal processing literature as the Baum–Welch re-estimation formula. We present here a slightly different version from the usual one, as in the original context Y was discrete and the conditional probability function itself to be estimated.

The maximum likelihood estimator of θ given Y is the maximiser of

$$
\begin{aligned}
L\left(\theta;Y\right) \\
&= \left. \sum_x f_{X,Y}\left(x,y;\theta\right) \right|_{y=Y} \\
&= \sum_x \prod_{j=0}^{T-1} f_{Y_j|X_j}\left(Y_j \mid x_j;\theta_1\right) f_{X_0}\left(x_0;\theta_2\right) \prod_{j=1}^{T-1} f_{X_j|X_{j-1}}\left(x_j \mid x_{j-1};\theta_2\right).
\end{aligned}
$$

This clearly computationally onerous task is reduced by using the EM algorithm. Suppose an estimator $\widehat{\theta}_k = \begin{bmatrix} \widehat{\theta}'_{k,1} & \widehat{\theta}'_{k,2} \end{bmatrix}'$ of θ is available. In particular, it is assumed that a good initial estimator $\widehat{\theta}_0$ may be found using other techniques. Then an estimator $\widehat{\theta}_{k+1}$, which increases $L\left(\theta;Y\right)$, is the value of θ which maximises

$$
\begin{aligned}
Q\left(\widehat{\theta}_k,\theta\right) &= \sum_x f_{X,Y}\left(x,Y;\widehat{\theta}_k\right) \log f_{X,Y}\left(x,Y;\theta\right) \\
&= \sum_x f_{X,Y}\left(x,Y;\widehat{\theta}_k\right) \sum_{j=0}^{T-1} \log f_{Y_j|X_j}\left(Y_j \mid x_j;\theta_1\right) \\
&\quad + \sum_x f_{X,Y}\left(x,Y;\widehat{\theta}_k\right) \log f_{X_0}\left(x_0;\theta_2\right) \\
&\quad + \sum_x f_{X,Y}\left(x,Y;\widehat{\theta}_k\right) \sum_{j=1}^{T-1} \log f_{X_j|X_{j-1}}\left(x_j \mid x_{j-1};\theta_2\right).
\end{aligned}
$$

We may thus obtain $\widehat{\theta}_{k+1,1}$ and $\widehat{\theta}_{k+1,2}$ separately: $\widehat{\theta}_{k+1,1}$ is just the value of

θ_1 which maximises

$$\sum_x f_{X,Y}\left(x, Y; \widehat{\theta}_k\right) \sum_{j=0}^{T-1} \log f_{Y_j|X_j}\left(Y_j \mid x_j; \theta_1\right)$$

while $\widehat{\theta}_{k+1,2}$ is the value of θ_2 which maximises

$$\sum_x f_{X,Y}\left(x, Y; \widehat{\theta}_k\right) \log f_{X_0}\left(x_0; \theta_2\right)$$

$$+ \sum_x f_{X,Y}\left(x, Y; \widehat{\theta}_k\right) \sum_{j=1}^{T-1} \log f_{X_j|X_{j-1}}\left(x_j \mid x_{j-1}; \theta_2\right).$$

Now,

$$\sum_x f_{X,Y}\left(x, Y; \widehat{\theta}_k\right) \sum_{j=1}^{T-1} \log f_{Y_j|X_j}\left(Y_j \mid x_j; \theta_1\right)$$

$$= \sum_{j=1}^{T-1} \sum_{x_j} f_{X_j,Y}\left(x_j, Y; \widehat{\theta}_k\right) \log f_{Y_j|X_j}\left(Y_j \mid x_j; \theta_1\right), \qquad (7.4)$$

$$\sum_x f_{X,Y}\left(x, Y; \widehat{\theta}_k\right) \log f_{X_0}\left(x_0; \theta_2\right)$$

$$= \sum_{x_0} \log f_{X_0}\left(x_0; \theta_2\right) f_{X_0,Y}\left(x_0, Y; \widehat{\theta}_k\right) \qquad (7.5)$$

and

$$\sum_x f_{X,Y}\left(x, Y; \widehat{\theta}_k\right) \sum_{j=1}^{T-1} \log f_{X_j|X_{j-1}}\left(x_j \mid x_{j-1}; \theta_2\right)$$

$$= \sum_{j=1}^{T-1} \sum_{x_{j-1},x_j} \log f_{X_j|X_{j-1}}\left(x_j \mid x_{j-1}; \theta_2\right) f_{X_{j-1},X_j,Y}\left(x_{j-1}, x_j, Y; \widehat{\theta}_k\right).$$

$$(7.6)$$

Often these computations are very simple, or approximations may be made which lead to simple computations. If, for example, the distribution of Y_t given X_t is Gaussian, $\widehat{\theta}_{k+1,1}$ will just be a weighted least squares estimator if θ_1 forms the parameter of a linear model. If θ_2 consists of an initial probability vector and the transition probability matrix, then $\widehat{\theta}_{k+1,2}$ is readily computed as weighted versions of terms formed from terms calculated in the forward–backward equations. In that case, let $\widehat{\pi}$ be the next estimator of the initial probability vector, and $\widehat{A}$ be that of the transition probability matrix. We obtain from (7.5)

$$\widehat{\pi}_j = \frac{f_{X_0,Y}\left(s_j, Y; \widehat{\theta}_k\right)}{\sum_l f_{X_0,Y}\left(s_l, Y; \widehat{\theta}_k\right)} \qquad (7.7)$$

and

$$\widehat{A}_{il} = \frac{\sum_{j=1}^{T-1} f_{X_{j-1},X_j,Y}\left(s_i, s_l, Y; \widehat{\theta}_k\right)}{\sum_{j=1}^{T-1}\sum_m f_{X_{j-1},X_j,Y}\left(s_i, s_m, Y; \widehat{\theta}_k\right)}. \qquad (7.8)$$

In the HMM implementation described in Section 7.4, we shall limit the amount of estimation to be done by modelling the transition probability matrix in terms of one parameter. While this is a little more complicated, it results in greater parsimony.

We now present the forward–backward equations, which enable us to compute the expressions needed in (7.4), (7.5) and (7.6). We shall omit reference to a particular value of θ, but it must be understood that each pass through these equations will use the successive iterates. Let

$$\alpha_t(i) = f_{X_t,Y_0,\dots,Y_t}(s_i, Y_0, \dots, Y_t)$$

and

$$\beta_t(i) = f_{Y_{t+1},\dots,Y_{T-1}|X_t}(Y_{t+1}, \dots, Y_{T-1} \mid s_i).$$

The following algorithm may be used to calculate these quantities efficiently.

Algorithm 10 *Forward–backward Equations*

(i) *Put* $\alpha_0(i) = f_{Y_0|X_0}(Y_0 \mid s_i) f_{X_0}(s_i) = b_{i0}\pi_i$ *and* $\beta_{T-1}(i) = 1$, *for all* i.

(ii) *For* $t = 1, \dots, T-1$, *calculate, for* $i = 1, \dots, d$

$$\alpha_t(i) = b_{it}\sum_j \alpha_{t-1}(j) A_{ji}. \qquad (7.9)$$

(iii) *For* $t = T-2, \dots, 0$, *calculate, for* $i = 1, \dots, d$

$$\beta_t(i) = \sum_j \beta_{t+1}(j) A_{ij} b_{j,t+1}. \qquad (7.10)$$

The above equations are easily verified using the Markovian nature of $\{X_t\}$ and the conditional independence assumption. The left side of (7.9) is equal to

$$\sum_j f_{X_{t-1},X_t,Y_0,...,Y_t} (s_j, s_i, Y_0, \ldots, Y_t)$$

$$= \sum_j f_{X_{t-1},Y_0,...,Y_{t-1}} (s_j, Y_0, \ldots, Y_{t-1}) f_{Y_t|X_t} (Y_t \mid s_i) f_{X_t|X_{t-1}} (s_i \mid s_j)$$

$$(7.11)$$

while the left side of (7.10) is

$$\sum_j f_{X_{t+1},Y_{t+1},...,Y_{T-1}|X_t} (s_j, Y_{t+1}, \ldots, Y_{T-1} \mid s_i)$$

$$= \sum_j f_{Y_{t+1}|X_{t+1}} (Y_{t+1} \mid s_j) f_{Y_{t+2},...,Y_{T-1}|X_{t+1}} (Y_{t+2}, \ldots, Y_{T-1} \mid s_j)$$

$$\times f_{X_{t+1}|X_t} (s_j \mid s_i) \,.$$

We then have the following results.

Lemma 6 *With $\alpha_t (i)$ and $\beta_t (i)$ defined above,*

$$f_{X_t,Y} (s_i, Y) = \alpha_t (i) \beta_t (i) \,,$$
$$f_{X_{t-1},X_t,Y} (s_i, s_j, Y) = \alpha_{t-1} (i) A_{ij} b_{jt} \beta_t (j)$$

and

$$f_Y (Y) = \sum_i \alpha_t (i) \beta_t (i) \,, \text{ for all } t.$$

Proof

$$f_{X_t,Y} (s_i, Y) = f_{X_t,Y_0,...,Y_t,Y_{t+1},...,Y_{T-1}} (s_i, Y_0, \ldots, Y_t, Y_{t+1}, \ldots, Y_{T-1})$$

$$= f_{Y_{t+1},...,Y_{T-1}|X_t,Y_0,...,Y_t} (Y_{t+1}, \ldots, Y_{T-1} \mid s_i, Y_0, \ldots, Y_t)$$

$$\times f_{X_t,Y_0,...,Y_t} (s_i, Y_0, \ldots, Y_t)$$

$$= f_{Y_{t+1},...,Y_{T-1}|X_t} (Y_{t+1}, \ldots, Y_{T-1} \mid s_i) f_{X_t,Y_0,...,Y_t} (s_i, Y_0, \ldots, Y_t)$$

$$= \beta_t (i) \alpha_t (i) \,,$$

$$f_Y (Y) = \sum_i f_{X_t,Y} (s_i, Y)$$

$$= \sum_i \beta_t (i) \alpha_t (i)$$

and

$$f_{X_{t-1},X_t,Y}\left(s_i,s_j,Y\right)$$

$$= f_{X_{t-1},X_t,Y_0,\dots,Y_t,Y_{t+1},\dots,Y_{T-1}}\left(s_i,s_j,Y_0,\dots,Y_t,Y_{t+1},\dots,Y_{T-1}\right)$$

$$= f_{Y_{t+1},\dots,Y_{T-1}|X_{t-1},X_t,Y_0,\dots,Y_t}\left(Y_{t+1},\dots,Y_{T-1}\mid s_i,s_j,Y_0,\dots,Y_t\right)$$
$$\times f_{X_{t-1},X_t,Y_0,\dots,Y_t}\left(s_i,s_j,Y_0,\dots,Y_t\right)$$

$$= f_{Y_{t+1},\dots,Y_{T-1}|X_t}\left(Y_{t+1},\dots,Y_{T-1}\mid s_j\right)f_{X_{t-1},X_t,Y_0,\dots,Y_t}\left(s_i,s_j,Y_0,\dots,Y_t\right)$$

$$= \beta_t\left(j\right)b_{jt}\alpha_{t-1}\left(i\right)A_{ij}$$

by the same reasoning as was used in (7.11). □

7.3.3 The scaled forward–backward equations

In practice, and especially for large values of T, the forward–backward equations must be scaled, for there will be large numbers of multiplications performed and underflow or overflow can be a problem. Such an approach has been used by Muñoz and Streit (1991), but the idea first appeared in Levinson, Rabiner and Sondhi (1983, p. 1051). Clearly, pre-scaling of the b_{jt} by the same constant does not alter any relative values, so that this may be done with impunity, and we need only worry about underflow. It is assumed in what follows that the b_{jt} have been scaled by, for example, their geometric mean. A scaled version of the forward–backward equations is given by the following algorithm.

Algorithm 11 *Forward–backward Equations (scaled)*

  (i) *Put $\xi_0 = \max_i\left(b_{i0}\pi_i\right)$, $\zeta_{T-1} = 1$, $\tilde{\alpha}_0\left(i\right) = \left(b_{i0}\pi_i\right)/\xi_0$, and $\tilde{\beta}_{T-1}\left(i\right) = 1$, for all i.*

  (ii) *For $t = 1,\dots,T-1$, calculate, for $i = 1,\dots,d$*

$$\tilde{\alpha}_t\left(i\right) = \left\{b_{it}\sum_j\tilde{\alpha}_{t-1}\left(j\right)A_{ji}\right\}\Big/\xi_t. \qquad (7.12)$$

  where $\xi_t = \max_i\left\{b_{it}\sum_j\tilde{\alpha}_{t-1}\left(j\right)A_{ji}\right\}$.

  (iii) *For $t = T-2,\dots,0$, calculate, for $i = 1,\dots,d$*

$$\tilde{\beta}_t\left(i\right) = \left\{\sum_j\tilde{\beta}_{t+1}\left(j\right)A_{ij}b_{j,t+1}\right\}\Big/\zeta_t,$$

  where $\zeta_t = \max_i\sum_j\tilde{\beta}_{t+1}\left(j\right)A_{ij}b_{j,t+1}$.

(iv) *Then* $\alpha_t(i) = \widetilde{\alpha}_t(i) \prod_{j=0}^{t} \xi_j$ *and* $\beta_t(i) = \widetilde{\beta}_t(i) \prod_{j=t}^{T-1} \zeta_j$.

The benefit from using the scaled version will accrue from using these in (7.4), (7.5) and (7.6). For example, if we were to compute

$$\sum_{j=0}^{T-1} \sum_{x_j} f_{X_j,Y}\left(x_j, Y; \widehat{\theta}_j\right) \log f_{Y_j|X_j}\left(Y_j \mid x_j; \theta_1\right)$$

using the unscaled quantities, we would obtain

$$\sum_{j=0}^{T-1} \sum_{k} \log\left\{f_{Y_j|X_j}\left(Y_j \mid s_k; \theta_1\right)\right\} \alpha_j(k) \beta_j(k),$$

with numerical problems associated with possible underflow in $\alpha_j(k)$ for large j and in $\beta_j(k)$ for small j. We could use the scaled quantities, however, and consider instead

$$\sum_{j=0}^{T-1} \nu_j \sum_{k} \log\left\{f_{Y_j|X_j}\left(Y_j \mid s_k; \theta_1\right)\right\} \widetilde{\alpha}_j(k) \widetilde{\beta}_j(k), \qquad (7.13)$$

where

$$\log \nu = \max_t \left(\sum_{k=0}^{t} \log \xi_k + \sum_{k=t}^{T-1} \log \zeta_k\right)$$

$$\log \nu_j = \sum_{k=0}^{j} \log \xi_k + \sum_{k=j}^{T-1} \log \zeta_k - \log \nu.$$

Similarly, we obtain

$$\xi_0 \left(\prod_{j=0}^{T-1} \zeta_j\right) \sum_{j} \log\{f_{X_0}(s_j; \theta_2)\} \widetilde{\alpha}_0(j) \widetilde{\beta}_0(j)$$

for

$$\sum_{x_0} f_{X_0,Y}\left(x_0, Y; \widehat{\theta}_j\right) \log f_{X_0}(x_0; \theta_2),$$

yielding

$$\widehat{\pi}_k = \frac{\widetilde{\alpha}_0(k) \widetilde{\beta}_0(k)}{\sum_j \widetilde{\alpha}_0(j) \widetilde{\beta}_0(j)}$$

for (7.7). Also, (7.8) is best calculated using

$$\widehat{A}_{il} = \frac{\sum_{j=1}^{T-1} \mu_j \widetilde{\alpha}_{j-1}(i) A_{il} b_{lj} \widetilde{\beta}_j(l)}{\sum_{j=1}^{T-1} \mu_j \widetilde{\alpha}_{j-1}(i) \sum_m A_{im} b_{mj} \widetilde{\beta}_j(m)},$$

where

$$\mu = \max_t \left(\sum_{k=0}^{t-1} \log \xi_k + \sum_{k=t}^{T-1} \log \zeta_k \right),$$

$$\mu_j = \sum_{k=0}^{j-1} \log \xi_k + \sum_{k=j}^{T-1} \log \zeta_k - \mu$$

and $\sum_{k=0}^{-1} c_k$ is 0 by definition. If θ_2 does not consist of the non-redundant elements of π and A, scaling still results in an algorithm which is numerically more stable than the unscaled version.

7.3.4 Other trackers

A track which is available very cheaply as a by-product of the forward–backward equations, known as the MAP (maximum *a priori*) track, is given by $\{x_0, \dots, x_{T-1}\}$ where

$$x_t = s_j$$

and

$$j = \arg\max_i \widetilde{\alpha}_t(i)\,\widetilde{\beta}_t(i) = \arg\max_i \alpha_t(i)\,\beta_t(i) = \arg\max_i f_{X_t,Y}(s_i, Y).$$

The estimated state at epoch t is thus that which maximises the conditional probability function of X_t given Y. As this is easily computed at each iteration of the EM algorithm, the track gives an excellent visual cue to the convergence of the EM algorithm, and often seems to be very close to the Viterbi track, even when the SNR is small.

Adaptive estimation techniques have been developed for the estimation of parameters in linear and nonlinear systems. The Kalman filter and extended Kalman filter (EKF) have been used successively in many circumstances. The common thread amongst these techniques is the use of current estimates of the parameters, together with a small-dimensional statistic, to update the estimates of the parameters and the small-dimensional statistic. The computational savings are often enormous, especially when the sample size is large, or the data is analysed online. When the parameters are expected to change with time, the algorithm may often be modified to incorporate 'forgetting factors', which limit the effect due to remote data. There cannot, however, be a completely adaptive implementation of the full HMM technique, since *all* of the data is needed for the Viterbi algorithm, and *all* of the data is needed to estimate the system parameters using the

EM algorithm. We can nevertheless estimate θ using a small fixed amount of data, and thenceforth use only the forward equations to estimate the track. The 'estimator' x_t of X_t will then be s_j where $j = \arg\max_i \tilde{\alpha}_t(i)$, where $\tilde{\alpha}_t(i)$ is given by (7.12) and the scaling factors ξ_t need not be stored. Although this will give a rapid track estimate, the track will certainly not be optimal, and a slow change in the system (represented by slow changes in θ) will result in the track wandering from the 'truth'. Nevertheless, when the SNR is moderately high, and the system parameters may be assumed to be stable, tracking accuracy is good, especially compared with competing techniques such as the Extended Kalman Filter.

7.4 HMM frequency tracking

The first frequency tracker formulated using HMM terminology was developed by Streit and Barrett (1990), who used the moduli of several Fourier coefficients through time to track frequency. They assume that the frequencies coincide with Fourier frequencies. An important feature of the tracker is its ability to *detect* as well as track, which is implemented via a *zero state*, which signifies *track lost*. The probabilities of transition from a real state to the zero state, and from the zero state to the real states, are extra system parameters which must be estimated. In practice, the tracking performance is very sensitive to changes in these transition probabilities. The combination of detection and tracking into a single algorithm is a key feature of the recent book by Stone, Barlow and Corwin (1999).

Barrett and Holdsworth (1993) used the moduli and relative phases between successive Fourier coefficients, and allowed the frequency states to differ from the Fourier frequencies. The problem with this approach is that the relative phases are *not* conditionally independent, given the frequency states, unless only every second relative phase is used, in which case important information is lost. Another problem with the implementation of the technique is the complexity of the joint probability density function of the amplitudes and relative phases. Quinn, Barrett and Searle (1994) avoided the problem by developing a technique which used the ratios of successive Fourier coefficients in batches of time blocks, and assumed that the states were constant within each batch. The joint asymptotic distributions in that case are then simple, but important information is again ignored. Xie and Evans (1991, 1993) have extended the HMM frequency tracker to cover multiple targets and/or multiple close frequencies.

What follows is a generalisation of the above work, which uses raw Fourier coefficients. Although the complexity of the Markov chain is an order higher, the distributions are much easier to work with (no modified Bessel functions or their integrals need be calculated) and no approximations need be made. It should be stressed from the outset that we do not actually believe that the frequencies form a Markov chain. In practice, however, we do not know how the frequencies change through time. Even if we do know the shape of the frequency variation, we may have to estimate unknown parameters and this may be very difficult in practice. For example, if there is a frequency shift due to the Doppler effect, it would be very difficult to implement a procedure which estimated the Doppler characteristics as well as the system parameters. *We shall suppose that the frequencies form a Markov chain only in order to impose constraints on the transition probabilities between frequencies.* The main problem associated with the development of a frequency tracker is easily seen from (7.1): let Y_q be formed from $\{z_{r,q}(m_T + j); r = 0, \ldots, R-1; j = -k, \ldots, l\}$, assume that $\{\delta_q\}$ forms a discrete time, discrete state space Markov chain, and let $X_q = \delta_q$. The requirements for a Hidden Markov Model are then not satisfied, since the distribution of Y_q given $\{X_j\}$ depends not only on X_q but on $\sum_{j=0}^{q} X_j$. The only exception to this is when the δ_q are integers. The following development can be greatly simplified if the variation in frequency is large and the accuracy needed in estimating the frequencies so low that allowing the δ_q to be integers would be acceptable. We shall assume from now on, however, that the δ_q are from the set $\left\{\frac{j}{M}; -Mk \le j \le Ml\right\}$. Then $\{X_q\}$ is also a Markov chain, where

$$X_q = \left[\begin{array}{cc} \delta_q & \sum_{j=0}^{q-1}\delta_j \end{array}\right]'.$$

The state space is, however, infinite and we can not use the algorithms of the previous section. Equation (7.1) shows, however, that the dependence on $\sum_{j=0}^{q-1}\delta_j$ is only through the fractional part of $R\sum_{j=0}^{q-1}\delta_j$. Let $e = \gcd(R, M)$, $R = eR'$ and $M = eM'$. Then the distinct values taken on by $R\sum_{j=0}^{q-1}\delta_j$ are the elements of $\left\{\frac{R'j}{M'}; j = 0, \ldots, M'-1\right\}$ or the fractions $\left\{\frac{j}{M'}; j = 0, \ldots, M'-1\right\}$. The sequence $\{X_q\}$, where

$$X_q = \left[\begin{array}{cc} \delta_q & M'\operatorname{fr}\left(R\sum_{j=0}^{q-1}\delta_j\right) \end{array}\right]',$$

and $\operatorname{fr}(x)$ denotes the fractional part of x, then also forms a Markov chain, but one that has a finite state space of dimension

$$M'\{M(k+l)+1\}.$$

Although this may be very large compared with $M(k+l)+1$, most transitions have probabilities which are 0, advantage of which may be taken when computing the EM estimators, the forward–backward equations and the Viterbi algorithm.

From (7.1), and using asymptotic distributions, the vector θ_1 consists of the real and imaginary parts of $D_T = T^{1/2}\frac{Ae^{i\phi}}{2}$ and the asymptotic variances of the real and imaginary parts of the $w_{x,r,q}\left(2\pi\frac{m_T+j}{T}\right)$, which we denote as σ^2 but which are really $\pi f_x(\lambda)$. Letting $\widetilde{A}_{ij} = \Pr\{\delta_q = s_j \mid \delta_{q-1} = s_i\}$, the transition probabilities of $\{X_q\}$ are obtained from

$$
\Pr\left\{ X_q = \begin{bmatrix} s_j & u_j \end{bmatrix}' \middle| X_{q-1} = \begin{bmatrix} s_i & u_i \end{bmatrix}' \right\}
$$
$$
= \begin{cases} \widetilde{A}_{ij} & ; \quad u_j = M'\,\mathrm{fr}\left(\frac{u_i}{M'} + Rs_i\right) \\ 0 & ; \quad \text{otherwise.} \end{cases}
$$

The implementation of the forward–backward equations is not straightforward, as there must be some ordering of the states. Moreover, a direct implementation would be computationally inefficient, because of the dimension of the state space. Rather than imposing an ordering, we shall rewrite the forward–backward equations, the EM algorithm and the Viterbi algorithm in a matrix form which is easily implemented by a matrix-oriented package such as MATLAB^{TM}. Let $U = M(k+l)+1$, and define the $M' \times U$ matrices b_q, α_q and β_q by

$$
\begin{aligned}
b_q &= \{b_q(i,j)\}_{i=1,\dots,M';j=1,\dots,U}\,, \\
\alpha_q &= \{\alpha_q(i,j)\}_{i=1,\dots,M';j=1,\dots,U}
\end{aligned}
$$

and

$$
\beta_q = \{\beta_q(i,j)\}_{i=1,\dots,M';j=1,\dots,U}
$$

where

$$
\begin{aligned}
b_q(i,j) &= f_{Y_q|X_q}\left(Y_q \middle| \begin{bmatrix} s_j & i-1 \end{bmatrix}'\right), \\
\alpha_q(i,j) &= f_{X_q,Y_0,\dots,Y_q}\left(\begin{bmatrix} s_j & i-1 \end{bmatrix}', Y_0,\dots,Y_q\right)
\end{aligned}
$$

and

$$
\beta_q(i,j) = f_{Y_{q+1},\dots,Y_{Q-1}|X_q}\left(Y_{q+1},\dots,Y_{Q-1} \middle| \begin{bmatrix} s_j & i-1 \end{bmatrix}'\right).
$$

Then the (unscaled) forward–backward equations may be written, letting $\odot$ denote the Hadamard or element-by-element product, as

Algorithm 12 *Frequency tracking forward–backward equations*

(i) *Let*

$$\alpha_0 = \left[\begin{array}{c} \pi' \\ 0_{(M'-1)\times U} \end{array}\right] \odot b_0, \quad \beta_{Q-1} = [1_{M'\times U}].$$

(ii) *For $q = 1, \ldots, Q-1$, let*

$$\alpha_q = (\widetilde{\alpha}_{q-1}A) \odot b_q \qquad (7.14)$$

where

$$\widetilde{\alpha}_{q-1}(i,j) = \alpha_{q-1}\left\{i'(i,j),j\right\}$$

and

$$i'(i,j) = M' \operatorname{fr}\left(\frac{i-1}{M'} - Rs_j\right) + 1.$$

(iii) *For $q = Q-2, \ldots, 0$, let*

$$\widetilde{\beta}_q = (\beta_{q+1} \odot b_{q+1})\, A'$$

and

$$\beta_q(i,j) = \widetilde{\beta}_q\left\{i''(i,j),j\right\} \qquad (7.15)$$

where

$$i''(i,j) = M' \operatorname{fr}\left(\frac{i-1}{M'} + Rs_j\right) + 1.$$

The algorithm seems a lot more complicated than the forward–backward equations for the usual HMM. However, in matrix-oriented computer languages such as MATLAB^{TM}, it may be programmed and run efficiently. For example, (7.14) may be implemented in MATLAB^{TM} by a single line of code

```
alp=(al(p1)*A).*b;
```

where b stands for b_q, al for α_{q-1}, A for A, alp for α_q and p1 for a matrix of indices which has the entries $\{1, \ldots, M'\}$ in each column, but permuted, and is calculated only once. The matrix p1 is also computed quickly with a single line of code, depends only on M', R and the s_j and is only calculated once. Similarly, β_q may be calculated by two lines of MATLAB^{TM} code

```
bep = (be.*b)*A';
bep = bep(p2);
```

where b stands for b_{q+1}, be for β_{q+1}, bep for β_q and p2 another permutation matrix which is only computed once. The implementation thus minimises storage and computational requirements. Of course, a scaled version will be used in practice as underflow problems are even more evident because of the increase in dimension of the state space.

There are similar computational savings to be made by rewriting the Viterbi algorithm:

Algorithm 13 *Frequency tracking Viterbi algorithm*

(i) *Let*

$$\Delta_0 (i, j) = \begin{cases} \log \pi_j + \log b_0 (i, j) & ; \quad i = 1 \\ -\infty & ; \quad i \neq 1. \end{cases}$$

(ii) *For $q = 1, \ldots, Q - 1$, and for each $i = 1, \ldots, M'$; $j = 1, \ldots, U$, put*

$$z_q (i, j) = \arg \max_k \left\{ \Delta_{q-1} \left(i' (i, k), k \right) + \log A_{kj} \right\}$$
$$\varpi_q (i, j) = i' \{i, z_q (i, j)\}$$

and

$$\Delta_q (i, j) = \Delta_{q-1} \{\varpi_q (i, j), z_q (i, j)\} + \log A_{z_q(i,j),j} + \log b_q (i, j),$$

where

$$i' (i, j) = M' \, \mathrm{fr} \left(\frac{i - 1}{M'} - R s_j \right).$$

(iii) *Let*

$$(\mu_{Q-1}, \nu_{Q-1}) = \arg \max_{i,j} \Delta_{Q-1} (i, j)$$

and, for $q = Q - 2, \ldots, 0$, put

$$\nu_q = z_{q+1} (\mu_{q+1}, \nu_{q+1})$$

and

$$\mu_q = \varpi_{q+1} (\mu_{q+1}, \nu_{q+1}).$$

The Viterbi track is then $\begin{bmatrix} s_{\nu_0} & \cdots & s_{\nu_{Q-1}} \end{bmatrix}'$. The algorithm may again be written in matrix form using permutation matrices, but the details are not given here.

The exact form of the EM algorithm will, of course, depend on the parametrisation. Whichever way we parametrise the transition probabilities, we have

$$\theta = \begin{bmatrix} \theta_1' & \theta_2' \end{bmatrix}'$$

where $\theta_1 = \begin{bmatrix} \operatorname{Re} D & \operatorname{Im} D & \sigma^2 \end{bmatrix}'$ and

$$\log f_{Y_q|X_q}\left(Y_q \mid x_q; \theta_1\right)$$
$$= -R\left(l + k + 1\right)\log\left(2\pi\sigma^2\right)$$
$$-\frac{1}{2\sigma^2}\sum_{r=0}^{R-1}\sum_{p=-k}^{l}\left|z_{r,q}\left(\widehat{m}_T + p\right) - De^{2\pi i\left\{R\sum_{s=0}^{q-1}\delta_s + \left(r+\frac{1}{2}\right)\delta_q\right\}}\frac{\sin\left(\pi\delta_q\right)}{\pi\left(\delta_q - p\right)}\right|^2.$$

Thus, with the α and β understood to be calculated using $\widehat{\theta}_j$,

$$\sum_x f_{X,Y}\left(x, Y; \widehat{\theta}_j\right)\sum_{q=0}^{Q-1}\log f_{Y_q|X_q}\left(Y_q \mid x_q; \theta_1\right)$$

$$= -RQ\left(l + k + 1\right)\log\left(2\pi\sigma^2\right)\sum_x f_{X,Y}\left(x, Y; \widehat{\theta}_j\right) - \frac{1}{2\sigma^2}\sum_x f_{X,Y}\left(x, Y; \widehat{\theta}_j\right)$$

$$\times\sum_{q=0}^{Q-1}\sum_{r=0}^{R-1}\sum_{p=-k}^{l}\left|z_{r,q}\left(\widehat{m}_T + p\right) - De^{2\pi i\left\{R\sum_{s=0}^{q-1}\delta_s + \left(r+\frac{1}{2}\right)\delta_q\right\}}\frac{\sin\left(\pi\delta_q\right)}{\pi\left(\delta_q - p\right)}\right|^2.$$

This is maximised with respect to D and σ^2 when

$$D = \widehat{D}_{j+1} = \frac{\varpi_1}{\varrho_1} \quad \text{and} \quad \sigma^2 = \widehat{\sigma}_{j+1}^2 = \frac{\varpi_2}{\varrho_2},$$

where

$$\varpi_1 = \sum_{q=0}^{Q-1}\sum_{r=0}^{R-1}\sum_{p=-k}^{l} z_{r,q}\left(\widehat{m}_T + p\right)\sum_u\frac{\sin\left(\pi s_u\right)}{\pi\left(s_u - p\right)}e^{-2\pi i\left(r+\frac{1}{2}\right)s_u}$$
$$\times\sum_v \alpha_q\left(u, v\right)\beta_q\left(u, v\right)e^{-2\pi i\left(\frac{v}{M'}\right)},$$

$$\varrho_1 = R\sum_{q=0}^{Q-1}\sum_{u=1}^{U}\sum_{p=-k}^{l}\left\{\frac{\sin\left(\pi s_u\right)}{\pi\left(s_u - p\right)}\right\}^2\sum_v \alpha_q\left(u, v\right)\beta_q\left(u, v\right),$$

$$\varpi_2 = \sum_{u,v}\alpha_{Q-1}\left(u, v\right)\sum_{q=0}^{Q-1}\sum_{r=0}^{R-1}\sum_{p=-k}^{l}\left|z_{r,q}\left(\widehat{m}_T + p\right)\right|^2$$
$$-\left|\widehat{D}_{j+1}\right|^2 R\sum_{q=0}^{Q-1}\sum_{u=1}^{U}\sum_{p=-k}^{l}\left\{\frac{\sin\left(\pi s_u\right)}{\pi\left(s_u - p\right)}\right\}^2\sum_v \alpha_q\left(u, v\right)\beta_q\left(u, v\right),$$

and

$$\varrho_2 = 2RQ\left(l + k + 1\right))\sum_{u,v}\alpha_{Q-1}\left(u, v\right).$$

These computations may again be implemented using matrix operations and are thus efficiently calculated using MATLAB^{TM}.

We next describe several parametrisations of θ_2 and their associated estimation procedures.

(i) Maximum likelihood. Here we just put $\pi' = \left[\begin{array}{ccc} \frac{1}{U} & \cdots & \frac{1}{U} \end{array} \right]$ and $A_{ij} = \frac{1}{U}$, for all i, j. We thus do not need to estimate θ_2, iterating the above estimation equations for θ_1 and the forward–backward equations until 'convergence' has occurred and using the Viterbi algorithm. This produces the maximum likelihood track discussed in Section 7.2.

(ii) Ignorance method. With π and A defined initially as above, we re-estimate π and A during each iteration, using generalisations of (7.7) and (7.8). It is easily seen that the estimate of π_k is (scaled version)

$$\widehat{\pi}_k = \frac{\widetilde{\alpha}_0 \left(1, k \right) \widetilde{\beta}_0 \left(1, k \right)}{\sum_j \widetilde{\alpha}_0 \left(1, j \right) \widetilde{\beta}_0 \left(1, j \right)}$$

while the estimate of A_{ij} is (unscaled version, for simplicity)

$$\widehat{A}_{kl} = \frac{\sum_{j=1}^{Q-1} \sum_{i=1}^{M'} \alpha_{j-1} \left(i, k \right) A_{kl} b_j \left\{ i'' \left(i, k \right), l \right\} \beta_j \left\{ i'' \left(i, k \right), l \right\}}{\sum_{j=1}^{Q-1} \sum_{i=1}^{M'} \sum_{m=1}^{U} \alpha_{j-1} \left(i, k \right) A_{km} b_j \left\{ i'' \left(i, k \right), m \right\} \beta_j \left\{ i'' \left(i, k \right), m \right\}}$$

where

$$i'' \left(i, k \right) = M' \, \mathrm{fr} \left(\frac{i-1}{M'} + R s_k \right).$$

(iii) Parametric method. If U is large, the above method results in too many parameters being estimated. We have had success in adapting the method used by Streit and Barrett, which assumes that the frequency tracks form a random walk with

$$\mathrm{Pr} \left\{ \delta_q - \delta_{q-1} = \frac{j}{M} \right\} \propto \exp \left(-\frac{j^2}{2M^2 \nu^2} \right). \tag{7.16}$$

In other words, the frequency increments are independent and identically distributed with mean zero, and have a distribution which is a discretised form of the normal distribution with variance ν^2. The initial distribution π is assumed unknown, and is re-estimated at each step, and only ν^2 needs to be re-estimated to obtain the next estimate of A. As long as M is large (in practice, larger than 10 will do), ν

small, and k and l both greater than 0,

$$\frac{1}{M} \sum_{j=-kM}^{lM} \exp\left(-\frac{j^2}{2M^2\nu^2}\right) \sim \int_{-k}^{l} \exp\left(-\frac{x^2}{2\nu^2}\right) dx$$

$$= \sqrt{2\pi\nu^2}\left[\Phi\left(\frac{l}{\nu}\right) - \Phi\left(\frac{-k}{\nu}\right)\right]$$

$$\sim \sqrt{2\pi\nu^2}.$$

Using this to approximate the constant of proportionality in (7.16), and thus A, produces reasonable results. The estimator of ν^2 is then calculated in a straightforward manner, but the details will not be reproduced here.

Finally, we indicate a procedure for computing initial estimates for the EM algorithm. For each $q = 0, \ldots, Q-1$, estimate δ_q by maximising (6.14) using Y_q. This is described in the following section. Initial estimators of D and σ^2 are then easily found by regression, an initial estimator of π is $\left[\begin{array}{ccc} \frac{1}{U} & \cdots & \frac{1}{U} \end{array}\right]'$ and, if using the parametric method, an estimator of ν^2 is given by the sample variance of the actual transitions in the estimated δ_q.

7.5 Real data example

Data was obtained from a single hydrophone (underwater microphone) while a quiet underwater object was passing at some distance. There was localised acoustic energy detected near a certain frequency. It was decided to Fourier transform the data in 412 small time-chunks, and store five Fourier coefficients near this frequency from each of these 412, at the same frequencies. All units have been normalised. Figure 7.1 shows the frequency (out of five) at which the periodogram is maximised.

This figure does not show sufficient detail to enable us even to eyeball a frequency track. Suppose we now choose $k = l = 2$, $M = 20$, $R = 1$ and $Q = 412$, with centre frequency represented by the middle frequency, and calculate $S_T(\delta)$ at $\delta = -\frac{40}{20}, -\frac{39}{20}, \ldots, \frac{39}{20}, \frac{40}{20}$ from (6.14) for each of the time blocks of length RT. Figure 7.2 is an intensity plot of $\log S_T$, which is similar to the usual *lofargram, sonogram* or *spectrogram*, but which has much higher resolution. Lighter shades represent larger values, and the maximising δ are represented by a '+'.

At least one track is apparent, starting near $\delta = 0$ (the value 3 on the frequency axis) and disappearing near the end of the data near $\delta = -0.5$. As the signal appears to be weak, we make the not unreasonable assumption

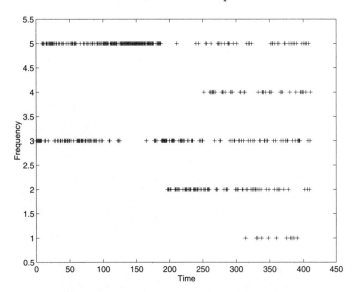

Fig. 7.1. Coarse periodogram maximiser

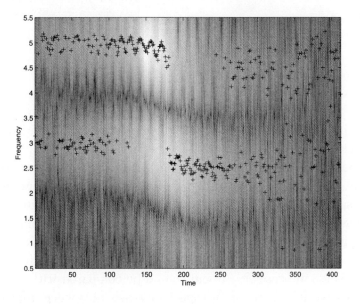

Fig. 7.2. High resolution spectrogram with $R = 1$

that the frequency variation is slow, and recompute $S_T(\delta)$ with $R = 5$, (and therefore $Q = 82$). The result is depicted in Figure 7.3. There is now clear evidence of two tracks, which represent two sinusoidal signals whose frequencies have Doppler shifted because of relative motion between source and receiver. In what follows, we shall refer to the estimates of δ obtained by maximising $S_T(\delta)$ defined in (6.14) as the FTI estimates, and the track formed from these as the FTI track.

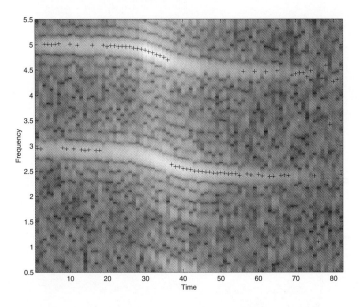

Fig. 7.3. High resolution spectrogram with $R = 5$

In order to obtain only the lower track, we reduce k and l to 1, and break the data up into three segments of 135 time periods. We carry out separate HMM trackers, using the parametric method, on each segment, changing the Fourier coefficients used (i.e. changing m) in the next segment if the track has deviated by half a bin or more from the central Fourier frequency. For segments subsequent to the first, initial estimators of the system parameters are provided by the final estimators of those parameters from the previous segment, speeding the tracking somewhat. Figure 7.4 represents the Viterbi and FTI tracks when $R = 1$, and should be compared with Figure 7.2, while Figure 7.5 is for the case $R = 5$ and should be compared with Figure 7.3.

The segmentation procedure is especially reasonable when it is considered that the amplitude of the signal changes substantially over the period, while the HMM procedure has been developed under the assumption that the amplitude is fixed.

Finally, the following figures display the Viterbi tracks using the maximum likelihood method; Figure 7.6 for the case where $R = 1$, which should be compared with Figures 7.2 and 7.4, and Figure 7.7 for $R = 5$, which should

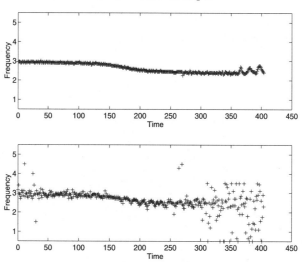

Fig. 7.4. Viterbi track (top) and FTI track (bottom) for $R = 1$

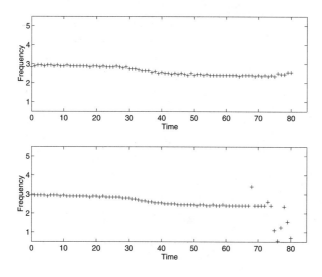

Fig. 7.5. Viterbi track (top) and FTI track (bottom) for $R = 5$

be compared with Figures 7.3 and 7.5. It is remarkable that the mere linking of phases between time blocks produces such an improvement, and that the Markovian mechanism makes so little difference to the results. It should be noted here that the reason for the poor tracking performance towards the end is that the signal has disappeared, which is evident from Figure 7.1. The seemingly better tracking performance of the Viterbi track is most likely due to the Markovian assumption, which prefers small deviations in frequency to large, even if this is not completely supported by the data.

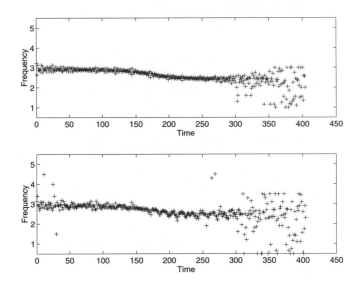

Fig. 7.6. ML track (top) and FTI track (bottom) for $R = 1$

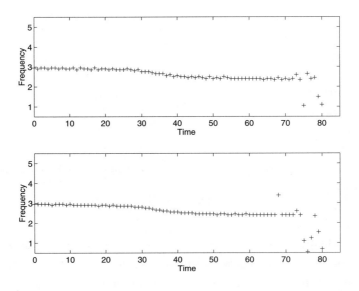

Fig. 7.7. ML track (top) and FTI track (bottom) for $R = 5$

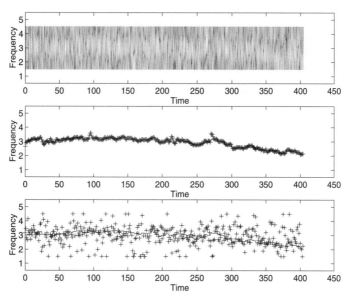

Fig. 7.8. Spectrogram, Viterbi track+truth, FTI track+truth for $R = 1$

7.6 Simulated example

Figures 7.8 and 7.9 below depict the Viterbi tracks from a single simulation, for which data was generated according to the (parametric) HMM model, and the parametric version of the algorithms was used to obtain the Viterbi tracks. A time series of length 414720 was generated, consisting of a single sinusoid with $\rho = 1$ and $\phi = 0$, with noise pseudo-Gaussian and independent and identically distributed with mean zero and variance $\frac{1}{2}10^{5/2} \sim 158.11$, and with frequency fixed for each segment of length $5T = 5120$ and frequency increments independent and identically distributed and Gaussian with mean zero and variance $0.01 \left(\frac{2\pi}{T}\right)^2$. The SNR was thus -25dB. As in the real data example, we divided the time series into three of length $QT = 135 \times 1024$ and constructed three Viterbi tracks, with the second and third tracks calculated along the lines of the previous section. Figure 7.8 represents the results for the case where $R = 1$, the top third representing the spectrogram, the middle showing the true frequencies and the Viterbi track and the bottom third showing the FTI track. Figure 7.9, which shows the same results but using $R = 5$, shows how much accuracy may be gained by 'aggregating' data.

Finally, we show the results of tracking using the Maximum Likelihood

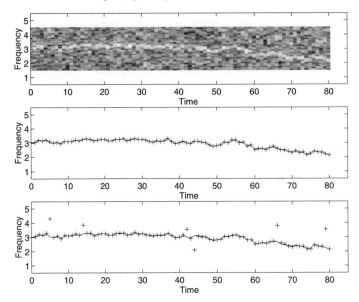

Fig. 7.9. Spectrogram, Viterbi track+truth, FTI track+truth for $R = 5$

HMM method. Figure 7.10 is for the case where $R = 1$, while Figure 7.11 is for $R = 5$.

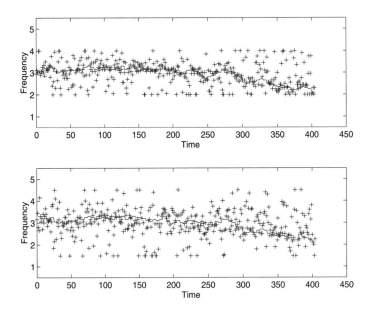

Fig. 7.10. ML track+truth, FTI track+truth for $R = 1$

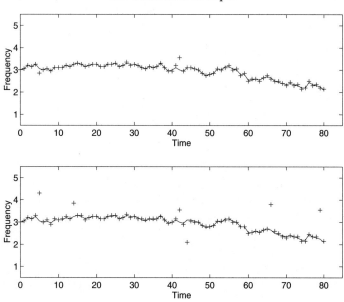

Fig. 7.11. ML track+truth, FTI track+truth for $R = 5$

Appendix 1

MATLAB^{TM} programs

Name ar2

Description The program 'ar2' computes the autoregressive fit of order 2 to a time series, and returns the maximiser of the corresponding spectral density.

Code

```
function w = ar2(x)
%usage w = ar2(x)
% x = time series
% w = ar2 frequency estimator
c = zeros(3,1);
T = length(x);
xb = mean(x);
for j = 0:2
    c(j+1) = (x(1:T-j)-xb)'*(x(j+1:T)-xb)/T;
end
b = - [c(1) c(2);c(2) c(1)]\[c(2);c(3)];
w = acos(-b(1)/(2*sqrt(b(2))));
```

Name arfit

Description The program 'arfit' used the Durbin–Levinson algorithm to compute autoregressive fits to a time series, and AIC, HQIC and BIC to estimate the order of the best autoregressive fit.

Code

```
function [y,ji,c,p] = arfit(x,k)
% Usage [y,ji,c,p] = arfit(x,k)
% x = time series
% k = maximum order
```

```
% y: rows correspond to different orders and contain
% autoregressive estimates followed by estimate of residual
% variance
% ji = orders (AIC HQIC BIC)
% c = autocovariances, lags 0 through k
% p =
n=length(x);
cr = [2/n 2*log(log(n))/n 2*log(n)/n];
es = [1 1 1];
xb=mean(x);
x=x-xb;
y=zeros(k+1,k+1);
c=zeros(k+1,1);
p=zeros(k+1,1);
b=zeros(k,1);
ac=zeros(k+1,3);
for i = 0:k
c(i+1) = x(i+1:n)'*x(1:n-i)/n;
end
s(1)=c(1);
ac(1,:) = log(s(1))*es;
b(1)=-c(2)/c(1);
p(1)=b(1);
s(2)=s(1)*(1-b(1)*b(1));
ac(2,:) = ac(1,:) + log(1-b(1)*b(1))*es + cr;
y(1,1:3)=[b(1) s(2) s(1)];
for i = 1:(k-1)
d=c(i+2)+b(1:i)'*c(i+1:-1:2);
e=-d/s(i+1);
b(1:i)=b(1:i)+e*b(i:-1:1);
b(i+1)=e;
p(i+1)=e;
s(i+2)=s(i+1)*(1-e*e);
ac(i+2,:) = ac(i+1,:) + log(1-e*e)*es + cr;
y(i+1,1:(i+2)) = [b(1:(i+1))' s(i+2)];
end
[max,ji]=min(ac);
ji=ji-1;
```

Name asvar, asvsc

Description The program 'asvar' computes the asymptotic variance of the generalised FTI estimator, relative to the asymptotic variance of the periodogram maximiser. The program 'asvsc' is a script file which uses 'asvar' to produce four subplots for various combinations of k, l and R.

Code

```
function y=asvar(k,l,R,N)
%usage y=asvar(k,l,R,N)
% k, l: Frequency bins are labelled -k to l
% R = number of time blocks
% N = number of values at which asvar is computed
d=(1:N)/(2*N);
kl=k+1;
dc=ones(kl,1)*d;
db=[(-k:-1)'; (1:l)']*ones(1,N);
dd=dc-db;
e1=dc./dd;
e2=db./(dd.*dd);
u1=diag(e1'*e1)+1;
u2=diag(e2'*e1);
u3=diag(e2'*e2);
s=pi*d;
s=s./sin(s);
s=s.*s;
den=pi*pi*(R*R-1)*u1+3*(u3-u2.*u2./u1);
y=pi*pi*R*R*s'./den;
t=sum(1./((1:k).^2))+sum(1./((1:l).^2));
ad=pi*pi*R*R/(pi*pi*(R*R-1)+3*t);
y=[ad;y];

%script for figure 6.2
N=50;
d=(0:N)'/(2*N);
y=zeros(N+1,5);
for j=1:5
y(:,j)=asvar(1,1,j,N);
end
subplot(2,2,1),plot(d,y)
xlabel('\delta')
ylabel('RAV')
```

```
title('k=l=1')
y=zeros(N+1,5);
for j=1:5
y(:,j)=asvar(2,2,j,N);
end
subplot(2,2,2),plot(d,y)
xlabel('\delta')
ylabel('RAV')
title('k=l=2')
y=zeros(N+1,5);
for j=1:5
y(:,j)=asvar(3,3,j,N);
end
subplot(2,2,3),plot(d,y)
xlabel('\delta')
ylabel('RAV')
title('k=l=3')
y=zeros(N+1,5);
for j=1:5
y(:,j)=asvar(4,4,j,N);
end
subplot(2,2,4),plot(d,y)
xlabel('\delta')
ylabel('RAV')
title('k=l=4')
```

Name freqest

Description The program 'freqest' uses the Quinn & Fernandes technique to estimate k frequencies, amplitudes and phases. A frequency is estimated and the associated sinusoid removed by regression. Another frequency is estimated from these residuals, etc, until k sinusoids have been estimated. Finally, the amplitudes and phases are estimated jointly from a regression on the cosine and sine regressors evaluated at the k frequency estimates. This program uses the programs 'qf', 'srg' and 'sinreg'.

Code

```
function [w,amp,ph,amp1,ph1,ress] = freqest(x,k)
%usage [w,amp,ph,amp1,ph1,ress] = freqest(x,k)
% This program naively estimates k sinusoids by
% using the Quinn & Fernandes technique to estimate
```

```
% a frequency, removing the sinusoid by regression,
% estimating a frequency using the residuals, etc.
% Finally, the estimated frequencies are used simultaneously
% to estimate the amplitudes and phases.
% x = time series
% k = number of sinusoids
% w = estimate of frequencies
% amp, ph = initial estimates of amplitudes and phases
% amp1, ph1 = final estimates of amplitudes and phases
% regss = regression sum of squares at final estimates
w=zeros(k,1);
amp=w;
ph=w;
y=x;
nu=(length(x)-1)/2;
for j=1:k
    wt=qf(y,0);
    [ress,y,b]=sinreg(wt,y);
    r=(b(2)-i*b(3))*exp(-i*wt*nu);
    amp(j)=abs(r);
    ph(j)=angle(r);
    w(j)=wt;
end
[ress,y,b]=sinreg(w,x);
r=(b(2:k+1)-i*b(k+2:2*k+1)).*exp(-i*w*nu);
amp1=abs(r);
ph1=angle(r);
```

Name freqest2

Description The program 'freqest2' uses the Quinn & Fernandes technique to estimate k frequencies, amplitudes and phases. A frequency is estimated and the associated sinusoid removed by regression. Another frequency is estimated from these residuals, etc., until k sinusoids have been estimated. The vector of frequency estimates is then used as the initial estimate in an iterative to maximise the regression sum of squares, using the MATLAB^{TM} function 'fmins'. This program uses the programs 'qf', 'srg' and 'sinreg'.

Code

```
function [w1,amp1,ph1,w2,amp2,ph2,ress] = freqest2(x,k)
%usage [w1,amp1,ph1,w2,amp2,ph2,ress] = freqest2(x,k)
```

```
% This program estimates k sinusoids by using the Q & F
% technique to estimate a frequency, removing the sinusoid
% by regression, estimating a frequency using the residuals,
% etc. The estimated frequencies are then used to
% initialise an iterative procedure which calculates the
% nonlinear regression estimates of all the parameters.
% x = time series, k = number of sinusoids,
% w1, w2 = initial, final estimates of frequencies
% amp1, amp2 = initial, final estimates of amplitiudes
% ph1, ph2 = initial, final estimates of phases
% regss = regression sum of squares at final estimates
w1=zeros(k,1);
amp1=w1;
ph1=w1;
y=x;
nu=(length(x)-1)/2;
for j=1:k
    wt=qf(y,0);
    [ress,y,b]=sinreg(wt,y);
    r=(b(2)-i*b(3))*exp(-i*wt*nu);
    amp1(j)=abs(r);
    ph1(j)=angle(r);
    w1(j)=wt;
end
w2=fmins('sinreg',w1,[],[],x)
[ress,y,b]=sinreg(w2,x);
r=(b(2:k+1)-i*b(k+2:2*k+1)).*exp(-i*w2*nu);
amp2=abs(r);
ph2=angle(r);
```

Name fti

Description The program 'fti' computes 5 different frequency estimates using 3 Fourier coefficients.

Code

```
function [y1,y2,y3,y4,y5]=fti(x)
%usage [y1,y2,y3,y4,y5]=fti(x)
% x = time series
% y1 = fti1, y2 = fti2, y3 = fti3
% y4 = fti3, with a slight twist
```

```
% y5 = average of two initial estimates, expected to do worst
T=length(x);
n=floor((T+1)/2);
y=fft(x);
z=y.*conj(y);
[a,b]=max(z(1:n));
m=b-1;
z1=real(y(b-1)./y(b));
z2=real(y(b+1)./y(b));
u1=z1/(1-z1);
u2=-z2/(1-z2);
z=u1;
if u1 > 0 & u2 > 0
    z=u2;
end
y1=2*pi*(m+z)/T;
y3=(u1+u2)/2+(u2-u1).*(3*z.^3+2*z)./(3*z.^4+6*z.^2+1);
y4=(u1+u2)+(u2.^2-u1.^2).*(3*z.^2+2)./(3*z.^4+6*z.^2+1);
y4=y4/2;
y5=(u1+u2)/2;
s6=sqrt(6)/24;
s2=sqrt(2/3);
y2=(u1+u2)/2+log(3*u2.^4+6*u2.^2+1)/4
y2=y2-s6*(log(u2.^2+1-s2)-log(u2.^2+1+s2));
y2=y2-log(3*u1.^4+6*u1.^2+1)/4
y2=y2+s6*(log(u1.^2+1-s2)-log(u1.^2+1+s2));
y2=2*pi*(m+y2)/T;
y3=2*pi*(m+y3)/T;
y4=2*pi*(m+y4)/T;
y5=2*pi*(m+y5)/T;
```

Name ftibl

Description The program 'ftibl' computes the maximiser of the likelihood given several Fourier coefficients in several consecutive time blocks. It optionally plots the regression sum of squares (block periodogram).

Code

```
function [y,w,wmax] = ftibl(x,T,k,l,S,pl)
%usage [y,w,wmax] = ftibl(x,T,k,l,S,pl)
% x = time series
```

```
% T = fft block length
% Uses Fourier coefficients at bins m-k to m+1
% where m is maximiser of periodogram block sums
% S = number of frequency values per bin
% pl = 1 if plot of block periodogram required
% y = block periodogram values
% w = frequencies corresponding to y
% wmax = block periodogram estimate of frequency
T1=length(x);
R=floor(T1/T);
n=floor((S+1)/2);
kl=k+l+1;
kv=(-k:l)';
u=zeros(T,R);
u(:)=x(1:R*T);
a=[(n-S:n-1)']/S;
del=kron(ones(kl,1),a)+kron(kv,ones(S,1));
aa=a*ones(1,kl);
z=fft(u);
z1=z.*conj(z);
y=sum(z1')';
[b,m]=max(y);
w=2*pi*(m-1+del)/T;
z=z(m-k:m+l,:).';
v=fft(z,S);
v=[v(n+1:S,:);v(1:n,:)];
y=zeros(kl*S,1);
oskl=ones(S,kl);
for p=-k:l
   e=find(kv ~= p);
   d=oskl;
   ab=aa(:,e);
   d(:,e)=ab./(ab+p*ones(S,kl-1)-ones(S,1)*kv(e)');
   num=sum((d.*v).')';
   num=num.*conj(num);
   den=sum((d.*d).')';
   z=num./den;
   ind=(p+k)*S+(1:S)';
   y(ind)=z;
end
```

```
[a,b]=max(y);
wmax=w(b);
if pl == 1
   plot(w,y)
   xlabel('Frequency (radians per unit time)')
   title('Block periodogram')
end
```

Name musicsp
Description The program 'musicsp' computes the music spectrum and opt-
ionally plots it.
Code

```
function [y,f]=musicsp(x,r,k,pad,pl)
%usage [y,f]=musicsp(x,r,k,pad,pl)
% x = time series
% r = number of sinusoids
% k = dimension of autocovariance matrix to use
% pad = padding factor. If pad is 1, then no padding
% y = vector of music 'spectrum' values
% peaks in y correspond to frequency estimates
% f = vector of frequencies
% If pl = 1 then music spectrum is plotted
c=zeros(k,1);
T=length(x);
n=floor((T*pad+1)/2);
f=2*pi*(0:n-1)'/(T*pad);
xb=mean(x);
% If k is large, the usual fft method of computing
% autocovariances might be faster
for j=0:k-1
   c(j+1)=(x(1:T-j)-xb)'*(x(j+1:T)-xb)/T;
end
C=toeplitz(c);
if isreal(x)
   r=2*r;
else
   C=conj(C);
end
[v,d]=eig(C);
```

```
[b,m]=sort(diag(d));
P=v(:,m(1:k-r));
z=fft(P,T*pad);
z=z(1:n,:);
z=z.*conj(z);
if k-r > 1
   z=sum(z')';
end
y=1./z;
if pl == 1
   plot(f,y)
   xlabel('Frequency (radians per unit time)')
   title('Music spectrum')
end
```

Name per

Description The program 'per' calculates and optimally displays the periodogram of x, zero padded so that it is calculated at $\lfloor (mT + 1)/2 \rfloor$ frequencies rather than $\lfloor (T + 1)/2 \rfloor$.

Code

```
function [y,w]=per(x,m,pl)
%usage [y,w]=per(x,m,pl)
% x = time series
% m = padding factor (1 means no padding)
% y = periodogram
% w = frequencies (radians per unit time)
% pl = 1 if plot required
[T,s]=size(x);
if T == 1
   x=x';
   T=s;
end
n=T*m;
n2=floor((n+1)/2);
w=(0:n2-1)*2*pi/n;
y=fft([x;zeros(n-T,1)]);
y=2*(y.*conj(y))/T;
y=y(1:n2);
if pl == 1
```

```
    plot(w,y)
    xlabel('Frequency (radians per unit time)')
    title('Periodogram')
end
```

Name pis

Description The program 'pis' computes the Pisarenko frequency estimates. The program automatically detects whether the time series is real
or complex.

Code

```
function w = pis(x,k)
%usage w = pis(x,k)
% x = time series (real or complex)
% k = number of sinusoids
% w = vector of frequency estimates
if isreal(x)
    k2=2*k+1;
else
    k2=k+1;
end
c=zeros(k2,1);
T=length(x);
xb=mean(x);
% If k is large, the usual fft method of computing
% autocovariances might be faster
for j=0:k2-1
    c(j+1)=(x(1:T-j)-xb)'*(x(j+1:T)-xb)/T;
end
C=toeplitz(c);
if ~isreal(x)
    C=conj(C);
end
[v,d]=eig(C);
[b,m]=sort(diag(d));
P=v(:,m(1));
w=roots(P);
w=angle(w);
if isreal(x)
    w=w(w > 0);
```

```
end
```

Name qf
Description The program 'qf' computes the Quinn & Fernandes frequency
estimator w. It optionally plots the filtered periodogram κ.
Code

```
function w = qf(x,pl)
%usage w = qf(x,pl)
% w = Q & F frequency estimate, x = time series
% pl = 1 if plot of filtered periodogram is required
[T,s]=size(x);
if T == 1
   x=x';
   T=s;
end
n=floor((T+1)/2);
xb=mean(x);
x=x-xb; %mean-corrected series
z=fft(x);
z=z(1:n);
z=z.*conj(z);
[y,k]=max(z);
a=2*cos(2*pi*(k-1)/T);
b=[1];
T1=T-1;
for j=0:1
   c=[1;-a;1];
   y = filter(b,c,x);
   v = x(2:T)'*y(1:T1)/(y(1:T1)'*y(1:T1));
   a = a+2*v;
end
w=acos(a/2);
if pl == 1
   z=fft([x;zeros(size(x))]);
   z=z.*conj(z);
   z=real(fft(z))/(2*T*T); % Autocovariances
   % Now compute kappa
   z=real(fft([0;z(2:T)./(1:T-1)']));
   plot(2*pi*(0:n-1)'/T,z(1:n))
```

```
    xlabel('Frequency (radians per unit time)')
    ylabel('\kappa')
end
```

Name qfpic, qfgen

Description The program 'qfpic' was used to generate the Figures 4.2 through 4.6. It produces four subplots, depicting the time series, its periodogram I_T, the windowed periodogram κ_T and its derivative. The program 'qfgen' is a script file, used to generate the data, to call 'qfpic', and to store the pictures as encapsulated postscript files.

Code

```
function [z,zi] = qfpic(x,pad)
%usage [z,zi] = qfpic(x,pad)
% x = time series
% pad = padding factor. If pad is 1, then no padding
% z = vector of $\kappa_T$ values
% zi = vector of $\kappa_T^{\prime} values
n=length(x);
subplot(221)
m=fix((n*pad+1)/2);
np=(0:m-1)'*2*pi/(n*pad);
xb=mean(x);
plot((1:n)',x)
xlabel('t')
ylabel('X_t')
title('Time Series')
x=x-xb;
if (q > 1)
z=fft([x;zeros(n*(q-1),1)]);
else
z=fft(x);
end
z=(z.*conj(z))*2/n;
subplot(222)
plot(np, z(1:m));
xlabel('\omega')
ylabel('I_T(\omega)')
title('Periodogram')
z=real(fft(z))/(2*q*n);
```

```
z=z(1:n);
if (q > 1)
zi=[0;z(2:n)./(1:n-1)';zeros((q-1)*n,1)];
z=imag(fft([z;zeros((q-1)*n,1)]));
else
zi =[0;z(2:n)./(1:n-1)'];
z=imag(fft(z));
end
z=z(1:m);
subplot(223)
plot(np,z,np,zeros(size(z)));
xlabel('\omega')
ylabel('\kappa_T^\prime(\omega)')
zi=real(fft(zi));
zi=zi(1:m);
subplot(224)
plot(np,zi);
xlabel('\omega')
ylabel('\kappa_T(\omega)')
title('Windowed Periodogram')%end of qfpic.m

%qfgen.m Script file for generating figures in Chapter 4
t1=(0:127)';
t2=(0:1023)';
x=cos(t1)+randn(128,1);
qfpic(x,5);
print -deps2 f2.eps
figure
x=cos(0.3*t1)+0.4*cos(0.5*t1)+0.5*randn(128,1);
qfpic(x,5);
print -deps2 f3.eps
figure
x=cos(0.3*t2)+0.4*cos(0.5*t2)+0.5*randn(1024,1);
qfpic(x,5);
print -deps2 f4.eps
e=filter([1],[1 0 0.81],randn(128,1));
x=sqrt(2)*cos(t1)+1/sqrt(2)*cos(2*t1)+e;
figure
qfpic(x,5);
print -deps2 f128c.eps
```

```
e=filter([1],[1 0 0.81],randn(1024,1));
x=sqrt(2)*cos(t2)+1/sqrt(2)*cos(2*t2)+e;
figure
qfpic(x,5);
print -deps2 f1024c.eps
```

Name quadint

Description The program 'quadint' computes the quadratic interpolator frequency estimate.

Code

```
function w=quadint(x)
%usage w = quadint(x)
% x = time series
% w = quadratic interpolation frequency estimate
T=length(x);
n=floor((T+1)/2);
y=fft(x);
y=y.*conj(y);
[a,b]=max(y(1:n));
del = (y(b+1)-y(b-1))/(2*y(b)-y(b-1)-y(b+1));
w=2*pi*(b-1+del)/T;
```

Name rifevinc

Description The program 'rifevinc' computes the Rife & Vincent frequency estimate.

Code

```
function w = rifevinc(x)
%usage w = rifevinc(x)
% x = time series
% w = Rife & Vincent's frequency estimate
T=length(x);
n=floor((T+1)/2);
y=fft(x);
y=abs(y);
[a,b]=max(y(1:n));
if y(b+1) > y(b-1)
   del = y(b+1)/(y(b)+y(b+1));
else
   del = -y(b-1)/(y(b)+y(b-1));
```

```
end
w=2*pi*(b-1+del)/T;
```

Name sinreg
Description The program 'sinreg' computes the sum of squares for regression of x on cosine and sine terms evaluated at the elements of the vector w.
Code

```
function [ress,y,b] = sinreg(w,x)
%usage [ress,y,b] = sinreg(w,x)
[T,s] = size(x);
if T == 1
   x=x';
   T=s;
end
p=ones(length(w),1);
n = (0:T-1)'-(T-1)/2;
v=[ones(size(x)) cos(n*w')];
m1 = w*p' - p*w';
m2 = w*p' + p*w';
m1=srg(m1,T);
m2=srg(m2,T);
m=(m1+m2)/2;
ss=srg(w,T);
mp=[T;ss];
mq = [ss' ;m];
mp=[mp mq];
np = [sum(x);cos(w*n')*x];
b1=mp\np;
ress=x'*x-np'*b1;
y=x-v*b1;
m=(m1-m2)/2;
np=sin(w*n')*x;
b2=m\np;
y=y-sin(n*w')*b2;
b=[b1;b2];
ress=ress-np'*b2;
```

Name sinsim
Description The program 'sinsim' simulates a sum of sinusoids in Gaussian white noise. The frequencies, amplitudes and phases must be entered as column vectors.
Code

```
function x=sinsim(T,w,amp,ph,s)
%usage x=sinsim(T,w,amp,ph,s)
% Simulates a sum of sinusoids in Gaussian noise
% T = sample size
% w = column vector of frequencies (radians per unit time)
% amp = column vector of amplitudes
% ph = column vector of phases (radians)
% s = standard deviation of noise
x=cos((0:T-1)'*w'+ones(T,1)*ph')*amp+s*randn(T,1);
```

Name srg
Description The program 'srg' computes the function

$$f(x) = \frac{\sin(nx/2)}{\sin(x/2)}$$

safely, at the vector of values x.
Code

```
function y = srg(x,n)
j=find(x);
y=n*eye(size(x));
y(j) = sin(x(j)*n/2)./sin(x(j)/2);
```

References

Ainsleigh, P.L., Kehtarnavaz, N. and Streit, R.L. (2000). Hidden Gauss–Markov models for signal classification, *preprint*.

An, H.Z., Chen, Z.G. and Hannan, E.J. (1983). The maximum of the periodogram, *J. Mult. Anal.* **13**, 383–400.

Anderson, B.D.O. and Moore, J.B. (1979). *Optimal Filtering* (Prentice Hall, Englewood Cliffs).

Barrett, R.F. and Holdsworth, D.A. (1993). Frequency tracking using HMM models with amplitude and phase information, *IEEE Trans. on SP* **41**, 2965–2976.

Bartlett, M.S. (1967). Inference and stochastic processes, *J. Roy. Statist. Soc. A* **130**, 457–477.

Billingsley, P. (1995). *Probability and Measure* (Wiley, New York).

Bloomfield, P. (1976). *Fourier Analysis of Time Series: an Introduction* (Second Edition 2000) (Wiley, New York).

de Boor, C. (1978). *A Practical Guide to Splines* (Springer-Verlag, New York).

Brillinger, D.R. (1974). *Time Series Data Analysis and Theory* (Expanded Edition 1981) (Holden-Day, San Francisco).

Brillinger, D.R. (1987). Fitting cosines: some procedures and some physical examples, in *Applied Statistics, Stochastic Processes and Sampling Theory*, ed. I.B. MacNeill and G.J. Umphrey (Reidel, Dordrecht).

Brunt, D. (1917). *The Combination of Observations* (Cambridge University Press).

Chan, Y.T., Lavoie, J.M.M. and Plant, J.B. (1981). A parameter estimation approach to estimation of frequencies of sinusoids, *IEEE Trans. on ASSP* **29**, 214–229.

Clarkson, V., Kootsookos, P.J. and Quinn, B.G. (1994). Analysis of the variance threshold of Kay's weighted linear predictor frequency estimator, *IEEE Trans. on SP* **42**, 2370–2379.

Cooley, J.W. and Tukey, J.W. (1965). An algorithm for the machine calculation of complex Fourier series, *Math. Comp.* **19**, 297–301.

Cox, D.R. and Hinkley, D.V. (1979). *Theoretical Statistics* (Chapman and Hall, London).

Cramér, H. (1946). *Mathematical Methods of Statistics* (Princeton University Press).

Daley, N. (2000). Problems in the Estimation of Frequency, University of London PhD. thesis.

David, F.N. and Kendall, M.G. (1949). Tables of Symmetric Functions, *Biometrika* **36**, 431–449.

Dieudonné, J. (1969). *Foundations of Modern Analysis* (Academic Press, New York).

Ferguson, B.G. and Quinn, B.G. (1994). Application of the short-time Fourier transform and the Wigner-Ville distribution to the acoustic localisation of aircraft, *J. Acoust. Soc. Amer.* **96:2 part 1**, 821–827.

Ferguson, J.D. (1980). Hidden Markov analysis: an Introduction, in *Proceedings of the Symposium on the Applications of Hidden Markov Models to Text and Speech*, 8–15. (IDA-CRD, Princeton).

Fernandes, J.M., Goodwin, G.C. and De Souza, C.E. (1987). Estimation of models for systems having deterministic and random disturbances, in *Proceedings of 10th World Congress on Automatic Control* **10**, 370–375. (Pergamon Press).

Fisher, R.A. (1929). Tests of significance in harmonic analysis, *Proc. R. Soc. London A* **125**, 54–59.

Gardner, W.A. (1988). *Statistical Spectral Analysis: a Nonprobabilistic Theory* (Prentice Hall, Englewood Cliffs).

Händel, P. (1995). On the performance of the weighted linear predictor frequency estimator, *IEEE Trans. on SP* **43**, 3070–3071.

Hannan, E.J. (1970). *Multiple Time Series* (Wiley, New York).

Hannan, E.J. (1973). The estimation of frequency, *J. App. Prob.* **10**, 510–519.

Hannan, E.J. (1979). The central limit theorem for time series regression, *Stoch. Proc. Appl.* **9**, 281–289.

Hannan, E.J. and Deistler, M. (1988). *The Statistical Theory of Linear Systems* (Wiley, New York).

Hannan, E.J. and Huang, D. (1993). On-line frequency estimation, *J. Time Series Anal.* **14**, 147–161.

Hannan, E.J. and Mackisack, M. (1986). A law of the iterated logarithm for an estimate of frequency, *Stoch. Proc. Appl.* **22**, 103–109.

Hannan, E.J. and Quinn, B.G. (1989). The resolution of closely adjacent spectral lines, *J. Time Series Anal.* **10**, 13–22.

Hannan, E.J. and Wahlberg, B. (1989). Convergence rates for inverse Toeplitz matrix forms, *J. Multivariate Anal.* **31**, 127–135.

Huang, D. (1996). On low and high frequency estimation, *J. Time Series Anal.* **17**, 351–365.

Huang, D. (2000). Approximate maximum likelihood algorithm for frequency estimation, *Statistica Sinica* **10**, 157–171.

Jennrich, R.I. (1969). Asymptotic properties of nonlinear least squares estimators, *Ann. Math. Statist.* **40**, 633–643.

Johnson, D.H. (1982). The application of spectral estimation methods to bearing estimation problems, *Proc. IEEE* **70**, 1018–1028.

Kavalieris, L. and Hannan, E.J. (1994). Determining the number of terms in a trigonometric regression, *J. Time Series Anal.* **15**, 613–625.

Kay, S.M. (1989). A fast and accurate single frequency estimator, *IEEE Trans. on ASSP* **37**, 1987–1990.

Kay, S.M. and Marple, S.L. (1981). Spectrum analysis – a modern perspective, *Proc. IEEE* **69**, 1380–1418.

La Scala, B.F., Bitmead, R.R. and Quinn, B.G. (1996). An extended Kalman filter frequency tracker for high-noise environments, *IEEE Trans. on SP* **44**, 431–434.

Levinson, S.E., Rabiner, L.R. and Sondhi, M.M. (1983). An introduction to the theory of probability functions of a Markov process to automatic speech recognition, *Bell Syst. Tech. J.* **62**, 1035–1074.

Luginbuhl, T.E. (1999). Estimation of General, Discrete-Time FM Processes, University of Connecticut PhD. thesis.

MacLeod, M.D. (1991). Fast DFT-domain algorithms for near-optimal tonal detection and estimation, *Proc. Inst. Acoust.* **13**, 102–109.

MacLeod, M.D. (1998). Fast nearly ML estimation of the parameters of real or complex single tones or resolved multiple tones, *IEEE Trans. on SP* **46**, 141–148.

McMahon, D.R.A. and Barrett, R.F. (1986). An efficient method for the estimation of the frequency of a single tone in noise from the phases of discrete Fourier transforms, *Sig. Proc.* **11**, 169–177.

Marple, S.L. (1987). *Digital Spectral Analysis with Applications* (Prentice Hall, Englewood Cliffs).

Molchan, G.M. (1990). Exact resolution thresholds for close frequencies, in *Proc. 5th International Vilnius Conf. on Probability and Statistics* **II**, 193–206.

Muñoz, J.L. and Streit, R.L. (1991). Connection machine implementation of hidden Markov models for frequency line tracking, in *Very Large Scale Computation in the 21st Century*, ed. J.P. Mesirov (SIAM, Philadelphia).

Nehorai, A. and Porat, B. (1986). Adaptive comb filtering for harmonic signal enhancement, *IEEE Trans. on ASSP* **34**, 1124–1138.

Pisarenko, V.F. (1973). The retrieval of harmonics from a covariance function, *Geophys. J.R. Astr. Soc.* **10**, 347–366.

Priestley, M.B. (1981). *Spectral Analysis and Time Series* (Academic Press, London).

Quinn, B.G. (1989). Estimating the number of terms in a sinusoidal regression, *J. Time Series Anal.* **10**, 71–75.

Quinn, B.G. (1992). Some new high-accuracy frequency estimators, in *Proceedings of the Third International Symposium on Signal Processing and its Applications, 16-21 August 1992, Gold Coast, Australia* **2**, 323–326.

Quinn, B.G. (1995). Doppler speed and range estimation using frequency and amplitude estimates, *J. Acoust. Soc. Amer.* **98:5 part 1**, 2560–2566.

Quinn, B.G., Barrett, R.F. and Searle, S.J. (1994). The estimation and HMM tracking of weak narrowband signals, in *Proceedings of 1994 International Conference on Acoustics, Speech and Signal Processing* **IV**, 341–344.

Quinn, B.G., Clarkson, V. and Kootsookos, P.J. (1998). Comments on "On the Performance of the Weighted Linear Predictor Frequency Estimator", *IEEE Trans. on SP* **46**, 526–527.

Quinn, B.G. and Fernandes, J.M. (1991). A fast efficient technique for the estimation of frequency, *Biometrika* **78**, 489–498.

Quinn, B.G. and Kootsookos, P.J. (1994). Threshold behaviour of the maximum likelihood estimator of frequency, *IEEE Trans. on SP* **42**, 3291–3294.

Quinn, B.G. and Thomson, P.J. (1991). Estimating the frequency of a periodic function, *Biometrika* **78**, 65–74.

Rabiner, L.R. and Juang, B.H. (1986). An introduction to hidden Markov models, *IEEE ASSP Magazine* **3**, 4–16.

Rice, J.A. and Rosenblatt, M. (1988). On frequency estimation, *Biometrika* **74**, 477–484.

Rife, D.C. and Boorstyn, R.R. (1974). Single tone parameter estimation from discrete-time observations, *IEEE Trans. Inf. Theor.* **20**, 591–598.

Rife, D.C. and Vincent, G.A. (1970). Use of the discrete Fourier transform in the measurement of frequencies and levels of tones, *Bell Syst. Tech. J.* **49**, 197–228.

Rissanen, J. (1989). *Stochastic Complexity in Statistical Inquiry* (World Scientific, New Jersey).

Sakai, H. (1984). Statistical analysis of Pisarenko's method for sinusoidal frequency estimation, *IEEE Trans. on ASSP* **32**, 95–101.

Schmidt, R.O. (1981). A Signal Subspace Approach to Multiple Emitter Location and Spectral Estimation, Stanford University PhD. thesis.

Schmidt, R.O. (1986). Multiple emitter location and signal parameter estimation, *IEEE Trans. Antennas Propag.* **34**, 276–280.

Shannon, C.E. (1948). A mathematical theory of communication, *Bell Syst. Tech. J.* **27**, 379–423 and 623–656.

Stoica, P. and Nehorai, A. (1989). MUSIC, maximum Likelihood, and the Cramer–Rao Bound, *IEEE Trans. on ASSP* **37**, 720–741.

Stone, L.R., Barlow, C.A. and Corwin, T.L. (1999). *Bayesian Multiple Target Tracking* (Artech House, Norwood MA).

Streit, R.L. and Barrett, R.F. (1990). Frequency line tracking using hidden Markov models, *IEEE Trans. on ASSP* **38**, 586–598.

Truong-Van, B. (1990). A new approach to frequency analysis with amplified harmonics, *J. Roy. Statist. Soc. B* **52**, 203–222.

Tukey, J.W. (1961). Discussion emphasising the connection between analysis of variance and spectrum analysis, *Technometrics* **3**, 191–219.

Turkman, K.F. and Walker, A.M. (1984). On the asymptotic maxima of trigonometric polynomials with random coefficients, *Adv. Appl. Prob.* **16**, 819–842.

Walker, A.M. (1971). On the estimation of a harmonic component in a time series with stationary independent residuals, *Biometrika* **58**, 21–36.

Wang, X. (1993). An AIC type estimator for the number of cosinusoids, *J. Time Series Anal.* **14**, 431–440.

Whittle, P. (1951). *Hypothesis Testing in Time Series Analysis* (Almqvist & Wiksell, Uppsala).

Whittle, P. (1952). The simultaneous estimation of a time series harmonic components and covariance structure, *Trab. Estadist.* **3**, 43–57.

Xie, X. and Evans, R. (1991). Multiple target tracking and multiple frequency line tracking using hidden Markov models, *IEEE Trans. on SP* **39**, 2659–2676.

Xie, X. and Evans, R. (1993). Multiple frequency line tracking with hidden Markov models – further results, *IEEE Trans. on SP* **41**, 334–343.

Yau, S.F. and Bresler, Y. (1992). Worst case Cramér–Rao bounds for parametric estimation of superimposed signals with applications, *IEEE Trans. on SP* **40**, 2973–2986.

Yule, G. U. (1927). On a method of investigating periodicities in disturbed series, with special reference to Wolfer's sunspot numbers, *Phil. Trans. Roy. Soc. London A* **226**, 267–298.

Author index

Subject index